Ammonia

Catalysis and Manufacture

With contributions by
K. Aika, L. J. Christiansen, I. Dybkjaer,
J. B. Hansen, P. E. Højlund Nielsen,
A. Nielsen, P. Stoltze, K. Tamaru

With 68 Figures and 23 Tables

Springer-Verlag
Berlin Heidelberg NewYork
London Paris Tokyo
Hong Kong Barcelona Budapest

Editor:

Anders Nielsen
Haldor Topsøe A/S
Nymøllevej 55, 2800 Lyngby/DK

ISBN 3-540-58335-1 Springer-Verlag Berlin Heidelberg NewYork
ISBN 0-387-58335-1 Springer-Verlag NewYork Berlin Heidelberg

Library of Congress Cataloging-in-Publication Data Ammonia: catalysis and manufacture/with contributions by K. Aika... [et al.; editor, Anders Nielsen]. p.cm. Includes bibliographical reference and index. ISBN 3-540-58335-1. -- ISBN 0-387-58335-1 (U.S.) 1. Ammonia. I. Aika, K. (Ken-ichi) II. Nielsen, Anders, 1934-. TP223.A453 1995. 661'.34--dc20 94-36677 CIP

Typesetting: Macmillan India Ltd., Bangalore-25
SPIN: 10077499 51/3020 - 5 4 3 2 1 0 - Printed on acid-free paper

Preface

This book owes its existence to Dr. Ekkehard Fluck, Director of the Gmelin-Institut. Dr. Fluck suggested that the individual chapters could be written by staff members of Haldor Topsoe A/S, and that emphasis should be given to the industrial manufacture of ammonia.

Upon careful consideration it was decided to ask two distinguished experts in catalysis, Prof. Kenzi Tamaru of the Science University of Tokyo and Prof. Ken-ichi Aika of the Tokyo Institute of Technology to write the chapter on ammonia synthesis on non-iron catalysts.

When I started to work in catalysis more than 50 years ago, first as a student in the Institute of Physical Chemistry with the late Prof. J. N. Brønsted and subsequently in the laboratories of Dr. Haldor Topsoe, the use of catalytic processes by industry and the literature on catalytic studies was still somewhat limited and allowed a single person to reasonably acquaint himself with the field. Today, catalysis is a step in the manufacture of most chemical products and most refinery streams undergo catalytic reactions. The number of physical tools applied to the study of catalysts is impressive and the literature on catalysis overpowering. A reasonably comprehensive volume on the topic of ammonia synthesis had to be a team effort. The first chapter entitled, "*Thermodynamic properties in Ammonia Synthesis*" is written by Dr. Lars J. Christiansen. It should be emphasized that this chapter does not contain the complete thermodynamics on ammonia, but concentrates on the thermodynamic properties used for the design and operation of ammonia synthesis units.

The second chapter "*Structure and Surface Chemistry of Industrial Ammonia Synthesis Catalysts*" is written by Dr. Per Stoltze. This chapter deals with the structure and surface chemistry of iron-based ammonia synthesis catalysts of the type used by industry. Certain studies of single crystal surfaces are included to the extent that they serve to add information to the main topic. This chapter includes a presentation of the unreduced catalyst; the reduction process and the bulk and surface structure of a reduced catalyst. A thorough discussion is given of the different states of sorption of nitrogen and of chemisorption of hydrogen, carbon oxides, ammonia and oxygen. The last part of the chapter gives a detailed account of the mechanism of ammonia synthesis on iron.

The third chapter is written by Profs. Ken-ichi Aika and Kenzi Tamaru, both of whom have contributed prominently to our knowledge of ammonia synthesis. Their chapter is entitled, "*Ammonia Synthesis over Non-Iron Catalysts and Related Phenomena*". It is recalled that osmium was indeed used in the early

Chapter 1

Thermodynamic Properties in Ammonia Synthesis

Lars J. Christiansen
Haldor Topsøe A/S Copenhagen, Denmark

Contents

Nielsen (Ed.), Ammonia

1.1 Introduction

The reaction of hydrogen and nitrogen, which is performed under pressure in the presence of a catalyst, is exothermic and reversible. The conversion to NH_3 is thus limited by chemical equilibrium. The ammonia is normally removed from the gas stream by cooling condensation and the unused reactants are recycled back to the inlet of the chemical reactor.

Design of energy efficient ammonia plants therefore requires an accurate knowledge of the following thermodynamic properties at the actual operating conditions:

1. PVT properties,
2. Thermodynamic Properties,
3. Phase equilibrium between condensed ammonia and gases,
4. Transport properties.

The thermodynamic properties or their derivatives must be continuous in order to ensure that there is no discontinuity in the derived thermodynamic property such as cooling condensation, etc.

An extensive review of the required data mentioned above has been given in Nielsen [45].

1.2 Equations of State (PVT Properties)

The pressure range utilized in ammonia plants implies that the ideal gas law cannot be used. This is corrected for by use of a compressibility factor Z., whereby

$$pV = ZRT \tag{1}$$

where p is the pressure, V the molar volume, T the temperature, and R the gas constant. The compressibility factor can be found from generalized charts, but it is more common to use an equation of state, where the pressure is given as a function of temperature, volume, and molecular parameters, such as the critical properties,

$$p = f(V,T, \text{parameters}) \tag{2}$$

By use of adequate mixing rules for the molecular parameters, the same type of equation can be used to calculate mixture properties. Numerous equations of state have been proposed and a review can be found in Reid et al. [50].

For detailed information on compressibility factors reference is given to the experimental data by Wiebe, Gaddy [56], by Bartlett et al. [3, 4]. These papers give values of the compressibility factors of hydrogen, nitrogen, and 3:1 mixtures of these gases at pressures up to 1000 atm and temperatures between − 70 and + 300 °C. The article by Wiebe and Gaddy also gives compressibility factors for three other hydrogen-nitrogen mixtures. Reference is also given to an extensive treatment by Sage et al. [52] of thermodynamic properties of two hydrogen-nitrogen mixtures, one of them the 76:24 two-component mixture and the other a mixture of these gases with CO_2, CO, and CH_4. Michels et al. [42] have calculated thermodynamic functions of two gas mixtures containing ammonia from compressibility isotherms, and Michels et al. [41] report compressibility and thermodynamic data of three ammonia-containing synthesis gas mixtures.

Thermodynamic properties of pure ammonia are shown in an extensive work by Haar, Gallagher [17]. They include densities of gaseous and liquid ammonia between − 50 and + 350 °C and from 0 to 1000 bar. Their work includes data from Circular of the National Bureau of Standard No. 142 [8], and they find excellent agreement between this work and newer data.

Properties of liquid ammonia from − 50 to + 65 °C and from saturation pressure to 370 bar are also reported by Zander, Thomas [62].

1.3 Thermodynamic Properties

1.3.1 Enthalpies of Pure Compounds and Mixtures

The calculation of the thermodynamic properties is accomplished in two steps. The first step is the calculation of the properties calculated for the ideal gas, the pure component heat capacities at constant pressure, and enthalpies and free energies of formation of the pure components. Ideal mixing rules are used. The second step is the calculation of the deviation from the ideal gas, which can be derived from an equation of state describing the mixture properties.

The mixture enthalpy H and the mixture heat capacity C_p are calculated as:

$$H = \sum_i y_i H_i^0 - H' \tag{3}$$

$$C_p = \sum_i y_i C_{pi}^0 - C_p' \tag{4}$$

where H_i^0 and C_{pi}^0 are the pure component ideal gas standard enthalpy and heat capacity, respectively, an y_i the mole fraction. The enthalpy departure from the ideal gas, H′, can be calculated using rigorous thermodynamics as:

$$H' = -\int_V^\infty \left[p - T\left(\frac{\partial p}{\partial T}\right)_{V,N} \right] dV - RT(z-1) \tag{5}$$

Insertion of the heat of reaction derived from equation (7) gives the following expression for the thermodynamic equilibrium constant as a function of temperature

$$R\ln K = a_0 - \frac{a_1}{T} + a_2 \ln T + a_3 T + \frac{a_4}{2} T^2 + \frac{a_5}{3} T^3 \tag{15}$$

The constant a_0 is fixed so that equation (14) gives the Gibbs free energy of formation at 25 °C. − 3915 kcal/kmol [24].

The equilibrium composition is then calculated in a two step procedure. The first step leads to the ideal gas composition as given by K_p and the second step is the correction for the non-ideal gas behavior by use of K_φ, which includes the influence of pressure and composition.

The most accurate experimental values appear to be those by Haber, Rossignol (18), by Schultz, Schaefer [53], by Haber et al. [21] for 30 atm pressure, by Larson, Dodge [32] for 10, 30, 50, and 100 atm, and by Larson [30] for pressures of 300, 600, and 1000 atm. Experimental and calculated data on ammonia synthesis equilibrium at pressures from 1000 atm to 3500 atm have been given by Winchester, Dodge [61].

The experimental data at high pressures have been analyzed by different equations of state to calculate the fugacity coefficients in K_φ. Gillespie [11] and Gillespie, Beattie [12] have used the Beattie-Bridgeman equation of state. The method by Newton [44] is a generalized method for calculation of fugacity coefficients.

In Figs. 1.1 to 1.3 the equilibrium % NH_3 are shown as a function of temperature at different pressures and at 0%, 10% and 20% inert (methane and argon) content, respectively. The fugacity coefficient ratio is calculated by the present author using the Martin-Hou equation of state, see [36]. Of course, other equations of state could have been used for calculating this coefficient.

It is interesting to note that because of the composition dependence of the fugacity coefficients the maximum yield of ammonia at high pressures may well exist at a hydrogen-nitrogen ratio different from 3. At 200 atm and 500 °C the maximum yield at equilibrium is to be expected for a ratio of 2.9 although the increase in only 0.01% NH_3.

1.5 Phase Equilibrium

The content of ammonia in the vapor phase and the amount of dissolved synthesis gases in the liquid ammonia phase are calculated from expressions for the β value, $\beta_i = y_i/x_i$, where y_i and x_i are the mole fractions in the gas and liquid phases, respectively.

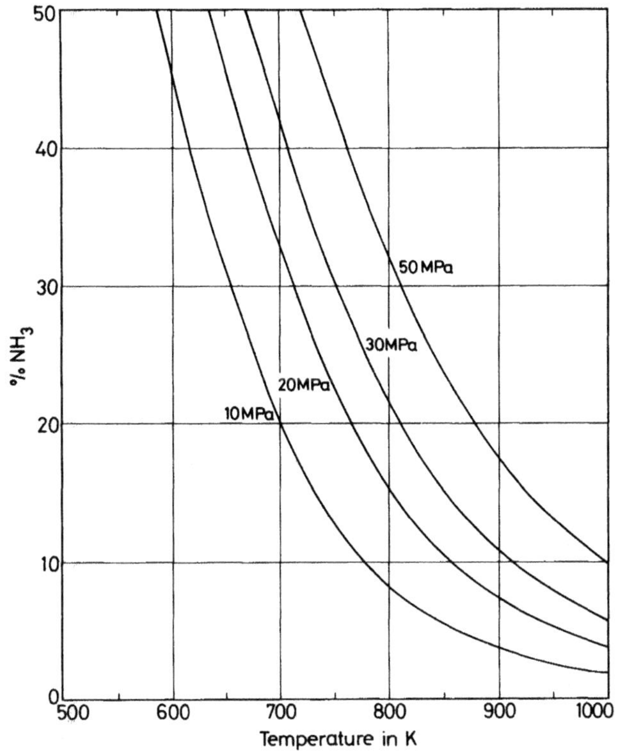

Fig. 1.1 Percentage of ammonia in equilibrium mixture (75% H_2, 25% N_2, 0% inerts).

Correlations for the β values are derived from the isofugacity criterion, which can be written as

$$\beta_i = \frac{\gamma_i f_i^0}{\varphi_i p} \tag{16}$$

where γ_i is the liquid phase activity coefficient, which is a function of temperature and composition, f_i^0 is the reference fugacity, and φ_i the gas phase fugacity coefficient, which is calculated from an equation of state. All parameters are given at the pressure of the mixture.

The reference fugacity for ammonia, which is the soluble component, is evaluated from the saturation fugacity of ammonia and a Poynting correction factor. For the dissolved gases, the reference fugacity is the Henry's constant referred to the saturation pressure of ammonia. For further details of the theory and the necessity of including the Poynting correction factor, see Reid et al. [50].

The activity coefficients and the fugacity coefficients consequently account for the deviation from Raoult's law for ammonia, and for the deviations from Henry's law for the dissolved gases. The validity of a published gas solubility for a binary system in a multicomponent system requires therefore that Henry's law

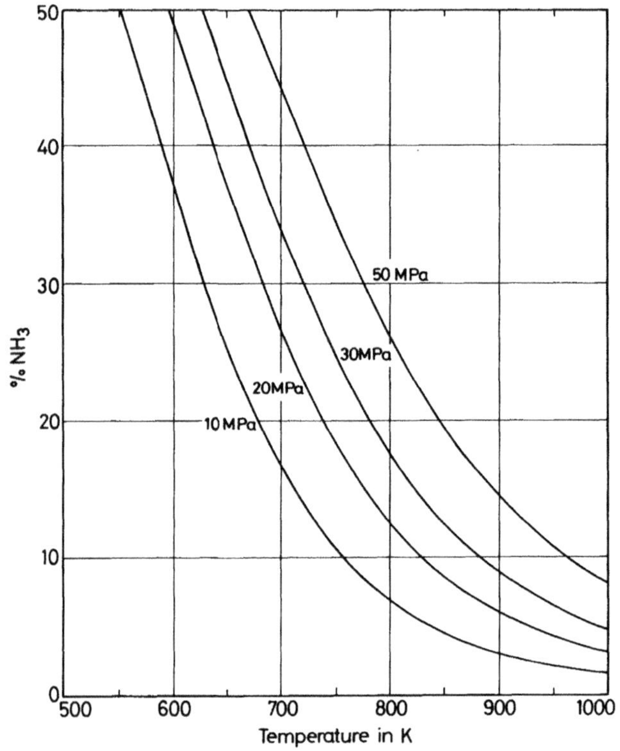

Fig. 1.2 Percentage of ammonia in equilibrium mixture (67.5% H_2, 22.5% N_2, 3% Ar, 7% CH_4).

is valid. This is often a good approximation in ammonia synthesis, where the amounts of dissolved gases are small.

The type of correlation as shown in equation (18) has been used by Alesandrini et al. [1] to evaluate phase equilibrium properties. They used the Redlich-Kwong equation of state and the van Laar type activity coefficients to correlate the experimental information for the binary systems.

1.5.1 Vapor Concentration of Ammonia

The activity coefficient for ammonia is greater than one and the fugacity coefficient decreases with increasing pressure thereby giving a higher concentration of ammonia in the gas phase than predicted by the Raoult's law alone.

Experimental measurements of the ammonia concentration in the vapor phase have been reported by Larson, Black [31] for the equilibrium with nitrogen and hydrogen from -22.5 to $+18.7\,°C$ and from 50 to 1000 atm. More recent data are reported by Michels et al. [40]. Their investigation

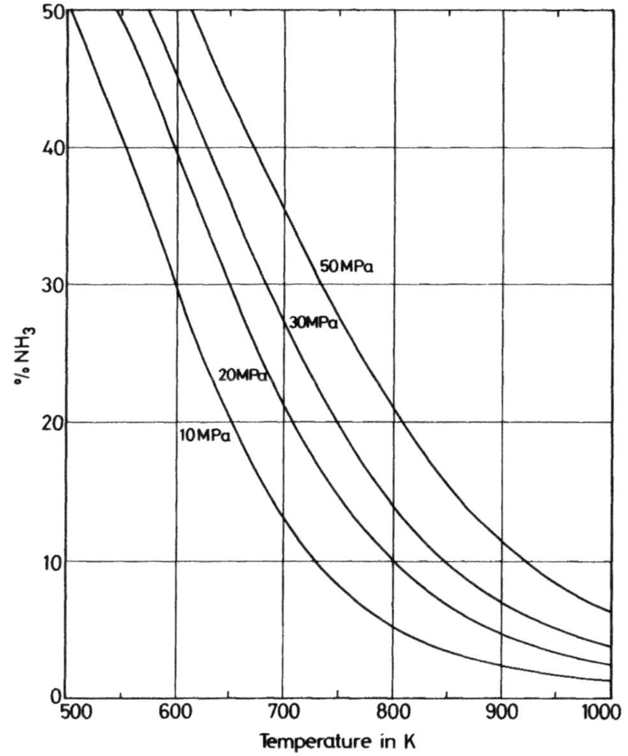

Fig. 1.3 Percentage of ammonia in equilibrium mixture (60% H_2, 20% N_2, 6% Ar, 14% CH_4).

covered the temperature interval from 0 to 121.8 °C and pressures from 25 atm to 785 atm. Michels et al. [38] later reported data in the same pressure range but at temperatures below 0 °C. Lefrancois, Vaniscotte [33] report data for the temperature interval − 70 to + 60 °C at pressures 300 and 500 kg/cm^2. These data agree well with the data by Michels et al. [38, 39]. Reddy, Husain [49] have given data for the ammonia vapor concentration in a mixture with gas phase mole ratios, H_2:N_2:Ar:CH_4, equal to 3:1:0.18:0.44.

The available data for the vapor concentration of ammonia in equilibrium with a 3:1 mixture of hydrogen and nitrogen are shown in Fig. 1.4 for pressures equal to 10, 20, 30, and 50 MPa. The data used are those published by Michels et al. These data are in good agreement with the data given by Lefrancois, Vaniscotte [33], by Heise [22] and by Zeininger [14] at 25 °C, whereas the data given by Larson, Black [31] give higher concentrations of ammonia, in particular at the high pressures.

In Fig. 1.4 the predicted values of the ammonia concentration are also shown assuming that Raoult's law is valid and neglecting the dissolved gases at 10 and 50 MPa, respectively. It is seen that Raoult's law is not valid.

The data presented by Reddy, Husain [49] also agree with Fig. 1.4 in spite of the content of inerts in their data. This is also the case for the data given by

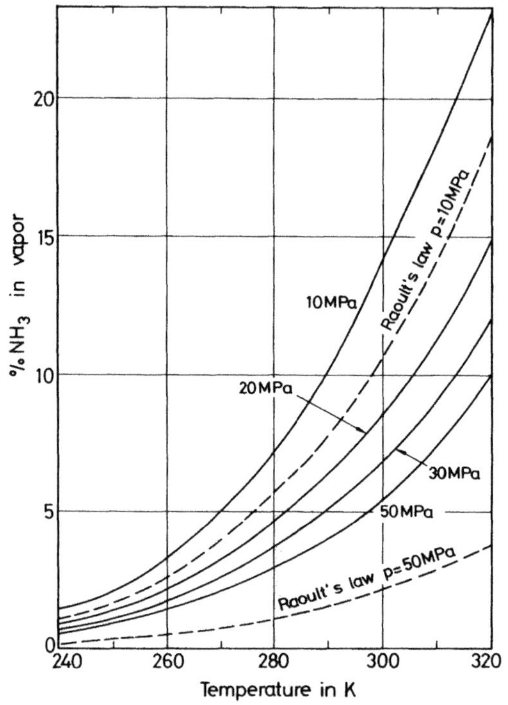

Fig. 1.4 Equilibrium vapor concentration of ammonia in 3:1 mixture of $H_2 + N_2$.

Zeininger [14]. There is hence some evidence that the pressure of inerts will not change the ammonia vapor concentration significantly although it will increase. More experimental information is needed in order to obtain more certainty about this point.

In the model given by Alesandrini et al. [1] ammonia concentrations are calculated which are too small at the higher pressures. The model has therefore been revised by Reddy, Husain [49].

1.5.2 Phase Equilibrium of Dissolved Gases

Solubilities of the gases hydrogen, nitrogen, argon, methane, and helium in the form of β values (definition see p. 6) are shown in the extensive treatment in Landolt-Börnstein [29]. The data cover the operating region normally used in ammonia synthesis.

Data for the solubility of hydrogen in ammonia can be found in Reamer, Sage [47], in Wiebe, Treamearne [57, 58]. In Heise [22], and in Zeininger [63]. Data for the solubility of nitrogen in ammonia can be found in Reamer, Sage [48], in Heise [22], and in Zeininger [63]. Data for the solubility of argon in ammonia can be found in Kaminishi [25], in Michels et al. [39], and in Heise [22]. Data for the solubility of methane in ammonia can be found in Kaminishi [25], and in Zeininger [63]. Data for the solubility of helium in ammonia can be found in Heise [22].

Data for the solubility of hydrogen-nitrogen in the ratio 3:1 in ammonia are shown in Larson, Black [31], in Michels et al. [40], in Lefrancois, Vaniscotte [34], in Michels et al. [38], and in Atroshchenko, Gavrya [2]. Data with methane can also be found in Zeininger [63] and in Konoki et al. [28].

β values for hydrogen, nitrogen, argon, and methane evaluated by using the method given in Alesandrini et al. [1] for a mixture with the 10% inert gases and $H_2/N_2 = 3$ are shown in Fig. 1.5. The parameters have been evaluated by use of the data above. β values are shown for 10, 20, and 30 MPa, respectively, as a function of temperature. The same figure also shows experimental data for the solubility of helium in the range given in Heise [22]. The β values decrease in the order helium, hydrogen, nitrogen, argon, and methane, which corresponds to an increase in solubility. The β values decrease with increase in temperature. The β values are not very sensitive to the content of inert gases, but they increase slightly (not significant in Fig. 1.5) with increasing inert gas level. The reason is that argon and methane have higher solubilities than hydrogen and nitrogen.

Very few data sets exist with all components present. Zeininger [63] gives data for a mixture with hydrogen, nitrogen and methane. The data agree well with those at high pressure shown in Fig. 1.5 whereas there is some divergence at low pressure. In this region, however, the solubility is very low.

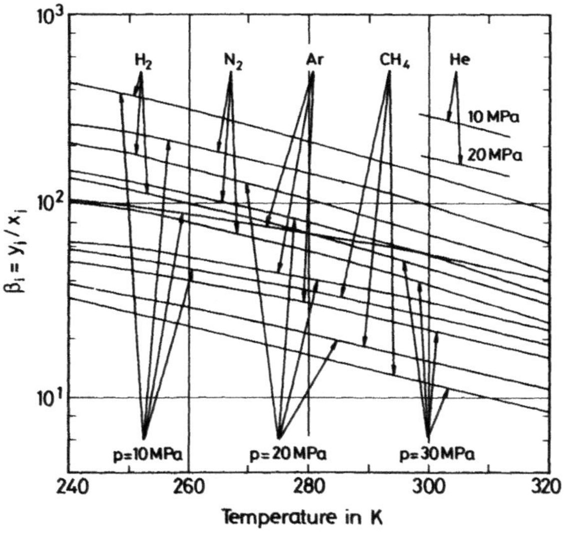

Fig. 1.5 Phase equilibrium β values for H_2-N_2-Ar-CH_4 mixtures (67.5% H_2, 22.5% N_2, 3% Ar, 7% CH_4).

1.6 Transport Properties

The transport properties to be discussed in the design of ammonia plants are the viscosity, the thermal conductivity, and the diffusion coefficients, which are used in calculation of transport rates in order to determine equipment sizes. The prediction of the properties for the gases are based on kinetic gas theory, whereas the properties of liquid ammonia are predicted by empirical correlations based on experimental data.

The properties of liquid ammonia with small amounts of dissolved gases are calculated by mixing the properties of the pure liquid NH_3 with the corresponding properties of the dissolved gases calculated by the kinetic gas theory as explained below. The mixing rule is the logarithmic mean.

The calculation of transport properties of gases can be divided into three parts similar to those used in the derivation of thermodynamic properties. The first part deals with the pure component low pressure values, which are functions of molecular parameters and temperature only. The second part is calculation of low pressure mixture properties using appropriate mixing rules, and the third part is the correction for influence of pressure. A detailed discussion of the kinetic gas theory is given in Hirschfelder et al. [23], and in Reid et al. [50]. The recommendations given in the latter reference are used closely in the following, since they give a good agreement with the available data for ammonia synthesis mixtures.

1.6.1 Low Pressure Properties

The pure component low pressure viscosity is predicted from the Chapman-Enskog equation which includes the effect of intermolecular forces. The equation for the viscosity η is:

$$\eta = 26.69 \frac{\sqrt{M_W T}}{\sigma^2 \Omega v} \tag{17}$$

where M_W is the molecular weight, T the temperature in K, σ the hard-sphere diameter in Å and Ωv the collision integral which is calculated from an intermolecular potential. Since ammonia is a polar gas it is necessary to use the so-called Stockmayer potential, which is identical with the Lennard-Jones potential but includes a term correcting for dipole-dipole interactions. The Stockmayer potential is then evaluated as a function of temperature. It contains two molecular parameters, the characteristic energy of interaction, and the dipole moment. The thermal conductivity of pure gases at low pressure can also be estimated from kinetic gas theory. The theory has been used by Bromley [6] to develop equations for the thermal conductivity of gases depending on the molecular structure. The equation gives the thermal conductivity λ, when the viscosity η is known in the form of the ratio $\lambda M_W/\eta$ which is equal to a function of the heat capacity and the various modes of vibration. The method of Bromley [6] gives acceptable results for the components present in ammonia synthesis gas.

1.6.2 Mixture Properties

The binary diffusion coefficients are also predicted using kinetic gas theory. The potential function used in evaluation of the collision integral is that given by Neufeld et al. [43] with the modification for polar gases given by Brokaw [5]. The low pressure mixture properties are calculated using an appropriate mixing rule, which also has been derived from kinetic gas theory. The formula includes a binary interaction term. For viscosity the interaction parameter given by Wilke [60] can be used. For thermal conductivity the form given by Lindsay, Bromley [35] for the binary interaction parameter can be used, and for the diffusion coefficients the formula given by Wilke [59] can be used.

1.6.3 Dependence on Pressure

The correction for the pressure dependence of transport properties of gases can be made by use of various correlations. These include the gas density which in turn is calculated from the compressibility factor. For viscosity the method proposed by Dean, Stiel [9] can be used, and for the thermal conductivity the method proposed by Stiel, Thodos [54] can be used. For the bulk diffusion coefficient it is normally assumed that it is inversely proportional to the density.

Viscosities of hydrogen-nitrogen and of hydrogen-ammonia mixtures have been reported in Pal, Barua [46]. Dembovskii [10] has reported viscosity data for mixtures of ammonia, hydrogen, and nitrogen. Thermal conductivities of nitrogen-hydrogen-ammonia mixtures are given by Golubev, Kiyashova [15]. Reference is also given to the work by Tsederberg [55]. Binary diffusion coefficients for hydrogen-ammonia and nitrogen-ammonia are given in Mason, Monchick [37]. The agreement between these experimental data and those calculated by the above-mentioned methods is acceptable for design purposes.

1.7 References

1. Alesandrini CG, Lynn S, Prausnitz JM (1972) Ind Eng Chem Process Design Develop 11: 253
2. Atroshchenko VI, Gavrya NA (1959) Zh Prikl Khim 32: 100; J Appl Chem [USSR] 32: 100
3. Bartlett EP, Cupples HL, Tremearne TH (1928) J Am Chem Soc 50: 1275

4. Bartlett EP, Hetherington HC, Kvalnes HM, Tremearne TH (1930) J Am Chem Soc 52: 1363
5. Brokaw RS (1969) Ind Eng Chem. Process Design Develop 8: 240
6. Bromley LA (1952) UCRL-1852: 1
7. Christiansen LJ, Kjær J (1982) Enthalpy Tables of Ideal Gases, Haldor Topsøe A/S, Copenhagen pp. 1
8. U. S. Department of Commerce (1923) Bur Std [U.S.] Circ No. 142: 1
9. Dean DE, Stiel LI (1965) Am Inst Chem Eng J 11: 526
10. Dembovskii VV (1968) Zavodsk Lab 34: 42; Ind Lab (USSR) 34: 52
11. Gillespie LJ (1925) J Math Phys 4: 84
12. Gillespie LJ, Beattie JA (1930) Phys Rev 36: 743
13. Gillespie LJ, Beattie JA (1930) Phys Rev 36: 1008
15. Golubev IF, Kiyashova VP (1979) Tr GIAP No. 52: 57
16. Haar L (1968) J Res Natl Bur Std A 72: 207
17. Haar L, Gallagher JS (1978) J Phys Chem Ref Data 7: 635
18. Haber F, Le Rossignol R (1907) Ber Bunsenges Physik Chem 40: 2144
19. Haber F, Tamaru S (1915) Z Electrochem 21: 191
20. Haber F, Tamaru S, Oeholm LW (1915) Z Electrochem 21: 206
21. Haber F, Tamaru S, Ponnaz C (1915) Z Electrochem 21: 89
22. Heise F (1972) Ber Bunsenges Physik Chem 76: 938
23. Hirschfelder JO, Curtis CF, Bird RB (1954) Molecular Theory of Gases and Liquids, Wiley, New York, pp. 1
24. Dow Chemical Corp (1971) JANAF Thermochemical Tables. 2nd Ed. NSRDS NBS-37: 1
25. Kaminishi G-I (1965) Intern Chem Eng 5: 749
26. Kazarnovskii YaS, (1945) Zh Fiz Khim 19: 392; C.A. (1946) 1727
27. Kazarnovskii YaS, Karapet'yants MK (1941) Zh Fiz Khim 15: 966; C.A. (1942) 6884
28. Konoki K, Takeuchi K, Kaminishi G-I, Toriumi T (1972) J Chem Eng Japan 5: 103
29. Landolt-Börnstein (1980) 6th Ed. Pt. 4C 2: 189
30. Larson AT (1924) J Am Chem Soc 46: 367
31. Larson AT, Black CA (1925) J Am Chem Soc 47: 1015
32. Larson AT, Dodge RL (1923) J Am Chem Soc 45: 2918
33. Lefrancois B, Vaniscotte C (1960) Chaleur Ind No. 419: 183
34. Lefrancois B, Vaniscotte C (1960) Genie Chim 83: 139
35. Lindsay AL, Bromley LA (1950) Ind Eng Chem 42: 1508
36. Martin JJ, Hou Y-C (1955) Am Inst Chem Eng J. 1: 142
37. Mason EA, Monchick L (1962) J Chem Phys 36: 2746
38. Michels A, Dumoulin E, Th. Van Dijk JJ (1959) Physica 25: 840
39. Michels A, Dumoulin E, Th. Van Dijk JJ (1961) Physica 27: 886
40. Michels A, Skelton GF, Dumoulin E (1950) Physica 16: 831
41. Michels A, Wassenaar T, Wolkers GJ, de Graaf W, Louwerse P (1953) Appl Sci Res A 3: 1
42. Michels A, Wassenaar T, Wolkers GJ, van Seventer W, Venteville AJ (1954) Appl Sci Res A 4: 180
43. Neufeld PD, Janzen AR, Aziz RA (1972) J Chem Phys 57: 1100
44. Newton RH (1935) Ind Eng Chem 27: 302
45. Nielsen A (1968) An Investigation on Promoted Iron Catalysts for the Synthesis of Ammonia. 3rd Ed., Gjellerup, Copenhagen, pp. 1
46. Pal AK, Barua AK (1967) J Chem Phys 47: 216
47. Reamer HH, Sage BH (1959) J Chem Eng Data 4: 152
48. Reamer HH, Sage BH (1959) J Chem Eng Data 4: 303
49. Reddy KV, Husain A (1980) Ind Eng Chem Process Design Develop 19: 580
50. Reid RC, Prausnitz, JM, Sherwood TK (1977) The Properties of Gases and Liquids 3rd Ed. McGraw-Hill, New York pp. 1
51. Rossini FD (1953) Selected Values of Physical and Thermodynamic Properties of Hydrocarbons and Related Compounds. Carnegie Press, Washington pp. 1
52. Sage BH, Olds RH, Lacey WN (1948) Ind Eng Chem 40: 1453
53. Schulz G, Schaefer H (1966) Ber Bunsenges Physik Chem 70: 21
54. Stiel LI, Thodos G (1964) Am Inst Chem Eng J 10: 26
55. Tsederberg NV (1965) Thermal Conductivity of Gases and Liquids, MIT Press, Cambridge, Mass., pp.1
56. Wiebe R, Gaddy VL (1938) J Am Chem Soc 60: 2300

57. Wiebe R, Treamearne JH (1933) J Am Chem Soc 55: 975
58. Wiebe R, Treamearne JH (1934) J Am Chem Soc 56: 2357
59. Wilke CR (1950) Chem Eng Progr 46: 95
60. Wilke CR (1950) J Chem Phys 18: 517
61. Winchester LJ, Dodge BF (1956) Am Inst Chem Eng J 2: 431
62. Zander, M. Thomas W (1979) J Chem Eng Data 24: 1
63. Zeininger H (1973) Chem Ing Tech 45: 1067

Chapter 2

Structure and Surface Chemistry of Industrial Ammonia Synthesis Catalysts

Per Stoltze
Haldor Topsøe Research Laboratories Lyngby, Denmark

Contents

Nielsen (Ed.), Ammonia

2.1 Introduction

This chapter deals with the structure and surface chemistry of industrial ammonia synthesis catalysts. Results on catalyst models and single crystal surfaces are included to the extent that they illuminate the behavior of industrial catalysts.

The industrial catalyst is prepared by fusion. The catalyst may be supplied in the unreduced state after crushing and screening to the desired particle size or the catalyst may be reduced and subsequently stabilized by controlled oxidation in the catalyst factory. Although the reduced catalyst is pyrophoric, the prereduced catalyst can be safely handled. In the ammonia synthesis plant the catalyst is activated by reduction with a mixture of hydrogen and nitrogen as the final step in the start-up procedure for the plant. The reduction of the prereduced catalyst is faster and simpler than the start-up of the unreduced catalyst.

A number of useful reviews on the structure and properties of ammonia synthesis catalysts [1–18] and on ultra-high vacuum investigations related to ammonia-synthesis [10, 19–25] have been published.

Catalyst constituents, which have little or no catalytic activity by themselves, but which increase the catalytic activity for the catalyst are referred to as *promoters*. Promoters which increase the catalytic activity primarily by increasing the active area of the sample are referred to as *structural* or *textural* promoters. Promoters which increase the activity of the catalyst primarily by increasing the reaction rate per area are referred to as *chemical* or *electronic* promoters. Constituents, which decrease the activity of the catalyst when present in small amounts are referred to as *poisons*.

Catalysts containing Fe, one structural promoter and no electronic promoter are commonly referred to as *singly promoted*. Catalysts containing Fe, one structural promoter and one electronic promoter are referred to as *double promoted*, while catalysts containing Fe, more than one structural promoter and one or more electronic promoters are referred to as *multiply promoted*.

In the following, the notation e.g. (Fe, Al, K) will indicate a sample containing the elements Fe, Al, K and possibly non-metallic elements. The sequence indicates the relative concentrations of the metals, the first metal being the most abundant. The asterisk ($*$) represents a surface site; $X*$ represents a species X adsorbed on a surface site.

2.2 The Unreduced Catalyst

The unreduced catalyst consists of oxides of iron with up to a few percent of Al, Ca and K. Other elements may be present in small amounts.

2.2.1 Structure

The unreduced catalyst is produced by melting a mixture containing the different elements. The main phase of the resulting product is magnetite [14, 15, 26–29]. The details of the preparation procedure [30, 31] and the homogeneity in the unreduced state [32–35] have been found to influence the properties. From an examination of the shapes of the magnetite grains [14, 28], it is apparent that this phase is the first to solidify during the cooling of the melt. The grain boundaries [36, 37] between the magnetite grains may contain several different phases which will be discussed below.

2.2.1.1 Magnetite

For an industrial catalyst, energy dispersive X-ray analysis, X-ray powder diffraction, optical microscopy and Mössbauer spectroscopy show that part of the Al and Ca atoms are dissolved in the magnetite lattice [15, 27, 28, 38–43]. The lattice constant of the magnetite phase of an industrial catalyst is 8.377 kX [15]. The lines in the X-ray powder diffractions diagram are broadened [15]; the broadening is independent of particle size [15].

From the Mössbauer spectrum [27, 44, 45] and the X-ray power diffraction diagram [39] of the unreduced catalyst, it has been estimated that 85% of the Al in the unreduced catalyst is dissolved in the magnetite. Evidence for the dissolution of K[41, 46], Mg[41, 46], V[41], Si[41], W[46], and Mo[46] in the magnetite has been reported. However, due to the large size of K^+, only a small amount of K is found in the magnetite phase of the industrial catalyst.

Additional information on the structure of the magnetite phase comes from the study of catalyst models, in particular of (Fe, Al) solid solutions. Mössbauer spectroscopic studies of Fe_3O_4 [47] and of unstochiometric Fe-spinels [44] have been reported.

For unreduced precipitated (Fe,Al) samples, solid solutions of Al in Fe-oxides and of Fe in Al-oxides may be observed [48, 49] depending on composition [48] and the preparation method [49]. In experimental (Fe,Al)-oxide samples, dissolution of Al in the magnetite has been shown by X-ray powder diffraction [50]. The lattice constant decreases from 8.413 Å to 8.365 Å for Fe_3O_4–Al_2O_3 up to 13 atom % Al_2O_3 and is then constant [51] indicating the formation of a saturated solid solution with segregation of excess Al_2O_3 as a separate phase.

2.2.1.2 Grain Boundaries

The grain boundary regions may constitute about 7% of the volume of the unreduced catalyst [46], and may contain much higher concentrations of promoter [14, 52] than the magnetite. The phases detected in the grain boundaries are small amounts of wustite [14, 15, 28, 40], calcium ferrites with dissolved promoters and a glass phase rich in silicon [26, 40, 41]).

2.2.1.3 Wustite

Based on X-ray powder diffraction studies, some authors have concluded that the structure of the wustite is that of natural wustite [15, 45]. From energy dispersive X-ray analysis and X-ray powder diffraction studies, others have found evidence for the dissolution of Al[28, 40], Ca[28], Mg[41], K[40], V[41], or Si[41] in the wustite. In fused (Fe,Al)-oxide samples the grain boundaries contain small but significant amounts of wustite [40].

2.2.1.4 Ferrites

Two calcium ferrites differing in their content of dissolved promoters may coexist in the grain boundaries [28, 41]. The dissolution of Al[28], K[28], Mg[41], V[41], or Si[41] in the calcium ferrites has been reported.

2.2.1.5 Glass Phases

The grain boundaries in the unreduced catalyst contain a glass phase. The main element of the glass phase is Si[26] or Fe[41]. The composition 17% SiO_2, 18% FeO, 9% CaO, 30%Al_2O_3, 24%K_2O[40] has been found from energy dispersive X-ray analysis.

2.2.2 Location of Promoters in the Unreduced Catalyst

While the composition of the solid solutions which constitute the unreduced catalyst is not known in detail, an extensive knowledge exists on the location of various additives both for catalyst models and for an industrial catalyst.

2.2.2.1 Aluminum

Evidence for the presence of Al dissolved in the magnetite phase has been found by X-ray powder diffraction [27, 28, 38–42, 53] and by chemical analysis of powders of varying particle size [54]. The solubility of Al in the $(Fe,Al)_3O_4$-phase has been determined to be 30 atom% Al[55] from measurement of the Curie temperature and 50 atom % Al[55], or 67 atom % Al[39] from measurement of the X-ray powder diffraction lattice constant. Other studies have indicated homogeneous solution of Al in magnetite, at least for small amounts of Al [56] and not too high temperatures [57].

In (Fe, Al) oxide catalyst models, some or all of the Al is dissolved in the Fe-oxide phases, as solid solutions between Fe-oxide and Al-oxide are readily formed [58, 59]. The solubility of Al in $(Fe, Al)_2O_3$ is < 15 atom % Al [58, 59] determined by X-ray powder diffraction, and < 9.2 atom % Al [58] determined by Mössbauer spectroscopy. In calcined (Fe,Co,Al) samples, Mössbauer spectroscopy shows αFe_2O_3 with dissolved Al_2O_3 for 0.10% Co and Fe-Co-spinel with a little αFe_2O_3 for 10–100% Co [60].

In the industrial catalyst, smaller amounts of Al are found dissolved in the calcium ferrites [28], and possibly in the wustite [28, 40] and in the glass phase [40].

2.2.2.2 *Calcium*

For an industrial catalyst, energy dispersive X-ray analysis, X-ray powder diffraction and optical microscopy indicate that Ca is found dissolved in the magnetite [28], in calcium ferrites [28, 41], possibly in the glass phase [40], and in the wustite [28]. Ca has been found in grain boundaries in a sintered (Fe,Ca) oxide catalyst [61].

2.2.2.3 *Potassium*

In the unreduced catalyst [11, 14, 40, 61, 62] and in (Fe,Mg) catalyst models [63], K is found in the grain boundaries. Additional amounts of K may be present as K-ferrites [64]. K has been reported to be associated with Si [54]. X-ray powder diffraction indicates that K is insoluble in magnetite [56].

The addition of Al makes the distribution of K more homogeneous [62], and the addition of Al or Zr decreases the volatility of K during preparation [65]. Both observations indicate the potential for K to react with acid oxides in the catalyst. CaO and SiO_2 decrease the water solubility of K [66]. Annealing increases the water solubility of K [66].

The amount of K which may be extracted by H_2O has been reported to increase [67, 68] or to decrease [66] by the addition of Si; the amount of K which may be extracted by H_2O is decreased [66] by the addition of Ca, and decreases [67, 68] or increases [66] by heating the unreduced catalyst. By scanning electron microscopy and energy dispersive X-ray analysis it was found that K segregates to the outer part of the catalyst particles with storage and prolonged use [69].

2.2.2.4 *Other Additives*

After sintering of a (Fe,Co,Al) based catalyst containing 15 atom % Co, Fe_3O_4 and $CoFe_2O_4$ were found [70]. In calcined (Fe,Co,Al) samples, Mössbauer spectroscopy shows αFe_2O_3 with dissolved Al_2O_3 for 0.10% Co, and Fe-Co-spinel with a little αFe_2O_3 for 10–100% Co [60]. The surface area of the unreduced sample decreases with an increasing Co content [71, 72]. For precipitated (Fe,Co,Al) the Fe(II):Fe(III) ratio increases [73] with increasing Co content.

In the unreduced catalyst, Mg is present dissolved in magnetite [63] and as an unidentified phase in grain boundaries [63].

The effect of Mn on the precursor has been studied [74],

In Mo containing samples, Mo has been detected as grains of K_2MoO_4 [75], $CaMoO_4$ [75], $FeMoO_4$ [75] $Fe_3(MoO_4)_3$ [76], or MoO_3 [76] depending on the composition and preparation procedure. Mo is soluble in magnetite [77].

For precipitated (Fe,Ni,Al), Ni increases the Fe(II):Fe(III) ratio [73].

In the unreduced catalyst, Si is found in the calcium ferrites [41] and in the glass phase [41, 26]. Smaller amounts may be present dissolved in the magnetite [41] or in the wustite [41]. Evidence for the association of Si with K has been reported [54]. Si is insoluble in magnetite in the absence of other promoters [56]. The addition of Si decreases the solubility of basic oxides in magnetite [56].

For W containing catalysts, W is found dissolved in the magnetite [46].

2.2.3 Texture

The size of the magnetite grains in the unreduced catalyst is rather variable [78]. The cross sectional area of magnetite in KM1 is $(1.28 \pm 0.14) \cdot 10^{-2} mm^2$ determined from planimetry [79]. The density is 4.8 g/cm^3 [14]. The porosity is negligible [14].

For the prereduced catalyst the density is 3.73 g/cm^3 [14]. The porosity is 0.11 cm^3/g, i.e., 41% [14], and electron microscopy [11, 14, 80] shows the presence of a well developed pore system.

Precipitated (Fe,Al) catalyst models in the unreduced state contain pores with a 19–20 Å radius and approximately 70 Å radius [48], while in the prereduced state these models have maxima in pore volume distribution at 20 Å and at 140 Å [48]. The 20 Å peak is thus unaffected by the reduction process.

The BET area is smaller for a catalyst prepared by sintering than for a similar catalyst prepared by precipitation [81]. The BET area decreases with increasing calcination temperature [58], and increases with increasing Al concentration [61] for precipitated catalysts. For samples calcined at 600 °C, the BET-area increases from 13 m^2/g at 0% Al_2O_3 to 215 m^2/g at 88% Al_2O_3 [48].

From a X-ray photoelectron spectroscopy study, the surface composition 3.2% Fe, 33.2% K, 8.4% Al, 3.9% Ca and 51.3% O (atomic %) was found [82] for the unreduced catalyst. The surface is enriched in K and Al compared to the bulk [52]. The prereduced catalyst shows more Fe in the surface by X-ray photoelectron spectroscopy than the unreduced catalyst [52]. Energy dispersive X-ray analysis of prereduced catalyst shows that the surface consists mainly of iron oxide [83]; Al, Ca and Si are inhomogeneously distributed in the surface [83].

2.2.4 Physical Properties

For unpromoted Fe-oxide the magnetization is maximum for Fe(II):Fe(III) = 0.50 (magnetite) [84]. The Curie temperature is constant for Fe(II):Fe(III) = 0.352 to 1.278 [84]. The Curie temperature decreases from 575 °C to 535 °C for Fe_3O_4 by the addition of up to 13 atom % Al_2O_3 [51]. The Curie temperature is 535 °C for higher Al-concentrations and is then constant [51]. For singly promoted samples containing Na, K, Cs, Ba, B, Al or Si, the Curie temperature is unchanged compared to unpromoted magnetite [84, 85]. The magnetization is reduced relatively to pure magnetite [84, 85]. For doubly promoted samples containing Al + Na, Al + K, B + K, Al + Ba, or Al + Si, both the Curie temperature and the magnetization is lower than for pure magnetite [84]. Addition of Li, Na, K, Mg, Ca, Ba or Si to (Fe, Al)-oxide samples either does not affect the Curie temperature, or decreases it slightly compared to (Fe, Al)-oxide [85].

The mechanical strength increases with increased Fe(II) content [86].

The heat capacity of an unreduced (Fe, Ca)-oxide sample displays a second order phase transition at 718 K [87].

The electrical conductivity of the unreduced catalyst has been studied [88].

2.3 The Reduction Process

The reduction of the industrial catalyst has been extensively studied, [89–92].

The reduced catalyst mainly consists of metallic Fe while the promoters remain in their oxidic state. The reduction process serves two purposes, firstly the surface of metallic iron is the active structure, and secondly the removal of the oxygen makes the material porous and increases the surface area by a large amount.

2.3.1 Reduction Temperature

During a temperature programmed reaction in hydrogen, an industrial catalyst is reduced in the range 500–900 °C with a maximum rate at 800 °C [73]. The prereduced catalyst reduces under the same experimental conditions at 523 °C [73]. The highest activity is reached by slow reduction in a specific temperature interval, this interval is 380–400 to 500–525 °C [93].

For $(Fe, Al)_2O_3$ with 5% Al, a peak at 950 °C is interpreted as the reduction of $FeAl_2O_3$ [73].

The reduction of 10% Fe supported on Al_2O_3 has been studied by Mössbauer spectroscopy [94]. At 460 °C Fe_2O_3 is reduced to Fe_3O_4, at 470 to 600 °C Fe_3O_4 is reduced to aluminates without consuming H_2, at 790 °C Fe(III) is reduced to Fe(0) and at 850 °C Fe(II) is reduced to Fe(0).

Cyanide based Fe catalysts reduce at 300–350 °C [95].

The reduction proceeds at a lower temperature [96] and higher catalytic activity is obtained for catalyst reduced as small grains.

For (Fe,Al) samples the reduction proceeds at higher temperatures and the activity is improved [12] or unchanged [19] after reduction in hydrogen + nitrogen mixtures compared to reduction in pure hydrogen. For unpromoted Fe the activity is improved [19] after reduction in mixtures of hydrogen and nitrogen compared to reduction in pure hydrogen

The kinetics of reduction for unpromoted magnetite in hydrogen is unaffected by the presence of water [97] while the reduction of an industrial catalyst is somewhat inhibited by water [97, 98]. The presence of water during the reduciton increases the average pore diameter [99] and decreases the activity of the catalyst. The activity of the catalyst increases with increasing space velocity during reduction [100]. The space velocity during reduction is more critical for samples reduced as small grains [100].

The rate of reduction increases with increasing Fe(II) content for (Fe,Al,K)-samples [101]. For multiply promoted samples [102, 103] the rate has also been reported to increase with increasing Fe(II) content. However, the rate does depend on the detailed structure of the catalyst.

2.3.2 Development of Properties During Reduction

2.3.2.1 Bulk Structure

The fully reduced catalyst consists of metallic Fe with little or no oxides of Fe present.

By the use of Mössbauer spectroscopy [27, 80, 104, 105] it has been found that wustite reduces before magnetite. The Fe formed from wustite has a more rounded shape than Fe formed from magnetite [28]. The reduction of the wustite leaves behind a system of pores with 600–2000 Å diameter [80]. For samples with an initially high concentration (27%) of wustite, the reaction zone is more diffuse than for samples of low initial wustite content.

The formation of wustite under conditions where wustite is metastable has been reported from magnetic measurements during reduction [106–108] and from X-ray powder diffraction [109] during reduction. Evidence against the formation of wustite has been reported from chemical analysis [110] and Mössbauer spectroscopic studies [27, 111].

For a (Fe,Al) based catalyst the presence of a paramagnetic phase under reduction has been detected by Mössbauer spectroscopy [112]; this phase was interpreted as $FeAl_2O_4$ inclusions in Fe [112]. Ca-ferrites are only partially reduced [40]. Mössbauer spectroscopy shows the formation Fe_3O_4 as an intermediate phase [111] during reduction of $(Fe,Al)_2O_3$ in H_2. A minimum in the reduction rate was observed [111] at a degree of reduction corresponding to the quantitative formation of Fe_3O_4.

2.3.2.2 Surface Properties

The outer shape of the grains is conserved during reduction and the porosity of the reduced sample thus has a simple relation to the amount of iron oxide in the unreduced sample.

For an industrial catalyst the pore volume [113, 114] increases during the entire reduction. The pore volume distribution of the reduced catalyst has a maximum at a pore radius of 100–120 Å and at a 260–430 Å [99, 115]. The smaller pores are formed by the reduction of magnetite [99]; the larger pores are formed by the reduction of wustite [99].

For an industrial catalyst the pore radius is constant at 50–100% reduction [114]. For (Fe,Mg), (Fe,Si), (Fe,Cr), (Fe,K) samples the pore radius is constant at 20–90% reduction and increases at 95–100% reduction [113, 116].

For an industrial catalyst the BET area increases during the entire reduction [114], increases at first and passes through a maximum at 90% reduction [113] or 95% reduction [117, 119].

After reduction, chemisorbed hydrogen has been detected by Laser Raman spectroscopy [120].

The CO chemisorption area increases slowly at 30–90% reduction and then increases rapidly at 96–100% reduction [113, 117–119, 121].

The oxygen chemisorption for an industrial catalyst is proportional to the degree of reduction [122] in the later stages of reduction.

The catalyst activity increases slowly at 30–90% reduction and then increases more rapidly than the CO-area at 96–100% reduction [117, 119]. This observation has been interpreted [117, 119] as the manifestation of the poisonous effect of the oxides on the surface.

Reduction at 1073 K increases both the NH_3 synthesis and the N_2-chemisorption rate; this has been interpreted as the reduction of Fe(II)-spinel [123].

2.3.3 Mechanism and Kinetic Models

The reduction of samples of an industrial catalyst with a small (4%) wustite content has been found by gravimetry [115, 124–126], Mössbauer spectroscopy [27, 127, 128], and electron microscopy [27, 80, 98, 115, 129, 130] to follow the core-and-shell model.

A kinetic model involving the rate of interface reaction, pore-diffusion and gas-film diffusion has been formulated for the core-and-shell reduction [115]. The enthalpy of adsorption of water on the reaction interface is 163.2 $\pm$ 45.2 kJ/mole in this model [115]. The core-and-shell model is invalid in the final stages of reduction for samples containing high (27%) concentrations of wustite [26, 27, 80] in the presence of small amounts of water [97, 131].

The reduction in a wet atmosphere has been found by scanning electron microscopy to follow the cracking-core model [98, 132].

As the main phase of the reduced catalyst is iron, the influence of a magnetic field on the reduction process has been studied [133].

2.3.3.1 Single Crystal Studies

The reduction of chemisorbed O* on Fe(100) by H_2 has been studied by Augeer electron spectroscopy and LEED at 473 and 673 K [134]. The reaction proceeds via the formation of H* [134]; no reaction is detected if the surface is completely covered by O* initially [134]. The apparent activation energy of the reaction is 59 $\pm$ 4 kJ/mole [134]. The transport of O to the surface is fast compared to the surface reaction [134].

2.3.4 Influence of Promoters on the Reduction Process

While the reduction of unpromoted Fe proceeds at a relatively low temperature and the reduced sample has a low porosity, the presence of the promoters leads to a much higher porosity for the reduced samples. However, the presence of the promoters also has a profound influence on the kinetics of the reduction process.

2.3.4.1 Aluminum

Al decreases the rate of reduction for (Fe, Al)-oxide samples [130], in particular in wet atmospheres [135, 136]. For Al_2O_3 supported catalyst models, the partial dissolution of Al during impregnation leads to a more difficult reduction [137].

2.3.4.2 Alkali

Mössbauer spectroscopic studies showed that alkali promotes the reduction of Fe_2O_3 to Fe at 300 °C [111, 138, 139]. Yet more direct measurements by temperature programmed reactions demonstrate that K decreases the rate of reduction for (Fe, K)-oxide samples [121] and for (Fe, Al, K)-oxide samples [121]. The rate of reduction increases through the sequence (Fe, Al, M), M = Li, Na, K, Rb, Cs for Fe_2O_3 based catalysts [139].

2.3.4.3 Other Elements

A number of elements, which are not present in an industrial catalyst, have been reported to affect the reduction. The elements which have been reported to increase the reduction temperature are Er [140], La [140], Mo [141–144], Pr [140], rare earths [145], Sc [140], and Sm [146–148], while the elements which have been reported to decrease the reduction temperature are Co [73], Cu [149], Ni [73, 150], Pd [151], and Re [152]. Ag [149] has been found to have no effect on the reduction.

2.4 Bulk Structure of the Reduced Catalyst

Some information on the structure and texture of the active catalyst may be inferred from studies of the spent catalyst [153], although the details of composition and structure may differ due to the violence of the reaction of the reduced catalyst with air and due to structural changes during the oxidation.

2.4.1 Iron

In the catalyst [14, 15, 27, 41, 112, 154] Fe is present in the reduced state mainly as the metal. The lattice constant of the iron is 2.8601 kX [15]. On the lean side of the gas phase equilibrium of the synthesis gas mixture Fe films do not form bulk nitrides [155]. On the rich side of the equilibrium Fe_4N may be formed [155].

In an industrial catalyst traces of unreduced Fe are detected by Mössbauer spectroscopy [27, 42]. These traces of Fe may be present in Ca ferrites with dissolved promoters [28, 41, 156] or in the glass phase [41, 156]. It has been suggested that the glass phase is inactive in the formation of the active catalyst [157].

For unpromoted Fe oxides, both Fe_3O_4 and FeO are completely reduced [40] at 550 °C in H_2.

For a number of (Fe,Al,K) and (Fe,Al,Cs) catalysts, in situ EXAFS and XANES [158, 159] indicate complete reduction of Fe in the reduced state of the catalyst. For an (Fe,Al)-oxide catalyst model containing 3% ASl_2O_3, Mössbauer spectroscopy indicates complete reduction of Fe [112] while for an (Fe,Al)-oxide catalyst model containing 10.2% Al_2O_3, traces of Fe(II) have been detected in the reduced state [154].

The degree of reduction for 0.05–15% Fe on Al_2O_3 is 77–97% after reduction in $3H_2 + N_2$ at 1 atm, 673 K [160]. This indicates that Fe supported on Al_2O_3 is more difficult to reduce that Fe promoted with Al_2O_3 and that (Fe,Al) oxide samples of high Al_2O_3 concentration may be poor models for the industrial catalyst.

The X-ray powder diffraction diagrams show line broadening for Fe due to particle size effects [14, 15, 27, 154]. The particle size for Fe is 300 Å [15]. Others have interpreted the in situ X-ray powder diffraction diagram as evidence for Fe being present as a metallic glass [161].

2.4.2 Aluminum

In the reduced catalyst Al_2O_3 have been shown by Auger electron spectroscopy [162] and X-ray photoelectron spectroscopy [52] to remain in its oxidic state. The Al-containing phases have been interpreted as Al_2O_3 segregated to the space between the Fe-crystallites based on evidence from chemisorption measurements [163] and X-ray powder diffraction studies [164]; as Al_2O_3 and $FeAl_2O_4$ based on evidence from EM and X-ray powder diffraction studies [48]; or as $FeAl_2O_4$ present as paracrystalline defects in the Fe crystallites based on evidence from secondary ion mass spectroscopy and Mössbauer studies [50, 154, 165, 166]. The presence of $FeAl_2O_4$ as paracrystalline defects has been challenged on the basis of Mössbauer spectroscopic studies [112, 165].

In reduced samples of $(Fe,Al)_2O_3$ with more than 2.4% Al_2O_3, Al_2O_3 is observed in the X-ray powder diffraction diagram [167].

2.4.3 Calcium

In the reduced catalyst CaO has been shown by Auger electron spectroscopy [162] and X-ray photoelectron spectroscopy [52] to remain in its oxidic state. Chemisorption measurements [163] and X-ray powder diffraction studies [164] show that during the reduction CaO segregate to the space between the Fe crystallites.

2.5 The Surface Structure of the Reduced Catalyst

The outer shape of the grains is conserved during reduction [28, 110, 113, 156, 168, 169], expands by up to 0.6% [168] or contracts by up to 0.5% [168] during reduction depending on the composition and reduction temperature.

2.5.1 The Pore System

The density of the reduced catalyst is 2.7–3.7 g/cm^3 [14]. The pore volume is 0.15 cm^3/g [67, 68], independent of the K concentration [[67, 68].

For the reduced catalyst, electron microscopy shows the presence of a well developed pore system [92, 170, 171]. An industrial catalyst with pores of 100 and 300 Å radius [14, 115] in the reduced state has been found to have 177 Å pores [172] after passivation. The reduction of promoted magnetite creates a system of pores with < 400 Å diameter [80]. The average pore radius increases with increasing K concentration from 50–200 Å (0% K) to 200–800 Å (0.78% K) after reducing at 600 °C [67, 68].

Precipitated (Fe, Al) catalyst models have pores with 19–20 Å radius and larger pores of 50–250 Å radius [48, 167].

2.5.2 Particle Size

The particle size has been studied by X-ray powder diffraction [173]. The cause of the line broadening in the X-ray powder diffraction diagram has been assigned to particle size effects [164], to the combined effect of particle size, defects and strain [174], or to the presence of paracrystalline defects [166, 173, 175].

If the line broadening is assigned to particle size effects alone, the calculated particle radius is 100–1000 Å [53], 175–250 Å [174], or 180 Å [164]. For catalyst models the particle radius is 305 ± 15 Å for (Fe, Al) [154] 125–155 Å for (Fe, Mg) and 305 Å for (Fe, Mg, K) [176].

If the line broadening in the X-ray powder diffraction diagram of an industrial catalyst in the reduced state is assigned to the presence of paracrystalline defects, the calculated particle radius is 200 Å [173, 175]. The calculated size of the paracrystalline defects is the same in both the unreduced and in the reduced states [173].

2.5.3 Surface Area

Physisorption measurements using N_2, CO, Ar, O_2 or CO_2 give the same area [177]; the areas determined by this method are 0.44–10.4 m^2/g [178]. The BET area for the reduced catalyst depends on the composition and structure prior to reduction and on the conditions during the reduction. Consequently very different values have been reported: 8 m^2/g [164], 11.6 m^2/g [179], 15.8 m^2/g [179], 15 m^2/g [93], or 20.9 m^2/g [180]. After passivation an area of 13.1 m^2/g found [172].

For an industrial catalyst the calculated area of the Fe crystallites based on the particle size obtained from X-ray powder diffraction is 2.6 times larger than the measured BET area of the sample [164].

The surface concentration of Fe in the reduced catalyst as determined by electron spectroscopy is 4% [162], 6.4% [82], or 15% [52].

The BET area of the catalyst increases with increasing Al content [48]. The increase is most pronounced at small Al concentrations. This has been attributed to the limited solubility of $FeAl_2O_4$ in Fe [175].

An increase of the wustite content in the unreduced sample has been reported to increase the BET area [181], to have no effect on the BET area [182] or to decrease the BET area [183, 184] of the reduced sample. Analogously an increase of the wustite content in the unreduced sample has been reported to increase [102, 180, 182, 184–187] the activity, to have no effect on the activity [180] or to decrease the activity [101, 102, 181, 188] of the reduced catalyst. The existence of a maximum in catalyst activity with respect to wustite content has been reported [188–190].

For (Fe, Al) catalyst models the BET area after reduction is 5.57 [191] or 5.86 m^2/g(Fe, Al) [191]. For unpromoted Fe the BET area after reduction is approx 1 m^2/g [192].

The stability of an industrial catalyst at temperatures higher than normal has been studied. High temperatures causes a decrease in the BET area [113, 193] and an increase in the average pore diameter [113]. The BET area and the CO chemisorption area remain proportional [193]. A phase rich in K, Al and Ca [28] seems to be formed.

The stability toward impure synthesis gas has been studied. Under exposure to water at synthesis temperatures the BET area [194] and the CO area [194] decreases.

2.5.4 Effect of Promoters

The cause of the structural promotion has been assigned to the segregation of refractory oxides to the surface [53, 154, 195, 196] or to the formation of paracrystalline defects [154, 166]. Theoretical considerations show that the structural promoters must be located near or on the surface to have any effect [197].

2.5.4.1 Aluminium

Al is a structural promoter [67, 68, 198–204]. The distribution of Al in the sample changes during reduction [40] and, in particular, the segregation of Al to the surface has been the subject of a large number of studies.

From studies by chemisorption measurements [163, 198, 199, 205–207], by X-ray powder diffraction studies [164], by electron microscopy [92, 170, 171], by X-ray photoelectron spectroscopy [205, 206, 208], and by scanning Auger electron spectroscopy [52, 209–211] it was concluded that Al_2O_3 is present as a thin surface layer in the reduced catalyst. Others have concluded that Al is segregated as both $FeOAl_2O_3$ and Al_2O_3 [53]. Al has been shown by Auger electron spectroscopy [162] and X-ray photoelectron spectroscopy [52] to remain in the oxidic Al(III)-state.

The content of Al in the catalyst is sufficient for a monolayer coverage [53] on the BET area. From chemisorption measurements for a (Fe, Al) sample it was concluded that 1% Al_2O_3 in the precursor leads to 35% total surface coverage [198, 199], while 10% Al_2O_3 in the precursor leads to 55% total surface coverage [198, 199].

For concentrations of Al in excess of the optimum, Al may be observed in bulk phases such as Al_2O_3 [167] by X-ray powder diffraction. Evidence for the existence of paracrystalline defects in reduced (Fe, Al) samples [212] has been found.

While an increase in Al content causes a monotonic increase in total area, the active area has been found to increase [198, 199] or decrease [67, 68]. The ratio of the catalytic activity to the BET area decreases [202, 203, 213] with increasing Al concentration. For this reason no simple relation exists between Al content and activity [65].

For (Fe, Al) catalysts the catalytic activity is maximum at 3.5–4 % Al_2O_3 [201], 3–4 % Al_2O_3 [214], 2.5–5 % Al_2O_3 [48], or 3 % Al_2O_3 [67, 68]. It has been suggested that the increase in activity with the Fe(II):Fe(III) ratio is caused by Al being more soluble in magnetite than in hematite [57, 215]. A maximum in BET area has been found by the addition of 2.5 % Al_2O_3 [67, 68] or 5–6 % $KAlO_2$ [204]. The addition of Al to multiply promoted samples decreases the pore radius [216]. Al increases the work function [217] for (Fe, Al, K) samples. These changes may be caused by changes in the K-coverage following the changes in Al content [217].

2.5.4.2 Calcium

Ca is a structural promoter [198, 199, 201]. The effect of Ca on the properties of the catalyst has been much less studied than the effect of Al.

Ca has been shown by Auger electron spectroscopy [162] and X-ray photoelectron spectroscopy [52] to remain in the oxidized state. Chemisorption measurements [163] and X-ray powder diffraction studies [164] show that CaO is segregated to the space between the Fe crystallites during the reduction. The segregation of Ca to the surface has been demonstrated by scanning Auger electron spectroscopy[52, 209–211].

Ca has been reported to increase the activity [218], to increase [198, 199] the surface area, and to increase the resistance toward impurities in the gas [67, 68].

The optimum Ca concentration in a (Fe, Al, K, Ca)-oxide catalyst is 2–2.5 % [201].

2.5.4.3 Potassium

K is an electronic promoter. K acts as an promoter both by impregnation and by addition to the melt [219].

The segregation of K to the surface has been demonstrated by chemisorption measurements [207, 220], by scanning Auger electron spectroscopy [52, 209–211], by X-ray photoelectron spectroscopy [208], and by electron microscopy [92]. Single crystal studies of K overlayers on Fe(110) demonstrate that K is not a structural promoter [221] and that K may even reduce the ability of Al to disperse Fe [221].

The migration of K to the surface of the reduced catalyst [40] has been demonstrated by energy dispersive X-ray analysis [28], by field iron mass spectroscopy [222, 223], by chemisorption of CO, CO_2, $N_2$and H_2 [207, 220] by scanning Auger electron spectroscopy [209–211], and by high-voltage electron microscopy [224].

For the catalyst it has been concluded that KH, KNH_2 and K_2O are less stable than KOH under NH_3 synthesis conditions [225, 226]. These considerations were based on the bulk phase thermodynamics.

For K adsorbed on an Al_2O_3 overlayer on Fe(100), the desorption temperature is increased to 600 °C [227], well above the typical reaction temperature [227].

On clean single crystal surfaces X-ray photoelectron spectroscopy and ultraviolet photoelectron spectroscopy show that K is present as the metal [228]. On Fe(100) [229, 230] chemisorbed K is disordered. On Fe(110) a hexagonal close packed structure is formed [228]. On Fe(111) K is disordered [230] or forms a (3×3) structure [231]. At room temperature it is unlikely that multilayer adsorption will occur [228] due to the low heat of vaporization for K.

K chemisorbed on Fe(100) desorbs under NH_3 synthesis at 20 atm. [232]. This is consistent with typical temperatures for NH_3 synthesis being above the desorption temperature for K. K∗/Fe(100) is stabilized by the presence of O∗ [232].

The presence of a K or a K + Al overlayer does not cause a recrystallization or an increase in catalytic activity after steaming and reduction in the NH_3 synthesis gas [233].

The migration of K on Fe/Al_2O_3 has been studied by Auger electron spectroscopy [234]. The migration is faster in H_2 than in O_2, and faster in moist gas than in dry. The kinetics of surface migration was found to be consistent with a surface diffusion mechanism [234].

Addition of small amount of K results in an increase in average pore diameter [67, 68, 235], no changes in the pore volume [67, 68], an decrease in the BET area [67, 68, 176, 236–241] a decrease of the active area [67, 68], an increase in average particle diameter [176], a decrease [176, 217, 237, 242–244] or an increase [243] in the work function. For single crystal surfaces at small K-coverages, the work function decreases with increasing K coverage [228, 229, 245].

The importance of K for the kinetics of NH_3 synthesis is discussed further in 7.1.2.

2.5.4.4 Other Promoters

Besides Al and Ca a number of elements have the ability to act as structural promoters, either alone or as a part of a mixture.

The following elements have been reported as structural promoters: Ba [246], Be [198, 199], Ce [200, 247, 248–250], Dy [200, 248–252], Er [252], Eu [250], mixtures of rare earths [132, 145, 253–256], Ho [252], Mg [56, 176, 198, 199, 202, 257–259], Mo [31, 76, 77, 141, 144, 260–270], Sc [262, 271, 272], Si [67, 68], Sm [147, 148, 273], U [274, 275], V [264], W [77, 141, 261, 263, 265, 270, 276], Y [277–280], Yb [277, 278, 280], and Zr [202, 239, 260, 278, 281].

Cr has been found to be structural promoter [198, 199, 260, 282, 283], while others have found that Cr has no effect on the catalytic activity [284] or acts to reduce the catalytic activity [285]. The large reactivity of Cr toward N_2 [286] and the considerable stability of nitrides of Cr [287] are complications in the deduction of the effect of small amounts of Cr added to the catalyst. Undoubtedly, this complication is the cause of much confusion in the study of the effects of trace amounts of a number of transition metals on the activity of ammonia synthesis catalysts.

The BET area increases through the sequence (Fe,Al), (Fe,Ti), (Fe,Cr), (Fe,Mg), (Fe,Mn), (Fe,Ca), (Fe,Si), (Fe,Be) for a small promoter content [198, 199]. The BET area increases with the promoter content [198, 199].

The alkali metals are electronic promoters, increasing the activity [139, 240, 288, 289], and decreasing the area [240]. The rate of synthesis increases through the sequence (Fe,Al,M), where M = Li, Na, K, Rb, Cs [139, 288, 290]. A decrease in the work function generally results in an increase in catalytic activity [291, 217, 242].

2.5.5 Models of Structural Promotion

Single crystals of Fe with evaporated overlayers of K and/or Al have been studied as models for the structure of the industrial catalyst. After heating of an Fe(110) crystal with an Al overlayer, the surface looks rough under the electron microscope [227]. The reaction leading to a dispersion of Al and Fe is primarily a reaction between oxidic phases [221]. The activity increases after reaction between the crystal and the Al-overlayer [227], possibly due to the formation of facets [227]. After heating of an Fe(110) crystal in 5 atm of N_2 to 450 C, no increase is found in the activity of the crystal [227]. The presence of Al stabilizes the Fe(111) and Fe(211) surfaces during NH_3 synthesis at 20 atm [221]. The heating of Fe(110) and Fe(100) with an Al overlayer causes a recrystallization of the Fe during heating in steam followed by reduction in NH_3 synthesis gas [233]. After reduction the surface is as active as F(111). The increase in activity is assigned to the formation of high index planes. For Fe(111) with an Al overlayer, the resulting increase in activity is very small [233].

Fe evaporated on a nonporous Al_2O_3 film has been studied as a model for structural promotion [292–296]. After heating in O_2, transmission electron microscopy shows a multilayer film of Fe-oxide coexisting with 3-dimensional crystallites of Fe [294]. After reduction of oxidic Fe/Al_2O_3 in H_2 at 500 °C with traces (1 ppm) of H_2O or O_2, torus-shaped crystallites are formed [295]. The composition of the tori approach the composition $FeAl_2O_4$ and/or $Al_2Fe_2O_6$. The core consists of less oxidic material [295]. After heating to higher temperatures the shapes are less distinct [295]. After prolonged heating the crystallites fragment, possibly due to mechanical fatigue [295]. Fe in oxidic form is present even after prolonged reduction at high temperatures [296]. The stabilization of oxidic Fe-compounds implies that the (Fe, Al) interfacila energy is small [296]. After heating in steam to 700 °C, no Fe particles are found by transmission electron microscopy [293], presumably because the Fe spreads as a thin film [293]. After reduction small Fe crystallites are found [293, 295]. The crystallites are smaller than the crystallites present before oxidation and reduction [293, 295]. The crystallites formed after oxidation and reduction are only smaller than the original crystallites if the reduction is of short duration [293, 295]. Sintering of the particles is caused by coalescence and by Ostwald ripening [296].

A large number of other catalyst models have been studied. The mostly frequently studied models are Fe/C [297–316]. Fe/Al_2O_3 [94, 108, 137, 317–321], Fe/MgO [108, 167, 319, 320, 322–330], Fe/SiO [319–321, 331, 339] Fe/TiO_2 [340, 342], FeTi intermetallics [343, 349] and FeZr intermetallics [350, 354].

2.5.6 Alloying with Transition Metals

A number of dilute alloys of transition metals in Fe are active for ammonia synthesis. The elements reported to form Fe alloys active for ammonia synthesis are Co [60, 123, 194, 355–363], Cu [338, 364], Ir [365], Ni [216, 268, 269, 356, 357, 363, 366–371], Os [372], Re [152, 264, 373], and Ru [264, 336, 363, 372, 374].

The information on the effect of alloy formation on the catalytic activity is generally conflicting. The effect of Sc during superheating of the catalyst is interpreted as an activity increase due to reduction of Sc_2O_3 to Sc followed by an activity decrease due to Sc + Fe alloy formation [271]. Small amounts of Co increases the rate of NH_3 synthesis and N_2-chemisorption [123]. Small amounts of Ni increases only the rate of NH_3 synthesis, not the rate of N_2 chemisorption [123].

A promising technique for the study of the catalytic activity of alloys is the study of chemisorption on iron overlayers on single crystals of other metals such as Fe/Ru [375], Fe/Re [376], and Fe/W [377].

2.6 Chemisorptive Properties of the Catalyst

An important step in heterogeneous catalysis is the adsorption and desorption of reactants and products of the reaction. Important information on the mechanism of ammonia synthesis has come from the study of the adsorption of H_2, N_2 and NH_3. The adsorption of H_2O and O_2 is interesting because of the role of H_2O as a poison for NH_3 synthesis.

As mentioned above, not all of the surface of the catalyst is active. The adsorption of other gases, such as CO and CO_2, is interesting because of these gases adsorb selectively on the catalyst surface.

The presence of adsorbed atoms on a metal surface may have consequences for subsequent adsorption of another gas. There are several possible outcomes of this procedure. The preadsorbed atoms may weaken or prevent the subsequent adsorption; the strength of adsorption for both species may increase; a compound may be formed or the preadsorbed atoms may be displaced to the bulk or to the gas phase.

While the reactions of the catalyst may be fairly complicated, studies of chemisorption on single crystals has resulted in a detailed understanding of the more important adsorption reactions.

2.6.1 Chemisorption of H

H_2 is a reactant in several reactions of the catalyst surface. The role of H_2 in the reduction of the surface has been treated above. The role of H_2 as a reactant in the synthesis of NH_3 will be treated below. The present section will treat the adsorption and desorption of H_2, the *ortho-para* conversion, and the $H_2 + D_2$ isotopic exchange.

2.6.1.1 Structure of Chemisorbed H

On single crystals a c(2 × 2) low-energy electron diffraction pattern is observed for H/Fe(1 1 0) at 140 K, 7 L [378] and a p(1 × 1) low-energy electron diffraction pattern is observed for H/Fe(1 1 0) at 140 K, 3500 L [378]. Disorder is observed for H/Fe(1 0 0), H/Fe(1 1 0) and H/Fe(1 1 1) [378] at higher temperatures. The complete $H_2 + D_2$ isotopic scrambling upon adsorption–desorption [378] indicates that H_2 is adsorbed as *atoms*. For coverages higher than the c(2 × 2) structure, adsorbate–adsorbate interactions are reflected in the low-energy electron diffraction structures observed at coverages between Fe:H = 1:0.25 and 1:1 [378].

High-resolution electron-energy-loss spectroscopy spectra of c(2 × 2) H/Fe(1 1 0) [191] are interpreted from selection rules, isotope effect, and wave number as $H*$ in the short bridge position [191].

By secondary ion mass spectroscopy from H_2 adsorbed on Fe the signal for H_2 is stronger than for H [379]. This is taken as evidence that H_2 does not dissociate [379]. However, this argument cannot be correct in view of the large difference in stability between H-atoms and H_2-molecules in the gas phase.

For the chemisorption of hydrogen on the catalyst, Emmett *et al* found a complex behavior [163, 380, 381]. Adsorption of H_2 on the catalyst was detected at −90 °C and above +100 °C [381]. Transients in the adsorption when the temperature is suddenly changed in the range 0–210 °C has also been observed by others [382]. Presumably this behavior is caused by two reactions where the low temperature reaction is a weakly exothermic equilibrium adsorption and the high temperature reaction is a reaction limited by a high activation energy.

2.6.1.2 Thermodynamics

The initial enthalpy of chemisorption has been determined for single crystal surfaces of Fe. For H/Fe(1 0 0) −86 kJ/mole was found by temperature programmed desorption [383]; for H/Fe(1 1 0) −109 kJ/mole [378] by temperature programmed desorption or −24.2 kcal/mole [384] by He-scattering; for H/Fe(1 1 1) −88 kJ/mole by temperature programmed desorption [378]. For D/Fe(1 1 0) −24.2 kcal/mole has been determined from He-scattering [384] and for D/Fe(1 1 1) −104 kJ/mole by calorimetry [385].

Adsorbate-adsorbate repulsion at high coverage is reflected in the low-energy electron diffractions patterns, in the occurrence of temperature programmed desorption peaks, and in the enthalpy of chemisorption.

For hydrogen adsorption on polycrystalline Fe, the enthalpy of chemisorption has been determined as −98 kJ/mole by calorimetry [385], −96 kJ/mole by calorimetry [386], −81.0 kJ/mole by electrical conductivity [387], −85.0 kJ/mole by volumetric chemisorption [388], −36 kcal/mole at zero coverage by volumetric chemisorption [389], −15 kcal/mole at 90% saturation [389], −20 kcal/mole at 140 K, zero coverage [390], −17.5 kcal/mole at 1% of saturation calculated from the equilibrium pressure [391], and −5 kcal/mole at 10% of saturation calculated from the equilibrium pressure [391].

For Fe/MgO the saturation coverage depends on particle diameter [328], i.e., the chemisorption appears to be structure sensitive. Structural sensitivity of NH_3 synthesis will be discussed further in Sect. 7.1.1.

The studies of the chemisorption of H_2 on the catalyst has been complemented by studies at low [391, 392] and at high pressures [393].

Calorimetry on a (Fe, Al, Ca, K) sample at 35 °C yields a Freundlich isotherm [394]. The decrease in heat of chemisorption is the same for promoted and unpromoted samples [391]. By volumetric chemisorption the adsorbed phase was found to behave as a 2 dimensional gas with a constant enthalpy of chemisorption, − 85.0 kJ/mole, up to 3.3 μmole/m^2 [388].

For an industrial catalyst the enthalpy of chemisorption is − 25 kcal/mole by calorimetry [395], − 28 kcal/mole at zero coverage from calorimetry [394], and − 12.3 kcal/mole calculated from calorimetry of NH_3 cracking [395].

From a thermodynamic model [396] of the chemisorption of H_2 we have found that under an equilibrium pressure of 1 atmosphere, the coverage by H∗ decreases from ~1 at 500 K to ~0.15 at 1000 K. By calculating the outcome of a volumetric chemisorption experiment from a model of the kinetics of ammonia synthesis [396], it is concluded that H∗ is by far too weakly adsorbed to be used in a titration of the number of active sites on the catalyst surface.

2.6.1.3 Adsorption Kinetics

For single crystal surfaces the sticking coefficient for H_2 into 2H∗ is 0.03 on Fe(1 0 0) at 250 K [383], 0.16 on Fe(1 1 0) at 140 K [378], and unity on Fe(pc) at 78–298 K [397]. The activation energy for Fe(1 1 0) is zero [384] or 0.70 kcal/mole [398].

The presence of H–H interactions [399] on the surface results in the formation of ordered domains at low temperatures. The kinetics of domain growth for H/Fe(1 1 0) has been studied by Monte Carlo simulations [400]. At $\theta = 0.500$ the domains grow according to an Allen-Cahn power law; the size of the domains is proportional to $t^{0.50}$ [400]. At $\theta = 0.667$ the rate is lower due to diffusion limitations [400]. The diffusion and reactions of H∗ at the surface has a number of similarities to the diffusion of H in the bulk [401].

For a film of Fe the rate of adsorption for H_2 at low temperature is proportional to the square root of the hydrogen pressure [402] and decreases exponentially with coverage [402]. The activation energy is 3–6 kcal/mole and increases with coverage [402]. The H_2 chemisorption is activated for small Fe particles on MgO [328], and for (Fe, Ir)/Al_2O_3 [365].

2.6.1.4 Desorption Kinetics

The temperature programmed desorption peak maxima at low coverage are H/Fe(1 0 0) 400 K [378], H/Fe(1 1 0) 480 K [378], H/Fe(1 1 1) 370 K [378] H/Fe(pc) 430 K [386]. For both H_2 and D_2 the desorption shows second-order kinetics [398, 384].

Additional peaks caused by adsorbate-adsorbate repulsion are seen for H_2 desorbing from single crystal surfaces at high initial coverages [378, 386].

The desorption temperature is 150 °C [403] or 120–170, 280–380 and 480–540 °C [404] for (Fe, Al) samples; 120–170, 280–380 and 480–540 °C for (Fe,

Al, Ca) samples [404]; and 120–170, 280–380 and 480–540 °C for (Fe, Al, Ca, K) samples [404].

Temperature programmed desorption of H_2 from the reduced catalyst has been reported to proceed at 90–240 °C [405] or to show 4 peaks [406]. The kinetics of thermal desorption of H_2 from the catalyst is second order [407]. Evidence that chemisorbed hydrogen remains on the surface after evacuation at 550 °C for 24 h has been reported [408].

2.6.1.5 Properties of Chemisorbed H

For H∗ on a Fe surface the ultra violet photoelectron spectroscopy (1s) peak is found at −5.6 eV [378].

High-resolution electron-energy-loss spectroscopy spectra of c(2 × 2) H/Fe(1 1 0) [191] are interpreted from selection rules, isotope effect, and wave number as H∗ in the short bridge position [191]. The 2 bands are interpreted as the symmetric Fe–H–Fe stretch at 1060 cm^{-1} and the asymmetric Fe–H–Fe stretch at 880 cm^{-1} [191].

Chemisorbed H decreases the work function for industrial catalysts [243, 279]. The effect of hydrogen on the work function has been reported to depend on the pressure [409]. There is evidence that the measured values of the work function change is distributed by the lack of mobility at 140 K [378]. Whereas if the temperature is increased, significant desorption takes place before equi-li-brium is established [378].

Magnetic measurements of hydrogen chemisorption on Fe/SiO_2 are reversible at 210 torr, 310 °C [410]. For 15 Å Fe/MgO, H∗ affects the magnetic moment below the superparamagnetic transition. Above this temperature no effect is found [326] for 80 Å Fe/MgO, H∗ does not affect the magnetic moment [326].

2.6.1.6 Effect of Promoters

For Fe single crystals the saturation coverages by H∗ is not effected by the presence of K [411], whereas for (Fe, Al) samples the amount of hydrogen chemisorption is increased by K-promotion [391, 412]. For Fe/Al_2O, tem-pera-ture programmed desorption shows that K increases the strength of adsorption for H [412].

At low coverages by K, preadsorbed K on Fe(1 0 0) increases the sticking coefficient for hydrogen [383]. The increase indicates that the sticking co-effi-cient is unity for K-promoted sites [383].

The heat of adsorption is increased by 6–10 kJ/mole for K promoted sites on Fe(1 0 0) [383, 411] and 8 kJ/mole for K-promoted sites on Fe(1 1 1) [411]. Even at low coverages by K, the temperature programmed desorption spectra of hydrogen desorbing from H/K/Fe(1 0 0) or from H/K/Fe(1 1 1) are not split into peaks assignable to promoted or unpromoted sites respectively [411].

The hydrogen chemisorption becomes activated when Al is present [193, 321]. A plausible explanation for this observation is that the structural promoters decorates the surface [321].

2.6.1.7 *Effect of Preadsorbed Species*

Preadsorbed O_2 inhibits H_2 chemisorption [207, 413] and decreases the rate of adsorption [383]. On p(1 × 1) O/Fe(1 0 0) the initial sticking coefficient for hydrogen is $1.0 \cdot 10^{-4}$ at 200 K [383].

The reaction between H_2 and preadsorbed O∗ is an important model for the reduction of the catalyst. For Fe(1 0 0) this reaction has been studied at 473 and 673 K at $2 \cdot 10^{-2}$ torr of H_2 [134]. The apparent activation energy is 59 ± 4 kJ/mole [134]. The reaction proceeds via the dissociative chemisorption of H_2 [134]. For this reason it is not unexpected that the reaction becomes quite inhibited by the presence of a monolayer of O∗ [134].

Preadsorbed S decreases the rate of adsorption. No H∗ is adsorbed on c(2 × 2) S/Fe(1 0 0) after 2000 L exposure at 200 K [383].

Preadsorbed C decreases the rate of adsorption. On c(2 × 2) C/Fe(1 0 0) the initial sticking coefficient for hydrogen is 10^{-3} at 200 K [383].

Preadsorbed CO inhibits H_2 chemisorption [207, 406], but has no effect on the enthalpy of chemisorption for H_2 [389].

Preadsorbed CO_2 inhibits H_2 adsorption at −78 °C [207].

For catalysts the results obtained when N_2 is adsorbed on preadsorbed H∗ and vice versa are not readily interpreted [380]. For single crystals the results are easier to interpret. When preadsorbed H∗ is exposed to N_2(g) at 77 K, N_2(g) is displaced [386, 414] and physisorbed N_2 may be formed on top of the H∗ layer [386].

By coadsorption of N_2 and H_2, FeN_2 is seen in secondary ion mass spectroscopy [379]. An Eley-Rideal mechanism has been suggested,

$$N_2{*} + H_2 = N_2H_2 \tag{1}$$

[379]. This conclusion is in disagreement with more direct studies of the chemisorption of N_2H_2 (Sect. 2.6.9) and of the mechanism of ammonia synthesis (Sect. 2.7).

Preadsorbed N∗ does not inhibit dissociative adsorption of H_2O [415]. Preadsorbed N∗ has been reported to decrease [207, 416, 406, 414] or to increase [417] the amount of chemisorbed hydrogen. Preadsorbed N∗ has been found to decrease [389], to increase [394] or to have no effect [414] on the heat of adsorption for H_2 [389]. Presumably, the results of the coadsorption of N∗ and H∗ depends sensitively on the experimental conditions as both displacement to the gas phase and the formation of NH_3 are realistic possibilities.

The presence of 0.1% N_2 in the reduction gas used for chemisorption measurements decreases the H_2 chemisorption by 25–30% [193, 380].

Preadsorbed N∗ is removed as NH_3 upon exposure to hydrogen [418, 419].

The synthesis of ammonia by hydrogenation of chemisorbed N∗ is faster than by the reaction of H_2 and N_2 over the same catalyst [420, 421]. This suggests that N_2 and $N*$ are not in equilibrium under ammonia synthesis conditions. This conclusion is in agreement with more direct studies of the mechanism of ammonia synthesis (Sect. 7).

^{15}N isotopic labeling shows that all chemisorbed N-atoms undergo hydrogenation with equal probability [422]. This suggest that all adsorbed $N*$ atoms have identical properties, i.e., the hypothesis that the surface should be strongly heterogeneous (Sect. 2.7.6) may be incorrect.

The rate of $N*$ removal from bulk Fe by reaction with H_2 is first order in $N*$ [423].

2.6.1.8 *The Hydrogen ortho-para Conversion*

The nuclear spin of H results in the existence of H_2 in two states, a singlet state called *ortho* and a triplet state called *para*. The interconversion of these two forms is a peculiar reaction as it requires the interaction with a magnetic substance but does not require the fission of the H–H bond.

The hydrogen *ortho-para* conversion is fast on (Fe, Al, K) and (Fe, Al, Si, Zr) samples at $-195\,°C$ [424]. The conversion over (Fe, Al, K) and (Fe, Al) is $\geq$ first order in H_2 [425].

The *ortho-para* conversion is not poisoned by preadsorbed $H*$ on an (Fe, Al, Si, Zr) sample [424] but has been found to be poisoned by preadsorbed H on an (Fe, Al, K) sample [424, 405].

2.6.1.9 *$H_2 + D_2$ Exchange*

The $H_2 + D_2$ isotopic exchange is interesting as one of the simplest chemical reactions of H_2 on the catalyst surface.

The activity of $H_2 + D_2$ exchange increases as Fe(pc) > Fe(1 1 0) > Fe(1 0 0) > Fe(1 1 1) [426].

Fast $H_2 + D_2$ isotopic scrambling has been found for an (Fe, Al, Si, Zr) sample at $-195\,°C$ [424], while the scrambling was slow on an (Fe, Al, K) sample at the same conditions [424].

2.6.2 Chemisorption of CO

The chemisorption of CO is interesting for the study of ammonia synthesis catalysts since this reaction provides a way of titrating the number of active sites on the surface. The reaction is complicated by the dissociation of CO.

2.6.2.1 *Structure of Chemisorbed CO*

The CO chemisorption is molecular at low temperature [207]. At higher temperatures the CO is dissociated. Evidence for the existence of more than one molecular species at low temperature has been found from X-ray photoelectron spectroscopy and ultra violet photoelectron spectroscopy [245, 427].

On single crystals a c(2 × 2) structure of CO/Fe(1 0 0) is formed at 373–400 K [427, 383]. For CO/Fe(1 1 0) at 300 K, c(2 × 4) is formed at low coverage [428, 429], p(1 × 2) is formed at high coverages [428]. By NEXAFS it has been found that for CO/Fe(1 0 0) the CO molecule is tilted $45 \pm 10°$ [430].

The disorder observed below room temperature [383, 427] [429] is caused by the lack of mobility [427]. Just below the dissociation temperature ordering is observed.

While molecular and dissociated CO have similar X-ray photoelectron spectroscopy spectra, O(1s) at 531 eV and C(1s) at 285 eV [245, 427, 431], the ultra violet photoelectron spectroscopy [245, 427] and laser Raman [432] spectra are different.

During sequential adsorption of ^{12}CO and ^{13}CO on Fe(1 0 0), isotopic exchange is observed only among the two strongest bound molecular states [433]. For the catalyst, isotopic scrambling between $^{13}C^{16}O$ and $^{12}C^{18}O$ is observed at −33 °C indicating the existence of dissociated CO at these temperatures. There is partial isotropic scrambling between the first and second exposure at −195 °C [434] and at −78 °C [435].

2.6.2.2 Thermodynamics

For single crystal surfaces the initial enthalpy of chemisorption for CO is −105 kJ/mole on Fe(1 0 0) by temperature programmed desorption [383], −96 kJ/mole on Fe(1 1 0) by temperature programmed desorption [245, 436], −91 kJ/mole by temperature programmed desorption on Fe(1 1 1) [437], and −155 kJ/mole by calorimetry on Fe(pc) [438].

The enthalpy of chemisorption for CO on Fe surfaces is −9.3 to −4.2 kcal/mole at 0 °C [439] on (Fe, Al), −23.1 to −16.9 kcal/mole at 0 °C [439] on (Fe, Al, K), −8.0 to −4.2 kJ/mole at −183 °C [439] on (Fe, Al, K), and −32 kcal/mole at 22 °C [389] on (Fe, Al, K). For Fe supported on Al_2O_3 a Freundlich isotherm has been found [160]. The enthalpy of adsorption is 12–30 kJ/mole at 37% of saturation [160].

For Fe/MgO the amount of CO chemisorption depends on particle diameter [328]. The ratio of the area determined from CO chemisorption to the area determined by the BET method decreases through the sequence Mg > Be > Al, Si, Cr > Mn > Ca > Ti [198, 199]. Infra-red spectroscopy and microcalorimetry show that the amount of weakly bound CO increases with the dispersion of Fe/MgO [328].

For multiply promoted samples the CO area is 5.6 m^2/g [180].

For the catalyst, pulse chemisorption and volumetric chemisorption give identical results at −78 °C and −196 °C [440]. For an industrial catalyst the CO chemisorption was initially found to be 0.7–1.4 m^2/g [14, 164], i.e., 9–18 μmole/g; in later studies the value 28–39 μmole/g was found [179]. Increa-sing the wustite content in the unreduced sample was reported to increase the CO area [181] or to decrease the CO area [183] of the reduced catalyst.

2.6.2.3 Adsorption Kinetics

The sticking coefficient is close to unity and remains constant up to 60% of a monolayer on Fe(1 0 0) [383] and up to 40% of a monolayer on Fe(1 1 1) [437] indicating the existence of a mobile precursor for the adsorption. The sticking coefficient decreases with temperature. For the catalyst the adsorption kinetics is first order [402].

2.6.2.4 Desorption Kinetics

At low initial coverage, temperature programmed desorption of CO from CO/Fe show peaks from *adsorbed molecules* at 430 K for Fe(1 0 0) [383], 400 K for Fe(1 1 0) [245, 436], 300 K [437] or 360 K [441] for Fe(1 1 1), and 350 K for Fe(pc) [438]. At low initial coverages, temperature programmed desorption of CO from *dissociated* CO on Fe show peaks at 800 K [383] for Fe(1 0 0), 710 K [441] for Fe(1 1 1) and 700 K [438] for Fe(pc). By sputtering of CO/Fe(1 1 1) the major desorption product is CO [442].

Isotopic labeling shows that physisorbed CO is desorbed in vacuum at − 195 C while chemisorbed CO is not [434]). Temperature programmed desorption of CO from the reduced catalyst yields 2 peaks, one peak at − 50 to + 180 °C, another peak at 650 °C [406]. Presumably, the peak at 650 °C is caused by the recombination of C∗ and O∗.

2.6.2.5 Dissociation Kinetics

Upon heating, the dissociation of adsorbed CO is complete at 390 K on Fe(1 1 0) [245]. The activation energy for dissociation of adsorbed molecules is 105 kJ/mole [383].

2.6.2.6 Properties of Chemisorbed CO

For single crystal surfaces the ultra violet photoelectron spectroscopy CO(4σ) peak is − 10.6 eV and the ultra violet photoelectron spectroscopy CO($\pi + 5\sigma$) peak is − 7.6 eV for CO/Fe(1 1 0) [245], while the ultra violet photoelectron spectroscopy C(2p) and the ultra violet photoelectron spectroscopy O(2p) peak are both − 5.6 eV for C∗ + O∗ on Fe(1 1 0) [245].

For CO/Fe(1 0 0) at exposures below 1 L at 350 K the CO stretch frequency is observed by high-resolution electron-energy loss-spectroscopy at 1180–1245 cm^{-1} [443]. At higher exposures at 110 K the CO stretch frequency is observed at 1900–2055 cm^{-1} [443]. This change in frequency is attributed to the transition from a *side-on* bonding at low coverages at a *end-on* bonding at higher coverages [443].

High-resolution electron-energy-loss spectroscopy spectra of CO/Fe(1 1 0) indicate that CO is positioned on-top with the C–O stretching frequency at 1890–1950 cm^{-1} [428]. The vibration frequency for the C–O stretch for

CO/Fe(1 1 0) is increased upon increasing exposure to CO [428] indicating adsorbate-adsorbate interactions.

On Fe(1 1 1) 3 different coordinations have been observed [437]. High-resolution electron-energy-loss spectroscopy spectra of C + O/Fe(1 1 1) show the Fe–C stretch at 420 cm^{-1} and the Fe-O stretch at 540 cm^{-1} [437].

At 100 K the work function increases for CO/Fe(1 0 0) until c(2 × 4) is complete at 3.8 L [436] and then remains constant due to depolarisation until p(1 × 2) is complete at 4.5 L [436]. Above room temperature the work function depends on pressure [436], presumably because adsorption is followed by dissociation. At small exposures the work function increases with CO coverage [428, 436, 438].

2.6.2.7 Effect of Promoters

A high-resolution electron-energy-loss spectroscopy study of CO/K/Fe(1 1 1) shows that up to one half monolayer of K, on-top, shallow-hollow and deep-hollow CO∗ are seen [437]. At higher K-coverages, on-top CO∗ is observed in high-resolution electron-energy-loss spectroscopy together with bands interpreted as CO∗ near K (1360–1435 cm^{-1}) and CO-bending modes (820 cm^{-1}) [437].

Temperature programmed desorption shows that the presence of K increases the heat of adsorption for CO [230, 444]. An increase of the heat of adsorption by 40–80 kJ/mole [230] has been deduced from the change in peak temperature.

The presence of K was found to increase the saturation coverage for CO [230, 245, 444]. For Fe(1 1 0) the saturation coverage is increased from 7.3 μmole/m^2 for the unpromoted surface to 11.7 μmole/m^2 for a surface precovered by 6.3 μmole of K per m^2 [245].

Other studies have found that K has no effect on that saturation coverage for CO on Fe(1 1 1) [437]. A possible explanation for the observed effects is that the adsorbate-adsorbate interactions limiting the CO adsorption on unpromoted Fe are screened by K. This will tend to increase the saturation coverage in the presence of K, while the blocking of sites by K will tend to decrease the saturation coverage by CO.

The initial sticking coefficient has been reported to increase [245], to remain constant [230] or to decrease [230] in the presence of preadsorbed K. Presumably conflicting results are obtained because the K blocks sites while increasing the strength of CO adsorption.

For single crystal surfaces the temperature programmed desorption peak maximum is 390 K for CO/K/Fe(1 0 0) [383], 420 K for CO/K/Fe(1 1 0) [245], and 500 K for CO/K/Fe(1 1 1) [437]. Temperature programmed desorption shows an extra peak at 700 K at small K coverages. This peak is due to the simultaneous desorption of K and dissociated CO [230]. Desorption of dissociated CO from K/Fe showed peaks at 800 K for Fe(1 0 0) [383], 700 and 820 K for Fe(1 1 1) [437].

The presence of adsorbed K leads to an apparently lower barrier to dissociation [444]. Presumably this is a reflection of the changed stability of the adsorbed molecule. If the adsorbed molecule is stabilized, the rate of desorption at a given temperature will be lower and this may appear as an increased tendency to dissociation.

For CO/K/Fe(1 1 0) with 8.2 μmole/m^2 of K, the ultra violet photoelectron spectroscopy (4σ) peak is -11.4 eV, and the ultra violet photoelectron spectroscopy CO($1\pi + 5\sigma$) peak is -7.8 eV [245].

The X-ray photoelectron spectroscopy O(1s) peak for CO/K/Fe(1 1 0) and respectively for C + O/K/Fe(1 1 0) is unaffected by the presence of K [245].

Preadsorbed K reduces the frequency of the CO stretch vibration [230].

2.6.2.8 Effect of Preadsorbed Species

Preadsorbed O* inhibits the chemisorption of CO [383, 207, 413], decreases the tendency for dissociation of CO* [383], and decreases the bond strength for CO* [445].

Preadsorbed S* partially blocks the adsorption of CO [446], decreases the initial sticking coefficient for CO [383], and decreases the tendency for dissociation of CO* [383].

Preadsorbed C* decreases the initial sticking coefficient for CO [383] and decreases the tendency for dissociation of CO* [383].

Preadsorbed N* partially inhibits CO chemisorption [207, 447, 448] and decreases the heat of adsorption for CO at 32 °C [389].

Preadsorbed CO_2 inhibits CO adsorption at -78 °C [207].

Preadsorbed Cl* blocks the adsorption of CO [449].

2.6.2.9 Correlation with Activity

Extensive chemisorption measurements on reduced (Fe, Al) and (Fe, Al, K) samples have been reported by Brunauer and Emmett [163]. From these studies it has been concluded that the Fe-area of the samples can be measured by low temperature chemisorption of CO [163, 193].

The active fraction of total area generally decreases with increasing promoter content [198, 199]. The activity is maximum when the ratio of the area determined by CO chemisorption to the area determined by the BET method is 0.3 [67, 68].

2.6.3 Chemisorption of CO_2

The chemisorption of CO_2 on the surface of the catalyst is primarily interesting as CO_2 appears to bind selectively to some parts of the surface of the catalyst.

2.6.3.1 *Structure of Chemisorbed CO_2*

For CO_2 adsorption on a stepped Fe(1 1 0) and on Fe(1 1 1) at 77 K, a linear and an unidentified species is formed [450]. At 140 K the unidentified species is dominating [450]. Above 140 K adsorbed CO_2 dissociates to CO*, O* and C* [450].

X-ray photoelectron spectroscopy and UPS for CO_2 on a Fe film shows that CO_2 is present at 80 K as a linear molecule and as a bent CO_2^- species [451]. For the catalyst, CO_2 is assumed mainly to bind to alkali at −78 C without dissociation [207]. Pure MgO, Cr_2O_3 and Al_2O_3 chemisorbs some CO_2 [198, 199].

2.6.3.2 *Thermodynamics*

For Fe(1 1 0) no adsorption of CO_2 is detectable in the range 77–340 K [450].

For a (Fe, Al) catalyst model the enthalpy of chemisorption is −8.7 to −6.5 kcal/mole at −78 °C [439] and −17.7 to −8.9 kcal/mole at −78 °C [439] for (Fe, Al, K).

Pulse chemisorption and volumetric chemisorption give identical results for the CO_2 chemisorption [440]. The enthalpy is difficult to determine because of the tendency to dissociate at higher temperatures.

2.6.3.3 *Dissociation Kinetics*

Above 140 K adsorbed CO_2 dissociates to CO*, O* and C* [450].

During passivation with CO_2 at 663–773 K, some carbide is formed in the industrial catalyst [452–457].

2.6.3.4 *Effect of Promoters*

The addition of basic promoters increases the CO_2 area [198, 199]. The ratio of the area determined by CO_2 chemisorption to the area determined by the BET method decreases through the sequence Ca > Mn, Mg, Si > Cr, Be > Al > Ti [198, 199]. For the industrial catalyst the CO_2 area is 59% of the BET area [164].

2.6.3.5 *Effect of Preadsorbed Species*

Preadsorbed O_2 does not inhibit CO_2 chemisorption [207, 413].

2.6.4 Physisorption of N_2

On Fe films at 78–273 K, three forms of N exists: physical adsorption, weak chemisorption and strong chemisorption [458, 459].

2.6.4.1 *Structure of Physisorbed N_2*

Based on detailed interpretation of the X-ray photoelectron spectroscopy spectrum the geometry of physisorbed N_2* is assigned to *end-on* [460, 461].

For the catalyst a phase transition in the physisorbed N_2 is observed at 5% of a monolayer at 79.8–90.3 K [164].

2.6.4.2 *Thermodynamics*

The saturation coverage is 5.8 ± 0.7 μmole/m^2 at 91 K for Fe(1 1 1) [462, 463].

The heat of physisorption is 5.2–3.1 kcal/mole [390] for Fe powder at 78–90 K.

For single crystal surfaces the heat of adsorption is 25–37 kJ/mole depending on the coverages by molecular and physisorbed N_2* [22, 462].

2.6.4.3 *Adsorption Kinetics*

The initial sticking coefficient for N_2(g) into the physisorbed N_2 state is 0.7 for Fe(1 1 1) at 85 K [462].

2.6.4.4 *Desorption Kinetics*

For Fe(1 1 1) a temperature programmed desorption peak is found at 84 K [22] or at 96 K [231]. Sequential adsorption of ^{14}N and ^{15}N shows that this peak is most likely a new N_2* species; possibly this peak is physisorbed N_2 upon N_2* [22].

2.6.4.5 *Kinetics of Conversion into αN_2*

The activation enthalpy for the conversion of physisorbed N_2* to chemisorbed N_2* is 18 kJ/mole for Fe(1 1 1) [462]. Estimation of the reaction rates indicates that the chemisorbed N_2* may be formed directly, rather than via physisorbed N_2*, at typical reaction conditions for NH_3 synthesis [464].

Below 90 K the equilibrium between the molecular chemisorbed state and the physisorbed state is not established [465].

2.6.4.6 *Properties of Physisorbed N_2*

Ultra violet photoelectron spectroscopy N(1s) peaks are found at -12 and -8 eV [462]. At 85 K X-ray photoelectron spectroscopy N(1s) peaks are found at 405.9 and 401.2 eV [462, 463]. High resolution electron energy loss spectroscopy shows a N–N stretch in γN_2 at 2100 cm^{-1} [231]. For N_2* the work function decreases linearly with the coverage up to saturation [22].

2.6.5 Molecular Chemisorption of N_2

Even if CO is isoelectronic with N_2, the bonding to the surface is different for the two molecules. CO adsorbs with a high sticking probability and high binding energy in end-on geometry, while N_2 adsorbs side-on in a more weakly adsorbed species. The dissociation of N_2* is much easier than the dissociation of CO*.

2.6.5.1 Structure of Chemisorbed N_2

A molecular, chemisorbed species, N_2*, is formed during the adsorption of N_2 on Fe [229, 466, 468].

The absence of ^{14}N–^{15}N exchange during adsorption-desorption [386, 469] and isotope labeling of N_2 in high-resolution electron-energy-loss spectroscopy demonstrates that this species is molecular [231].

The geometry of N_2* is assigned to side-on based on detailed interpretation of the X-ray photoelectron spectroscopy spectrum [460, 461] and of the high-resolution electron-energy-loss spectroscopy spectrum [470].

2.6.5.2 Thermodynamics

The saturation coverage for N_2* on Fe(1 1 1) is 1.16 μmole/m^2 [462, 463]. This is significantly less than for the physisorbed state.

The enthalpy of chemisorption for N_2* is -31 kJ/mole for N_2/Fe(1 0 0) [466], [229], -31 kJ/mole for N_2/Fe(1 1 0) [467], and -21 kJ/mole for Fe(pc) [469] from temperature programmed desorption studies.

2.6.5.3 Adsorption Kinetics

The sticking coefficient for N_2 into N_2* is 10^{-2} for Fe(1 1 1) [466]. This is lower than the sticking coefficient into the physisorbed state.

2.6.5.4 Desorption Kinetics

Temperature programmed desorption of N_2 from N_2*/Fe is 160 K for N_2/Fe(1 0 0) [466], 160 K for N_2/Fe(1 1 0) [467], 260 K for N_2/Fe(1 1 1) [469, 471], and 290 K for N_2*/Fe(pc) [471]. Others have reported desorption of N_2 from N_2*/Fe(pc) at 77, 100, 200–250 and 350 K [386].

2.6.5.5 Properties of Chemisorbed N_2

High-resolution electron-energy-loss spectroscopy spectra of N_2*/Fe(1 1 1) show a N–N stretch at 1490 cm^{-1} [463, 470, 472] indicating a considerable weakening of the N–N bond compared to the stretch found at 435 cm^{-1} [463, 472].

Laser Raman during NH_3 synthesis shows peaks at 2040, 1940, 423, 443 cm^{-1} [473]. These peaks have been assigned to N_2* [473].

The adsorption of N_2* increases the work function on Fe [22, 474]. The dipole moment of N_2* is 0.4 D [467].

2.6.5.6 *Effect of Promoters*

K increases the stability of N_2* [466]. The maximum stability is 10.5 kcal/mole [466]. This effect is strongest on Fe(1 0 0) [466]; this erases the difference in sticking coefficient between low index planes [466].

The decrease in N_2* adsorption at high coverages by K is consistent with N_2* adsorbing on Fe adjacent to K* rather than adsorption on top of K* [466].

The enthalpy of chemisorption has been determined from temperature programmed desorption and from equilibrium coverages. K increases the stability of N_2* [466]. The maximum stability is 10.5 kcal/mole [466]. The enthalpy of chemisorption is -46 kJ/mole for N_2/K/Fe(1 0 0) [466] and -43.9 kJ/mole for N_2/K/Fe(1 1 1) [466]; the heat of adsorption is increased by 3 kcal/mole by the presence of K [229].

From theoretical considerations it has been argued that the effect of K^+ is a local effect extending at most to the next neighbor position.

On exposing K/Fe(1 1 1) to N_2, promoted and unpromoted sites are simultaneously filled [466]. This indicates that the sticking coefficient into the N_2* is approximately identical for both promoted and unpromoted sites.

The temperature programmed desorption peak maximum for N_2*/K/Fe(1 1 1) is 210 K [466]. At small coverages by K, both promoted and unpromoted sites are seen in temperature programmed desorption [466]. At large K-coverages a short range N_2*–K* interaction is detected [231]. The temperature programmed desorption peak maximum is increased to 210 K. The stabilization is 16.2 kJ/mole [231].

High-resolution electron-energy-loss spectroscopy of N_2*/K/Fe(1 1 1) shows a N–N stretch at 1390 cm^{-1} [470] or 1370 cm^{-1} [470]. The reduction of the N–N stretching frequency in the presence of K [231, 472] does not represent a significant weakening of the N–N bond [475].

2.6.5.7 *Effect of Preadsorbed Species*

Preadsorption of O* inhibits the formation of N_2* [470, 476, 477].

2.6.6 Dissociative Chemisorption of N

Above room temperature the exposure of Fe to N_2 leads to the formation of N* through dissociative chemisorption, $N_2(g) + 2* = 2N*$. A number of reviews of ultra-high vacuum single-crystal studies of the kinetics and mechanism of N_2 adsorption on Fe are available [10, 20–23, 478].

2.6.6.1 Structure of Chemisorbed N

The exposure of Fe to N_2 at low and moderate pressures does not lead to the formation of bulk nitrides.

The adsorbate is atomic as $^{28}N + {}^{30}N$ are completely equilibrated after adsorption-desorption on Fe(1 1 0) at 680 K [468], on Fe(1 1 1) at 140–1000 K [468], on Fe(pc) [479, 480] and on the catalyst [481]. Complete isotopic scrambling indicates that the adsorption is dissociative. Others have concluded that a significant part of the adsorbed nitrogen at 380–400 °C is not dissociated [379, 482, 483]. These latter conclusions are based on less direct evidence.

On single crystals, N_2 adsorbing on Fe(1 0 0) forms c(2 × 2) [229, 466, 474, 484]. Detailed interpretation of the low-energy electron diffraction patterns for N/Fe(1 0 0) shows that N∗ is located in sites with C_4 symmetry 0.27 Å above the outermost plane of Fe-atoms [485]. The distance between the first and the second layer of Fe atoms is expanded by 7.7% compared to the bulk lattice of Fe [485].

Fe(1 1 0) [468] and Fe(1 1 1) [468, 474] reconstruct upon N_2 adsorption. Fe(1 1 0) forms c(2 × 3) [468], while Fe(1 1 1) forms (3 × 3), $(\sqrt{19} x \sqrt{19})$R23.4 °, $(\sqrt{21} x \sqrt{21})$R10.9 °, $(\sqrt{27} x \sqrt{27})$R30 °, (2 × 2) [474]. The c(2 × 2) pattern on Fe(1 1 1) is only observed when N segregates from the bulk [474]. N_2^+ implantation in Fe(1 1 1) yields p(1 × 1) [486].

The observed structures have been rationalized [468, 474]. The c(2 × 2) N/Fe(1 0 0) pattern is structurally very similar to the (0 0 2) plane of Fe_4N while recrystallisations of N/Fe(1 1 0) and N/Fe(1 1 1) can be interpreted as stepwise transitions toward the Fe_4N(1 1 1) structure. The sequence of patterns can be seen as an attempt to keep as many Fe-atoms fixed while accommodating an increasing number of N-atoms.

On Fe(1 1 0) the original surface structure is recreated during desorption in N∗ [468]. For Fe(1 0 0) partial desorption of N∗ results in disorder [474]; accommodating at 620 K recreates the order [474]. N_2 chemisorption on Fe(12, 1, 0) at 750 K leads to an increase in the density of steps and to facetting [487].

The chemisorption of N_2 has been studied on catalysts and catalysts models [488, 489]. The studies include high pressures [393] and high temperatures [490].

2.6.6.2 Thermodynamics

From a study of N_2 adsorption by volumetric chemisorption it was concluded that the chemisorption of N_2 on a (Fe, Al, K, Si) catalyst follows the Freundlich isotherm [491, 492]. The enthalpy of chemisorption is − 38 kcal/mole log(θ) [492].

For polycrystalline samples the enthalpy of chemisorption for N_2, i.e. for $N_2(g) + 2* = 2N*$, has been determined in a number of ways [493]. For a (Fe, Al) sample

$$H = -208.2 + 446.5\theta \text{ kJ/mole} \tag{2}$$

for converges $0.00 < \theta < 0.22$ [494], and

$$H = -136 + 122.2\theta \text{ kJ/mole} \tag{3}$$

for $0.22 < \theta < 0.90$ [495] has been determined from gravimetric studies.

For an industrial catalyst, -209 ± 16 kJ/mole at 400 °C [493], 30 ± 5 kcal/mole at 197 °C [395], and 32 kcal/mole [496] were obtained from direct calorimetric measurement; -70 kcal/mole at $\theta = 0$, and -16 kcal/mole at $\theta = 0.18$ [389] from volumetric chemisorption; -56 to -19 kcal/mole [497] from temperature programmed desorption; -175 kJ/mole [498] from the solubility of N in Fe; and -126 ± 21 kJ/mole at 197 °C [499] from a calorimetric study of NH_3 adsorption.

2.6.6.3 *Adsorption Kinetics*

Emmett and Brunauer pointed out that the rates of N_2 chemisorption and of NH_3 synthesis are comparable and slow [163].

The sticking coefficient, defined as the probability that one $N_2(g)$ molecule hitting the surface will chemisorb as $2N*$, is small. The sticking coefficient for N_2 determined by Auger electron spectroscopy is on Fe(1 0 0) 10^{-7} at 383 K [474, 484], 10^{-7} at 400 K [229], $1.4 \cdot 10^{-7}$ at 430 K [466], $4.0 \cdot 10^{-7}$ at 508 K [484]; on Fe(1 1 0) 1^{-7} at 583–733 K [468]; and on Fe(1 1 1) 10^{-7} to 10^{-6} [468] or $6 \cdot 10^{-6}$ at 583–733 K [467, 474]. On polycrystalline Fe, the sticking coefficient is less than 10^{-6} at 290 K [471], $5 \cdot 10^{-6}$ at 430 K [466], $4 \cdot 10^{-6}$ at 423 K [500] determined by Auger electron spectroscopy, and $1.6 \cdot 10^{-7}$ at 273 K, [479] and $2.5 \cdot 10^{-7}$ at 348 K [479] determined by volumetric chemisorption.

For Fe(1 0 0) the activation enthalpy for the sticking coefficient is independent of coverage [229], while for Fe(1 1 0) [229, 466, 468, 474, 484] and Fe(1 1 1) [474] the activation enthalpy becomes more positive with increasing coverage. The activation energy is approximately 5 kcal/mole [484] for Fe(pc). From the low value of the activaiton energy for the sticking coefficient, it is apparent that the small sticking coefficient is due to the unusually low value for the preexponential factor.

For a molecular beam of N_2 directed on Fe(1 1 0) crystal at 550 K, N_2 translational energy 0.176 and 0.647 eV, no diffuse scattering was observed [501]. This is not unexpected in view of the low sticking probability.

The dynamics of N_2 chemisorption on Fe(1 1 1) has been studied by molecular beam techniques using kinetic energies in the range 0.3 to 0.6 eV [502, 503, 504] and observation of the sticking by electron spectroscopy. The sticking probability increases from 10^{-6} at 0.09 eV to 10^{-1} at 4.3 eV [503]. The sticking coefficient increases at low surface temperatures [503]. This demonstrates that the barrier to dissociation is found in the entrance channel [502–504]. Consequently this barrier is more easily passed if energy is supplied to the molecule in the form of kinetic rather than vibrational energy. Vibrational energy is about half as effective as kinetic energy [502–504].

The high activity on the more open planes has been assigned to the existence of C_7 sites on such planes [505].

For (Fe, Al) samples the rate of N_2 chemisorption has been studied by gravimetry [494, 495]. The activation energy of chemisorption was found to increase, $22.0 + 324\theta$ kJ/mole up to $\theta = 0.22$, and then to remain constant [191, 495]. The rate of adsorption was still significant after 8 h at 201 torr and 250 °C [494]. The activation energy for desorption decreases with converge [191, 495]. A change in the preexponential factor in the rate of adsorption [494, 495] and for desorption [495] is interpreted as a compensation effect or as a change in the properties of the molecular precusor with coverage [494].

The entropy of the adsorbate is calculated from transition state theory. For $\theta < 0.10$ the adsorbate is immobile; for $0.22 < \theta < 0.30$ the adsorbate is immobile and dissociated [191].

2.6.6.4 Desorption Kinetics

On single crystal surfaces the desorption temperatures are 980 K [474, 484] for N/Fe(1 0 0), 920 K [468] for N/Fe(1 1 0), and 860 K [474] for N/Fe(1 1 1).

N_2 desorption from N$*$ chemisorbed on a Fe-overlayer on Ru(0 0 1) shows a temperature programmed desorption peak at 850–950 K [506].

The desorption of N_2 from the catalyst has been studied [406, 497].

2.6.6.5 Hydrogenation of Chemisorbed N

The hydrogenation of adsorbed N_2* and N$*$ has been studied [418, 507]. Preadsorbed N$*$ is removed exclusively as NH_3 at 218–313 °C [418]. The rate of reaction of N$*$ with H_2 is first order in H_2 [508], the rate increases with increasing θ_{N*} [508] and the rate is higher for D_2 than for H_2 [508]. ^{15}N isotopic labeling shows that all chemisorbed N-atoms undergo hydrogenation with equal probability [422]. By exposing N$*$ adsorbed on single crystal Fe surfaces to an increasing pressure of H_2(g) at 580 K, the disappearance of N$*$ is observed at approximately 100 torr of H_2 [414]. For catalysts the rate of hydrogenation of preadsorbed N$*$ depends on the adsorption temperature [509, 419].

By field mass spectroscopy of $N_2 + H_2$ interacting on an Fe tip at room temperature H, H_2 [510], N_2 [510], N [510] N_2H [510] and NH_3 [511] have been detected. By coadsorption of N_2 and H_2, FeN_2 is seen [379].

2.6.6.6 Properties of Chemisorbed N

The N(1s) ultra violet photoelectron spectroscopy peaks are −5.0 and −1.8 eV [474] for Fe(1 0 0) and −5.4 and −1.8 eV [474] for Fe(1 1 1). Ultra violet photoelectron spectroscopy of N_2 adsorbed on Fe(1 1 1) at 140–1000 K shows peaks at 5 eV, interpreted at N* [468].

For N_2 adsorbed on single crystal surfaces, the X-ray photoelectron spectroscopy N(1s) peaks are 380eV for Fe(1 0 0) [474] and 397 eV for Fe(1 1 1) [462, 463]. X-ray photoelectron spectroscopy of N(1s) in N_2/Fe(pc) adsorbed at −80 °C shows two peaks at 405.3 eV and 400.2 eV [471], while X-ray photoelectron spectroscopy of N(1s) in N/Fe(pc) adsorbed at 290 °C shows one peak at 397.2 eV [471]. The absence of a shift in core level energy for Fe upon N* adsorption shows that the Fe–N bond in N* is covalent [52].

High-resolution electron-energy-loss spectroscopy of N*/Fe(1 1 1) shows N–Fe at 450 cm^{-1} [463, 470].

For catalyst models infrared spectroscopy after N_2 adsorption shows a band at 1820 cm^{-1} [419] for Fe/SiO_2. This band disappears upon exposure to H_2 [419]. On Fe/MgO bands are observed at 2200 and 2050 cm^{-1} [512].

On single crystal surfaces the work function increases smoothly with coverage [474] on Fe(1 0 0), while for the Fe(1 1 1) surface the work function behavior is more complicated [474]. The change in work function at saturation is +0.33 eV for N*/Fe(1 0 0) [474] and +0.25 eV for N*/Fe(1 1 1) [468, 474].

The work function during N_2 adsorption at −80 °C on Fe-films goes through a minimum [513]. The shape but not the absolute value is reproducible [513].

On catalysts N_2 chemisorption has been reported to decrease the work function [243, 514] or to have no effect on the work function [243, 279].

For N_2 chemisorption on Fe/SiO_2 at 400 °C little change in the magneti-zation is observed [410].

2.6.6.7 *Effect of Promoters*

The reported structures of N*/K/Fe(1 0 0) are c(2 × 2) [229, 466] and (3 × 3) for N*/K/Fe(1 1 1) [500]. With up to 30% of the saturation coverage by K, the presence of K does not affect the saturation coverage of N* [229].

K increases the strength of N_2 chemisorption [412]. At small coverages by K, the number of promoted sites is proportional to the coverage by K, each K forming 1–2 promoted sites [229, 466]. At higher K-coverage, the number of promoted sites decreases with increasing coverage by K due to surface blocking by K [466]

K increases the rate of chemisorption for N_2 [231, 466, 515]. The sticking coefficient is $1.4 \cdot 10^{-7}$ for Fe(1 0 0) at 430 K, 0 μmole K/m^2 [466]; $3.9 \cdot 10^{-5}$ for Fe(1 0 0) at 430 K, 2.5 μmole K/m^2 [229, 466]; $5 \cdot 10^{-6}$ for Fe(1 1 1) at 430 K 0 μmole K/m^2 [466], $4 \cdot 10^{-5}$ for Fe(1 1 1) at 430 K, 3.3 μmole K/m^2 [466]; $4 \cdot 10^{-6}$ for Fe(pc) at 423 K, 0 μmole K/m^2 [500], and $2.5 \cdot 10^{-5}$ for Fe(pc) at 423 K, 2.2 μmole K/m^2 [500]. A maximum sticking coefficient is seen for 3.3 μmole/m^2 [229] of K atoms.

The activation energy for N_2 adsorption on K/Fe(1 0 0) is approximately 0 [229].

Thermal desorption of K from N*/K/Fe shows peaks at 570–810 K [466, 500]. Temperature programmed desorption shows that N* increases the stability of K* [466].

From electron spectroscopy studies it was found that N* and O* interacts with K [515]. On the surface K is associated with 2 or 3 chemisorbed O-atoms [515].

Al decreases the N_2 chemisorption on an area basis [210, 211, 516].

2.6.6.8 *Effect of Preadsorbed Species*

Preadsorption of O* inhibits the formation of N* [476, 477, 490]. One O* blocks the formation of one N* [477].

In a study [517] of N_2 adsorption on Fe(1 1 1) contaminated with 1.2 μmole O_2 per m^2, the initial sticking coefficient for N_2 into 2N* was 10^{-7} at 420 K and $2 \cdot 10^{-7}$ at 470 K.

For (Fe, Al) samples the rate of adsorption for N_2 is increased by the presence of H_2 [518, 519]. Chemisorption on an industrial catalyst give evidence that N and H chemisorb on identical sites [496].

Cl blocks the adsorption of N_2 [449].

2.6.6.9 *Isotopic Exchange*

The rate of ^{14}N-^{15}N exchange becomes measurable at typical reaction tempera-tures or NH_3 synthesis [520].

The N_2 isotopic exchange does not occur on Fe films [488], but does occur on fused Fe [490]. These findings may be related to the low surface area of these materials.

The kinetics of N-isotopic exchange was interpreted as N–N bond breaking being rate limiting [482, 520–523].

The activation energy for N_2 isotopic exchange is 36 kcal/mole [482], 33 kcal/mole [522] or 58 kcal/mole [524, 525].

N-isotopic marking show that under NH_3 synthesis, NH_3(g) and N* are in equilibrium [526, 528], while N* and N_2(g) are not in equilibrium [527, 528].

The $N_2 + N_2$ isotopic exchange, $NH_3 + N_2$ isotopic exchange is much faster [529, 530].

The rate of isotopic exchange measured for the gas phase in the presence of a catalyst is the same as the rate of exchange between the gas and chemisorbed N* [482, 521], indicating that the isotopic exchange predominantly proceeds on the surface of the catalyst.

The rate of isotopic exchange has been found to correlate with the catalytic activity for a series of catalysts [531] or even to be the same as the rate of ammonia synthesis [521]. Model calculations show that the coverage by N* is high during NH_3 synthesis [396]. This is not necessarily the case during a N_2 isotopic scrambling experiment. This may explain why the rate of NH_3 synthesis does not always correlate perfectly with the rate of N_2 isotopic scrambling. These results give evidence for the dissociative mechanism of ammonia synthesis [520].

Addition of H_2 to N_2 has been reported to increase the rate of $^{14}N + {}^{15}N$ isotopic scrambling [288, 481, 488, 520, 522, 532, 533], to have no effect of the rate [534] or to decrease the rate [288]. The observation of an increase in the rate of isotopic exchange in the presence of H has been interpreted as H removing an oxide from the surface [481, 533].

The presence of K has been found to be important for the effect of H [288].

N_2 isotopic exchange over fused Fe is inhibited by O_2 [490]. The activation energy for nitrogen isotope exchange is higher in the presence of chemisorbed oxygen [534].

2.6.7 Kinetic Models of N_2 Chemisorption

The numerical models of the kinetics of NH_3 synthesis will be discussed in Sect. 2.7.3 All contain a model of the kinetics of N_2 adsorption

$$N_2(g) + * \leftrightarrows N_2* \tag{4}$$

$$N_2* + * \leftrightarrows 2N* \tag{5}$$

as a special case. From the kinetic models the coverage of N* may be calculated as a function of exposure. These calculations, shown in Figure 2.1, are in good agreement with experiments at low and moderate exposures [396]. The calculated peak temperatures, see Figure 2.2, as well as the peak shapes are in reasonable agreement with experiments [396].

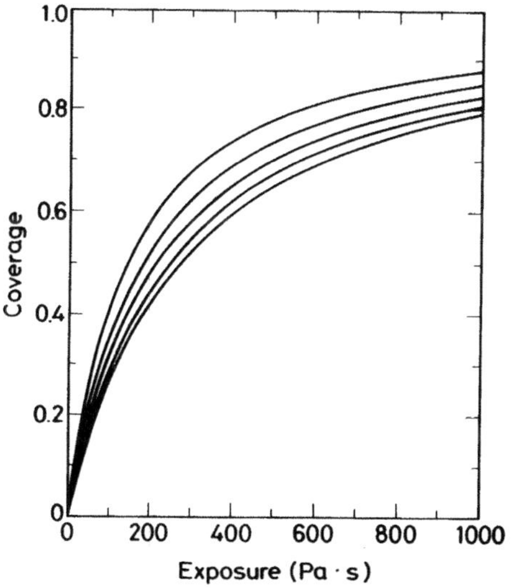

Fig. 2.1. Calculated coverages by N* vs exposure for N_2 adsorption on Fe. Pressure 10^{-6} torr. Temperature 300 K (upper curve), 350 K, 400 K and 500 K (lower curve). Reproduced from [396]

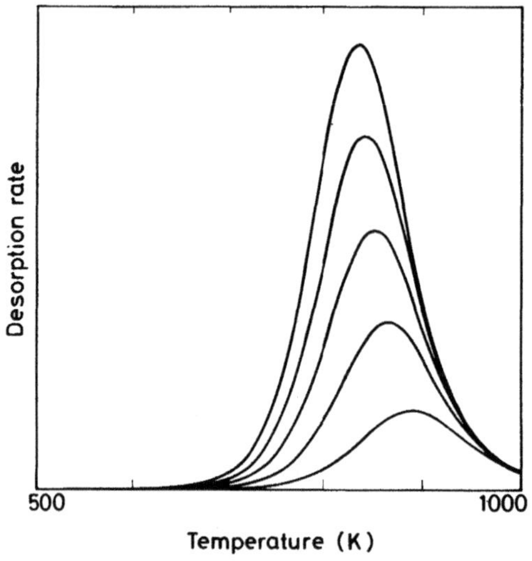

Fig. 2.2. Calculated TPD curves for N_2 desorbing from Fe. Heating rate 10 K/s. Initial coverage 0.20 (lower curve), 0.40, 0.60, 0.80 and 1.00 (upper curve). Reproduced from [396]

N_2* is weakly adsorbed on the surface of Fe. One model [396] calculates that the equilibrium coverages of N_2* under 101 kPa of N_2 is 0.27 for Fe and 0.30 for K/Fe at 500 K. The coverage is even smaller at higher temperatures [396]. These calculations are made based on the assumption that N_2* does not dissociate to form $N*$. $N*$ is much more strongly adsorbed than N_2*. Under 101 kPa of N_2, the equilibrium coverage by $N*$ is close to unity at 500–1000 K [396].

The activation energy for the sticking coefficient is small and the low value of the sticking coefficient must be ascribed to an unfavorable activation entropy. The negative activation energy found for adsorption on Fe(1 1 1) and K/Fe may be attributed to the top of the activation barrier for N_2* being below the level for N_2(g) [396]. An equivalent description may be developed from the statement that the decrease in sticking coefficient at higher temperatures is caused by the shorter lifetime of N_2* at higher temperatures [396].

Estimates of the sticking probability based on *transition state theory* are not particularly successful. This may be caused by incomplete equilibration of the transition state [396], by an underestimate of non-adiabatic quantum effects, or because the one-dimensional potential energy surface is an oversimplification of the problem.

If a Fe surface is given a brief but large exposure to N_2 at low temperature, the surface will be covered by N_2* [396]. The dissociation to $N*$ is thermodynamically favorable, but does not proceed as virtually no free sites are left [396]. At a higher temperature the coverage by N_2* will be less complete and there will be a sudden transition to complete coverage by $N*$ [396]. At even higher temperatures the coverage by $N*$ will be less than unity [396].

The thermal desorption of N_2 has been treated by Bowker, Parker and Waugh and by Stoltze and Nørskov. In their first paper [535], Bowker, Parker and Waugh used a conventional prefactor for the rate of $N*$ recombination.

This resulted in a somewhat too high binding energy for N* and led to the conclusions that the enthaply measured at high coverages by N* might be more appropriate [536]. Later the prefactors were adjusted [537] resulting is a better value for the stability of N*. Using the data of Stoltze and Nørskov, Bowker, Parker and Waugh [537] found a broad peak for the thermal desorption of N_2. This discrepancy was resolved by Trivino and Dumesic, who concluded that both the data-set of Stoltze and Nørskov and the revised data set of Bowker, Parker and Waugh reproduce the experimental data [474].

2.6.8 Chemisorption of NH_3

2.6.8.1 Structure of Chemisorbed NH_3

NH_3 may be adsorbed on Fe without decomposition at low temperatures; at higher temperatures a number of species are formed [538]. For NH_3 chemisorption on a stepped Fe(1 0 0) surface at low temperatures, ESDIAD shows that NH_3 is bonded to the surface through the N-atom [539] and that the molecule has free rotation around the nitrogen-surface bond [539]. Under increasing exposure at 110 K, chemisorbed NH_3 in *on-top* geometry and in multicoordinated geometry; adsorption on top of the chemisorbed layer and finally multilayer chemisorption is observed [540]. Above 155 K only chemisorbed NH_3 is formed [540]. Based on an interpretation of the ultra violet photoelectron spectroscopy spectra, the species formed upon adsorption of NH_3 on Fe is assigned to NH_3 * at 120–300 K [541], to NH* at 350 K [541] and to N* at 500 K [541].

NH_3* + D* isotopic scrambling is observable at low temperature for NH_3/Fe(1 1 1) [397] but not for NH_3/Fe(1 1 0) [541].

Low-energy electron diffraction for NH_3/Fe(1 0 0) [542] and NH_3/Fe(1 1 1) [542], [429] shows disorder. Low-energy electron diffraction for NH_3/Fe(1 1 0) shows disordered c(2 × 3) at 120 K [541], (2 × 2) at 280 K [429] and at 340 K [541]. The (2 × 2) has been interpreted as NH* [541]. By NH_3 adsorption on Fe(1 1 1) followed by flashing to 310 K, a (3 × 3) structure is observed [543]. More complex patterns are observed on stepped surfaces [543].

Facetting of Fe(1 0 0) after high temperature exposure to NH_3 has been observed [542].

By exposure of a Fe film to NH_3, several layers of nitride may be formed [544]. By reaction of this nitride layer with D_2, the product is mostly NH_3 indicating that the film contains H [544].

2.6.8.2 Thermodynamics

The enthalpy of chemisorption is −71 kJ/mole [541] for NH_3/Fe(1 1 0) and −84 kJ/mole [397] for NH_3/Fe(1 1 1) from temperature programmed desorption.

2.6.8.3 Adsorption Kinetics

At 120 K the initial sticking coefficient is 0.16 on Fe(1 1 0) [541].

2.6.8.4 Desorption Kinetics

At low coverage NH_3 desorbs from Fe(1 1 0) at 255 K [541], from Fe(1 1 1) at 330 K [397] or 380 K and 435 K [543]. Desorption of N_2 is observed at 900 K for NH_3/Fe(1 1 1) [397] and at 900 K from NH_3/Fe(pc) [229]. Desorption of H_2 is observed at 390 K from NH_3/Fe(1 1 1) [397].

2.6.8.5 Dissociation

For NH_3/Fe(1 0 0) [542] and NH_3/Fe(1 1 1) [397, 542] the dissociation becomes observable at 160 K and is complete at 320 K. For NH_3/Fe(1 1 0) the dissociation temperature is 260–290 K [540].

The dissociation of NH_3* on Fe(1 1 0) was reported to give N∗ and H∗ as the only reaction products [540]. Others have found evidence for the intermediate formation of NH∗ [545].

By field ion mass spectroscopy of NH_3 on Fe at room temperature, $4 \cdot 10^{-4}$ torr, N_xH_y species and FeN_xH_y species are detected [510]. Secondary ion mass-spectroscopy of NH_3/Fe(1 1 0) at 130 K shows $Fe_nNH_3^{m+}$ with n, m = 1, 2, NH^{n+} with n = 0, 1, 2, 3, 4, H^+, H_2^+ and Fe^+ [545]. The spectrum is interpreted as fragments of molecularily adsorbed NH_3* [545]. When the temperature is increased, $FeNH_3^+$ and $FeNH_3^{2+}$ decrease smoothly, while Fe^+ increases smoothly, illustrating the decrease in surface coverage and the reaction of NH_3(g) with the surface at higher temperatures [545]. The intensity of NH_3^+ with NH_2^+ decreases steeply above 300 K consistent with the temperature programmed desorption studies [545]. Both ions are probably formed from adsorbed species [545]. The intensity of NH^+ decreases more slowly than NH_2^+ and NH_3^+; the reason is probably that NH^+ is formed from both NH_3(g) and NH∗ [545].

2.6.8.6 Properties of Chemisorbed NH_3

Ultra violet photoelectron spectroscopy peaks for NH_3* are −7.4 and −11.8 eV on Fe(1 0 0) [542], −6.4 and −11.2 eV on Fe(1 1 0) [541], −7.2 and −11.6 eV on Fe(1 1 1) [397, 542]. NH∗/Fe(1 1 0) formed by thermal decomposition of NH_3*/Fe(1 1 0) shows ultra-violet photoelectron spectroscopy peaks at −5.2 and −8.4 eV [541].

The X-ray photoelectron spectroscopy N(1s) spectrum of NH_3/Fe(pc) shows one peak at 400 eV at 85 K [471] and two peaks at 397.2 and 400 eV at 290 K [471]. X-ray photoelectron spectroscopy and ultra violet photoelectron spectroscopy of NH_3/Fe(pc) are interpreted as Fe, after $2 \cdot 10^7$ L at 670 K, $5 \cdot 10^{-4}$ torr [229] and Fe_xN, with x approximately equal to 2, after $5.4 \cdot 10^{10}$ L at 670 K, 1 torr [229].

Auger electron spectroscopy shows that the core-level energy for Fe is independent of NH_3 coverage [229] indicating that the N–Fe bond is covalent.

High-resolution electron-energy-loss spectroscopy spectra of NH_3/Fe(1 1 0) at 120–315 K were reported and interpreted [540, 546]. The geometry of the adsorbed molecule is C_{3v}, i.e., the symmetry of the surface has a negligible effect on the adsorbed molecule. The high-resolution electron-energy-loss spectroscopy spectra are interpreted [540] as Fe–N stretch at 420 cm^{-1} for NH_3*, 400 cm^{-1} for ND_3*/Fe, symmetrical NH_3 stretch at 1170 cm^{-1} for NH_3*, 905 cm^{-1} for ND_3*, symmetrical NH_3 stretch at 3310 cm^{-1} for NH_3*, 2410 cm^{-1} for ND_{3+}*.

The change in work function at saturation is −1.98 eV for NH_3/Fe(1 0 0) [542], −2.4 eV for NH_3/Fe(1 1 0) [541] and −2.05 eV for NH_3/Fe(1 1 1) [542]. For NH_3/Fe(1 1 0) the work function decreases with exposure and passes through a minimum, −2.4 eV, at 15 L, 120 K [541].

2.6.8.7 Effect of Preadsorbed Species

The adsorption of NH_3 on Fe overlayers on Ru(0 0 1) has been studied [506].

Preadsorbed N* decreases the initial sticking coefficient and the enthalpy of chemisorption for NH_3 [541]. The saturation coverage for NH_3 is unchanged [541] or decreased [547].

The tendency to dissociate is reduced by preadsorbed c(2 × 2) N/Fe(1 0 0) [542] and (3 × 3) N/Fe(1 1 1) [397].

For a Fe(1 1 1) surface contaminated with 1.2 μmole O/m^2, the initial sticking coefficient is 0.03 at 300 K [517].

2.6.9 Adsorption of N_2H_4

Fe is an active catalyst for N_2H_4 decomposition at 26 and 365 °C [548]. Labeling with ^{15}N shows that the N_2 bond is not split [548]. Exposing Fe to N_2H_4 at 243 K leads to the formation of NH_3, H_2 and N_2 [549].

During adsorption of N_2H_4 on Fe(1 1 1), the tendency of dissociate is great. At 126 K ultra violet photoelectron spectroscopy shows the presence of N_2H_4* after 0.5 L exposure [550]; condensation is detected at 80 L, 126 K [550]. At 220 K ultra violet photoelectron spectroscopy and X-ray photoelectron spectroscopy detect N_2H_4* and NH_x* after 3 l of exposure [550]. A 550 K N* is formed during exposure [550].

2.6.10 Chemisorption of O_2

The chemisorption of O_2 on Fe has been the subject of a number of studies. This reaction is important for the catalyst during NH_3 synthesis and for the passivation of the catalyst to prevent an uncontrolled oxidation upon exposure to air [89, 551].

2.6.10.1 Structure of Chemisorbed O

The chemisorption of O_2 is complicated, as chemisorption is not clearly distinct from bulk oxidation.

On single crystal surfaces a c(2 × 2) low-energy electron diffraction pattern is formed on Fe(1 0 0) [552] and one Fe(1 1 0) [553, 555]. A c(3 × 1) [553, 555] and a split c (3 × 1) [554] patterns are seen on Fe(1 1 0) at intermediate coverage. A p(1 × 1) pattern is seen for Fe:O = 1:1 on Fe(1 0 0) [552, 556, 557] and on Fe(1 1 0) [553].

Only at low temperatures will the p(1 × 1) surface structure be reasonably completed before the bulk oxidation starts [558, 559]. The chemisorbed layer may be amorphous if the exposure is made at too low a temperature [557]. At room temperature, exposure of the c(2 × 2) structure to O_2 results in the growth of a 3 dimensional oxide [552, 554].

Auger electron spectroscopy and ultra violet photoelectron spectroscopy of O_2 chemisorption at 77 K on Fe have shown that the formation of oxide may be followed by the formation of molecularily adsorbed O_2 [476]. Evidence for the adsorption molecular O_2, even in the absence of oxide, has come from X-ray photoelectron spectroscopy spectra for O_2 adsorbed on single crystal surfaces, where a peak at − 533.6 eV is interpreted as the molecular precursor [560].

In suitable temperature ranges FeO can be observed to grow epitaxially on Fe(1 0 0) [561] and on Fe(1 1 0) [553, 555, 562]. With further exposure the sticking coefficient increases and the low-energy electron diffraction picture disappears [561]. This is interpreted as the nucleation and growth of FeO, which finally results in the formation of p(1 × 1) oxide on Fe(1 0 0) [552] and on Fe(1 1 0) [553].

For (Fe, Al) samples an unstochiometric magnetite [563–566] or Fe(III) oxide [563, 564, 567] may be formed. Passivated (Fe, Co, Al) samples consists of a Fe–Co alloy core; the oxide skin is enriched in Co [60]. The surface is covered with small particles of Fe(III) [566].

Mössbauer spectroscopy of the industrial catalyst shows that some Fe_3O_4 is formed [452]. Oxygen may be present in more than one oxide phase [568]. The chemisorption of oxygen on the surface of the catalyst is not always sufficient to prevent the reaction with air [569], possibly because metallic Fe may migrate through the oxide layer during storage [569, 570].

2.6.10.2 Thermodynamics

The amount of O_2 taken up under passivation in 0.75–1.0% O_2 in N_2 is 3–4% by weight [571]. The oxide film formed is approximately 30 Å [122] or 6–12 atomic layers of oxide [572].

The enthalpy of chemisorption is 0–120 kcal/mole for (Fe, Al) samples [439] at $-183\,^{\circ}C$ and 0-100 kcal/mole for (Fe, Al, K) samples [439] at the same conditions.

The enthalpy of chemisorption for an industrial catalyst is 420 kJ/mole up to 0.25 monolayer of O_2, then the enthalpy drops to 34 kJ/mole [573]. The strength of adsorption and the adsorption capacity for O_2 on the catalyst increases with temperature [574]. The optimum temperature range for passivation is 673–773 K [575].

By pulse chemisorption of O_2 on the catalyst, the O_2 chemisorption area is 1.7 m^2/g at $-78\,^{\circ}C$ and 1.9 m^2/g at $20\,^{\circ}C$ for a sample with a BET area of 10 m^2/g [576].

In the later stages of reduction the amount of oxygen adsorption is proportional to the degree of reduction [122].

2.6.10.3 Adsorption Kinetics

The initial sticking coefficient at room temperature for O_2 on Fe is 0.20 ± 0.01 on Fe(1 1 0) measured by molecular beam techniques [577], 0.13 [554] or 0.20 [577] measured by Auger electron spectroscopy and near unity measured by high-resolution electron-energy-loss spectroscopy and LEED [555]. Below room temperature the sticking coefficient remains constant almost to saturation [578], indicating the existence of a weakly adsorbed precursor. At room temperature the sticking coefficient decreases as expected for site blocking until the $c(2 \times 2)$ structure is complete [552–554].

Following an interruption in exposure for Fe films an increase in sticking coefficient is observed [577]. This is interpreted as the diffusion of Fe out through the oxide layer [573, 577].

For the catalyst the rate of chemisorption of O_2 decreases to 1% of the initial rate at 1 monolayer of $O*$ and to 10^{-6} times the initial rate at 50 monolayers [570].

2.6.10.4 Desorption Kinetics

During temperature programmed desorption of the passivated catalyst, H_2 and N_2, but no O_2 is observed [579].

2.6.10.5 Properties of Chemisorbed O

For the $c(2 \times 2)$ O/Fe(1 0 0) structure the ultra violet photoelectron spectroscopy O(1s) peak is -5.5 eV [552, 580]. This feature is interpreted as chemisorbed $O*$ [552]. At higher coverages peaks are observed at -1.5, -2.5 and -5.0eV [552, 580]. X-ray photoelectron spectroscopy shows peaks at 531.7 eV for chemisorbed oxygen [552, 578] and for oxide at 530.3 eV [560, 578]. X-ray

2.6.11 Adsorption of H_2O

At 77 K H_2O is chemisorbed without dissociation [586, 587]. The chemisorption energy of H_2O is 54.6–66 kJ/mole [586, 587]. Sputtering of the surface does not destroy the passivating effect of the adsorbed H_2O [588], while the electron beam during a Auger electron spectroscopy experiment leads to oxidation and destruction of the passivation [588].

2.6.12 Adsorption of H_2S

The chemisorption of S on Fe(1 0 0) by exposure to sulfur vapors leads to a number of structures [589]. At a Fe:S-ratio of 0.5 the structure is $c(2 \times 2)$ [589]. There is some evidence of a $p(2 \times 2)$ structure at lower coverages [589].

The rate limiting step for the formation of chemisorbed S* from H_2S on Fe is the dissociation of H_2S [423].

H_2S poisons the adsorption of N_2 on Fe-films [513]. When a (Fe, Al) sample is poisoned by H_2S, FeS is observed in the X-ray power diffraction diagram if the $H_2S:H_2$ ratio is sufficiently high [590]. At smaller H_2S concentrations a monolayer of S is formed [590].

After poisoning with 1.6 ppm H_2S in $3H_2 + N_2$ at 303 °C, a monolayer (i.e. 0.4–0.5 mg/m^2) is adsorbed [590]. The catalyst activity is lost a 0.2 mg S/m^2 [590].

Chemisorbed H_2S may not be removed with $3H_2 + N_2$ below 620 °C [590].

The addition of K and Al to Fe increases the resistance to poisoning by H_2S at 400–500 °C, 1 atm [591].

2.7 The Mechanism of NH_3 Synthesis

The key step in the synthesis of NH_3 form $N_2 + 3H_2$ is to dissociate the N–N triple bond in the N_2 molecule. The direct gas phase reaction would involve extremely endothermic and exothermic reactions. The resulting activation energies would be prohibitively high according to the principle of Sabatier.

All imaginable reaction mechanisms can be separated into two main cases. In the associative mechanism H is added to the N_2 molecule before dissociation of the N–N bond, e.g.

$$N_2 + * \leftrightarrows N_2* \tag{6}$$

$$N_2* + 2H* \leftrightarrows N_2H_2* + 2* \tag{7}$$

$$N_2H_2* + * \leftrightarrows 2NH* \tag{8}$$

Very endothermic or exothermic reaction steps are avoided if the N–N bond is broken synchronously with the addition of H to the N-atoms.

In the dissociative mechanism the N–N bond is dissociated before any N–H bond is formed, e.g.

$$N_2 + * \leftrightarrows N_2* \tag{9}$$

$$N_2* + * \leftrightarrows 2N* \tag{10}$$

$$N* + H* \leftrightarrows NH* \tag{11}$$

Very endothermic or exothermic reaction steps are avoided if the N–N bond is broken synchronously with the formation of the N$*$ surface bond.

The number of distinct mechanisms is further varied by the number of intermediate steps and the number of sites involved in the bonding of each intermediate. A large number of mechanisms result in kinetics expressions which are indistinguishable from an experimental point of view.

Catalysis may be understood at several levels. In recent years the understanding of ammonia synthesis has been taken to the level where the high pressure reaction has been treated in terms of numerical models based on a description of the reactants at the atomic level. In the present section we will first address the questions on the nature of the catalytically active structure in the catalysts and the information on the mechanism of ammonia synthesis available from chemisorption studies. We will then describe some models of ammonia synthesis in some detail and then proceed to discuss the remaining aspects of the mechanism of ammonia synthesis based on these models.

2.7.1 Nature of the Active Structure

Experimentally it has been verified that the synthesis of ammonia takes place on the surface of the catalyst [592] rather than in the bulk. However, isotope labeling experiments seems to indicate that small amounts of ammonia may be formed in the bulk at very high temperatures [593]. X-ray photoelectron spectroscopy demonstrates that Fe is present in the surface of the reduced catalyst as Fe^0 and not as Fe oxide [52, 208].

Under ammonia synthesis conditions, ultra-high vacuum data demonstrate that a bulk nitride is not formed [594] whereas during NH_3 decomposition the kinetics indicate that the reaction may proceed on a completely nitrided surface [595]. The increase in reaction rate observed by the addition of H_2 is interpreted as the reaction being faster on Fe metal than on Fe-nitride [595].

The coverage of the total surface by catalytically inactive structural promoters has been determined by a number of techniques. The coverage by catalytically inactive material is 60%, calculated from kinetic data for NH_3 synthesis [596], 55% [191, 207], 60% [207], or 45% [195] from chemisorption measurements, 45% from oxygen exchange with ^{18}O-labeled water [195], and 26% from D_2/H_2 isotopic exchange [191]. The active fraction of the total area generally decreases with increased promoter content [198, 199]. The active area decreases through the sequence Mg > Be > Al, Si, Cr > Mn > Ca > Ti [198, 199] for a fixed promoter concentration. Reduction at a very high temperature (1073 K) increases the NH_3 synthesis and N_2-chemisorption rate; this has been interpreted as reduction of a Fe(II)-spinel [123].

Calculated from the activation entropy, the density of active sites is 10^{12} per cm^2 for a porous Fe-catalyst and $2.5 \cdot 10^{16}$ per cm^2 for a Fe-film [597].

For (Fe, Al) samples large N_2 chemisorption correlates with high catalyst activity [524, 525]. In some cases no correlation between N_2 chemisorption and catalytic activity has been found. This lack of correlation has been ascribed to heterogeneity of the surface [598] or to structural sensitivity for Fe/MgO [328].

While the catalytic activity for the catalyst does not correlate with BET-area [599], the catalytic activity has been found to be proportional with the Fe area as determined by CO-chemisorption [53, 600]. Others have found no correlations or a complex relationship [164, 601], probably because promotion is important.

2.7.1.1 Structural Sensitivity

If we conclude that the active structure is the metallic iron surface, an important question is whether all basal planes have the same catalytic activity [602]. From the results for the chemisorption of N_2 on single crystals, the answer is that the rate of chemisorption *does* differ among the basal planes. Also the presence of chemisorbed K changes the rate of chemisorption significantly. For the catalyst differences in catalytic activity between the basal planes would be reflected in a dependence of the catalytic activity on the particle size, since smaller particles expose a relatively large number of atoms as high index planes or edge atoms.

The rate of NH_3 synthesis has been measured at 20 atm on single crystals of Fe [603, 604] in a high pressure microreactor. The ratio of activities of NH_3 synthesis on a Fe-single crystal is (1 1 1):(1 0 0):(1 1 0) = 418:25 1 [604].

The catalytic activity for Fe/MgO depends on the particle size of Fe; the cause for this has been assigned to the structure sensitivity [322–324, 328]. The increase in synthesis rate measured for Fe/MgO after reduction in pure ammonia is interpreted as surface reconstruction [324].

The high activity of the more open planes of Fe for both N_2 chemisorption and NH_3 synthesis [605] has been assigned to the existence of C_7 sites [505]. The high activity of C_7 sites has been explained in geometric terms [505]. Based on X-ray photoelectron spectroscopy studies of active clusters in (Ni, Fe)-alloys with 0–20% Fe it was concluded that an active site for NH_3 synthesis consists of approx 7 Fe atoms [606].

The investigation whether the catalytic activity differs among the basal planes of the iron is complicated by the recrystallization of the Fe surface during NH_3 synthesis. This has been observed both for unpromoted Fe [607], for (Fe,

Al) catalyst models [608] and for Fe single crystals with evaporated overlayers of Al and K [232].

2.7.1.2 The Effect of K

The segregation of K to the surface has been demonstrated by chemisorption measurements [207, 220], by scanning Auger electron spectroscopy [52, 209–211], and by X-ray photoelectron spectroscopy [208]. Chemisorption measurements [207] and spectroscopic studies using a Fe film in a microreactor [609] have shown that K is in close contact with Fe on the active surface.

Addition of a small amount of K results in an increase in the catalytic activity for ammonia synthesis [139, 217, 240, 242, 244, 248, 249, 288, 289, 609–611] and an increase in activity for nitrogen isotopic exchange [611]. For ammonia synthesis the acitivity effect depends on the pressure [563, 564].

Addition of too much K increases the activation energy for ammonia synthesis [176, 242]. The optimum concentration of K [67, 68] depends on the nature and concentration of the structural promoters [291], and on the preparation method [612].

Kinetic and spectroscopic studies of an Fe film after NH_3 synthesis suggest that the absorbed K atoms are associated not only with N but also with O on the surface [609]

While it is fairly obvious that Al and Ca are structural promoters and increase the activity of the catalyst by increasing the specific area, it is also obvious that K does increase the catalytic activity, but does not do so by increasing the specific area. While the effects of K on the kinetics of NH_3 synthesis may fairly easily be observed in a high pressure reaction experiment, the cause of the effects are very hard to deduce from the observed kinetics.

Evidence for more complex changes in the kinetics [613] has been presented. A complication in the interpretation of the effect of K is that the observed changes in the kinetics depend on the operating conditions for NH_3 synthesis [563, 564, 614, 615] and decomposition [236].

The synthesis of ammonia has been studied in microreactors over single crystal surfaces. At 20 atm the equilibrium coverage by K is 0.15 [232]. K increases the activity of Fe(1 0 0) and Fe(1 1 1). At 0.3% conversion the rate is increase by a factor of 2 in the presence of K [232]. The effect of K is larger at higher conversions [232]. Fe(1 1 0) and K/Fe(1 1 0) do not show any activity for ammonia synthesis [232]. The results for NH_3 synthesis over single crystals thus show that a significant difference in activity between the basal planes persists in the presence of K [616]. However, the results obtained for NH_3 synthesis over single crystals are limited to small conversions and thus the results may be more directly related to the case of a catalyst operating under unusual conditions than to the case of typical reaction conditions.

A detailed study of the rate of the catalytic reaction over Fe films precovered with K has been performed in a microreactor [609]. This study clearly demonstrated the promoting effect of K and the existence of an optimum coverage of K.

Based on studies of the catalyst, the origin of the promoting affect of K has been assigned to a decrease of the work function [217], to changes in the population of reaction intermediates [614, 617–619], to a shift in rate limiting step [620], to a decrease in the dipole moment of the transition state [621, 622], or to a destabilization of NH_3* [221, 232]. Others have speculated that the binding energy for intermediates may be too high on the surface of small Fe particles and that the primary effect of K is to induce particle growth [623, 624].

The numerical models show that the coverage by NH_3* during ammonia synthesis is small [396, 625]. This makes it difficult to understand why a destabilization of NH_3* by K should be responsible for the promoting effect of K [232, 227].

The most direct information on the origin of the effects comes from quantum mechanical calculations and single crystal chemisorption studies. From single crystal studies it is concluded that the effect of K is to increase the stability of N_2* [466]. High-resolution electron-energy-loss spectroscopy spectra of N_2* Fe(1 1 1) and N_2*/K/Fe(1 1 1) show that K does not promote the dissociation of N_2* by weakening the N–N bond in N_2* [475].

Based on quantum-mechanical calculations, the stabilization is expected to be mainly electrostatic and of short range [626–629]. From studies of the Xe(5p) photo emission from Xe/K/Ru(0 0 1) it has been concluded that K decreases the work function for the neighbor Ru-sites [630] and that this effect extends approximately 6 Å from the K nucleus [630].

The idea that the promoting effect of K is due to a stabilization of N_2* has been examined within the model of NH_3 synthesis by Stoltze and Nørskov. The stabilization of N_2* in the presence of K is 12 kJ/mole for Fe(1 1 1) [396, 625] as deduced from the changes in the temperature programmed desorption peak temperatures [231, 466]. Although it is a complication that the different basal planes exposed in the catalyst have very different catalytic activity in the absence of K, the results of the calculations strongly support the electrostatic picture of the promoting effect of K [396, 625].

2.7.2 Results from Chemisorption Studies

The ultra-high vacuum studies show that during NH_3 synthesis $N*$ is formed [414]. As the desorption temperature for $N*$ is well above normal synthesis temperature, the desorption of $N*$ as $N_2(g)$ will be slow at synthesis conditions [414]. If $N*$ were not consumed by the reaction, it would soon inhibit the synthesis due to blockage of active sites [414]. These observations are consistent with $N_2(g) + 2* = 2N*$ being the rate limiting step for NH_3 synthesis [414].

Temperature programmed desorption of N_2 from the industrial catalyst resembles temperature programmed desorption of N_2 in the range 85 to 220 K

[631]. This means that physisorbed N_2* and chemisorbed N_2* on the catalyst have the same properties of K/Fe(1 1 1) as on the surface on the industrial catalyst [631]. To the extent that the desorption of N_2 from $N*$ near 850 K may be followed by temperature programmed desorption, it appears that $N*$ also has the same properties on K/Fe(1 1 1) as on the surface of the industrial catalyst [631].

Chemisorption of N_2 [286, 632–640] and H_2 [634] on Fe in relation to NH_3 synthesis has been the subject of quantum-mechanical calculations. The dynamics of N_2 chemisorption has been simulated using a semiclassical wave packet technique [641]. The simulations agree with the molecular beam experiments in the conclusion that vibrational excitation is of some importance.

2.7.3 Kinetic Models of NH_3 Synthesis

Data for the kinetics measured on single crystal surfaces are important for the study of catalytic reactions as the surface of single crystals are approximations to the more complicated surface of catalysts [642]. This is supported by the agreement between rate measurements for Fe single crystals [603, 604] and the rate measured for an industrial catalyst at 1 atm [643].

The purpose of developing models of the kinetics and mechanism of catalytic reactions starting with a description of the reactants at the atomic level is to understand the kinetic phenomena, rather than to give a very accurate description of a few phenomena. While a model of a single phenomenon may give an accurate description of this phenomenon, this description may not be unique and extrapolation may lead to ambiguities for situations where no experimental data exist. However, if all aspects of interest are described using one model and this model is in reasonable agreement with available data, this model may also be used with some confidence in situations where experimental data are sparse or ambiguous.

2.7.3.1 Formulation of the Models

A number of models have been developed. The model by Bowker, Parker, and Waugh is based on the reaction sequence

$$N_2(g) + * \leftrightarrows N_2* \tag{12}$$

$$N_2* + * \leftrightarrows 2N* \tag{13}$$

$$H_2(g) + * \leftrightarrows H_2* \tag{14}$$

$$H_2* + * \leftrightarrows 2H* \tag{15}$$

$$N* + H* \leftrightarrows NH* + * \tag{16}$$

$$NH* + H* \leftrightarrows NH_2* + * \tag{17}$$

$$NH_2* + H* \leftrightharpoons NH_3* + * \tag{18}$$

$$NH_3* \leftrightharpoons NH_3(g) + * \tag{19}$$

Bowker, Parker, and Waugh proceed by using Arrhenius expressions with known or estimated values for all prefactors and activations energies.

The model by Stoltze and Nørskov [396, 625, 644, 645] is based on the reaction sequence

$$N_2(g) + * \leftrightharpoons N_2* \tag{20}$$

$$N_2* + * \leftrightharpoons 2N* \tag{21}$$

$$N* + H* \leftrightharpoons NH* + * \tag{22}$$

$$NH* + H* \leftrightharpoons NH_2* + * \tag{23}$$

$$NH_2* + H* \leftrightharpoons NH_3* + * \tag{24}$$

$$NH_3* \leftrightharpoons NH_3(g) + * \tag{25}$$

$$H_2(g) + 2* \leftrightharpoons 2H* \tag{26}$$

with the explicit assumption that the rate limiting step is the dissociation of N_2*

$$N_2* + * \leftrightharpoons 2N* \tag{27}$$

Stoltze and Nørskov proceed by applying statistical mechanical methods to this sequence, essentially expressing the thermodynamic properties of reactants and intermediates in terms of spectroscopic properties. Further they treat within the same model, the kinetics of adsorption of N_2 as well as the thermodynamics of adsorption for H_2 and NH_3.

Trivino and Dumesic have considered both of these reaction sequences and compared the results and the differences in approach.

While the models of Bowker, Parker, and Waugh and by Trivino and Dumesic do not make *a priori* assumptions on the nature of the rate limiting step, Stoltze and Nørskov make the explicit assumption that the dissociation of N_2* is rate limiting. The assumption leads to a considerable simplification in the further treatment of their model. While Bowker, Parker, and Waugh, and Trivino and Dumesic must calculate reaction rates iteratively, the model by Stoltze and Nørskov allows the derivation of explicit solutions for the coverages and reaction rates. Further, a number of aspects of the kinetics of ammonia synthesis, such as the activation enthalpy and the reaction orders may be investigated analytically in the model by Stoltze and Nørskov.

The solution of the models involves a number of approximations.

The reaction sites are treated as identical [396, 625, 645]; the reactants, intermediates and products chemisorb competitively on these sites. The competition for the sites important kinetic consequences. The identity of the sites is justified by the results from temperature programmed desorption from single crystals and by the result from quantum mechanical calculations.

The adsorption is assumed to be random. This is justified by the experimental measurements of the temperatures of ordering for the reaction intermediates. The binding energy on a given site is treated as independent of the occupation of the neighbor sites [396].

The gas phase is treated as ideal [396, 625, 645].

For calculations for comparison with laboratory data measured in plug-flow reactors [14, 617], the reactors are treated as isothermal, plugflow reactors [396]. The diffusion limitations are negligible [617, 646].

2.7.3.2 Input Parameters

The input parameters in the model by Stoltze and Nørskov are taken from experimental results. Vibration frequencies are taken from spectroscopic data on single crystals [396, 625, 645]. The rate data [396, 625, 645] for the dissociation of N_2* are taken from the sticking coefficient for N_2 and its activation energy. The binding energy for the intermediates is taken from temperature programmed desorption spectra. For the desorption of $N*$ as N_2 it is important that the low sticking coefficient for N_2 be taken into account [396]. Treatment of the desorption using a "normal" prefactor amounts to the implicit assumption of a different mechanism for desorption than for adsorption and the deduced binding energy for $N*$ is too high.

The number of sites found by CO chemisorption is used for the number of active sites for the catalyst [396]. Quantum mechanical calculations show that the effect of K is to stabilize N_2*. Experimentally the differences in sticking coefficient among the basal planes of Fe is small in the presence of K. The activity of all sites are assumed to be that of K/Fe(1 1 1) [396, 625, 645]. This may be further justified by the close similarity of the experimental temperature programmed desorption spectra of N_2* and $N*$ for the industrial catalyst and for K/Fe(1 1 1) [631].

2.7.3.3 Test

By construction the model reproduces the thermodynamics of adsorption for the intermediates and the kinetics of adsorption ad desorption for $N*$. As no measured data for the catalytic activity have been used in the determination of the input parameters, the model may be tested by a comparison between the calculated and experimental rates of ammonia synthesis. The test is made by calculating the exit ammonia concentration from the input composition and operating conditions for the reactor.

Bowker, Parker, and Waugh in their first paper [535] did not appreciate that the uniquely low sticking coefficient for N_2 must also be reflected in a uniquely low prefactor for the recombination of $N*$. The somewhat too high value for the binding energy of $N*$ resulted in a reaction rate for NH_3 synthesis too low by a factor of 10^5 [535]. Bowker, Parker, and Waugh suggested the use of binding energies appropriate for high coverages by $N*$ available from gravimetric

photoelectron spectroscopy of $Fe(2p_{3/2})$ shows that for $Fe:O < 1:1.5$ the oxidation state for Fe is $+3$ [578] and not $+2$ which has been suggested earlier.

At small coverages the work function increases with the coverage until a maximum is reached at the $c(2 \times 2)$ structure [552]. At higher coverage the work function decreases and goes through a minimum [552, 581].

The complicated behavior of the work function is not reflected in the bulk electronic structure. The electrical resistance of a Fe film increases smoothly with O_2 exposure [582].

2.6.10.6 Oxygen Isotopic Exchange

For an industrial catalyst passivated in industry no $^{18}O + ^{17}O$ exchange is detected, while catalyst passivated in the laboratory catalyses this reaction [583]. After evacuation, the surface of the prereduced catalyst is active in the isotopic exchange for molecular oxygen [584]. The occurrence of O-isotopic exchange was interpreted as some O being loosely associated with metallic sites [569]. This reaction is poisoned by CO and water [584].

2.6.10.7 Effect of Promoters

Scanning electron microscopy shows that Al diffuses into the bulk during oxidation [585].

Low-energy electron diffraction shows the formation of a $c(2 \times 4)$ structure on Fe(1 1 0) [54]. There is no evidence for the formation of K-oxide at K-coverages less than one monolayer [554]. The bulk oxidation starts approximately at $Fe:O = 1:0.5$ independent of K coverage [554]. N$*$ and O$*$ interacts with K [515].

For O_2 chemisorption on K/Fe(1 1 0) the sticking coefficient is unity [554]. The desorption temperature for K from K/Fe is increased from 670 K to 810 K in the presence of coadsorbed O$*$ [500].

The O(1s) X-ray photoelectron spectroscopy peak at 530.0 eV is independent of K coverage and of O-coverage [554]. At small coverages by O$*$ the work function is decreased for O/K/Fe due to depolarization of K–K interactions [500].

2.6.10.8 Effect of Preadsorbed Species

Preadsorbed N$*$ does not inhibit O_2 chemisorption [415] since the preadsorbed N$*$ is displaced to the bulk Fe upon O_2 exposure at 300 K [477, 486]. The presence of N in the bulk decreases the tendency for oxidation of the Fe [477]. The N dissolved in the bulk can only be removed by desorption after the O$*$ layer has been removed by reduction [477].

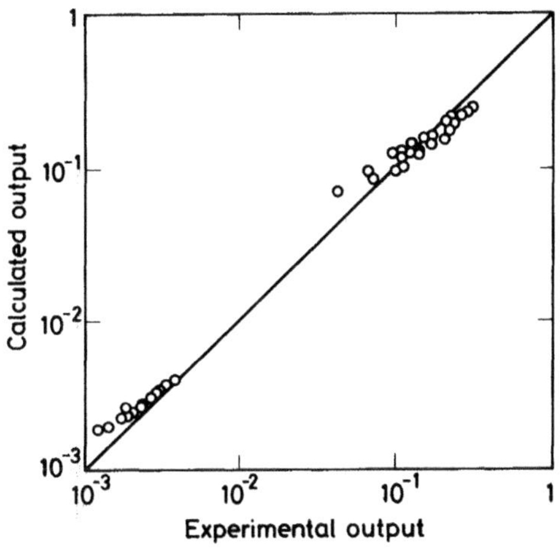

Fig. 2.3. Comparison of the calculated and experimental ammonia production over a Topsøe KM1 catalyst. The data set spans a broad range of pressures (1–300 atm), temperature (375–500 °C) and gas velocities. The synthesis gas is a 1:3 mixture of N_2 and H_2. Reproduced from [724]

isotherms [191]. However, it has been pointed out that these data are not consistent with measurements for the sticking coefficient over single crystals. Bowker, Parker, and Waugh later adjusted their data and obtained agreement with experimental rates within an order of magnitude.

Using a number of data sets at 1, 150 and 300 atm, Stoltze and Nørskov found that the calculated exit concentrations are in good agreement with the experimental results [396, 625, 645], see Fig. 2.3. The differences between calculation and experiment corresponds to an error in the rate of less than a factor of 1.5 [396, 645]. For (Fe, Al) the calculated rates are too high by a factor of about 3. This is clearly a consequence of assuming that all sites have the activity of Fe(1 1 1).

Trivino and Dumesic concluded [330] that the remaining discrepancy between the results by Bowker, Parker, and Waugh and by Stoltze and Nørskov area caused by an unusual small number of active sites assumed in the model by Bowker, Parker, and Waugh.

Not all the input parameters are equally important for the success of the model. Stoltze and Nørskov found that the critical parameters are the prefactor and the activation energy for the sticking coefficient, the ground state energy for N_2* and the ground state energy for $N*$ [396, 645]. These parameters are all rather accurately known from experimental data. The reason why these are the critical parameters is that the first three parameters determine the rate constant, while the groundstate energy for $N*$ determines the number of free sites [396, 645].

2.7.4 The Nature of Reaction Intermediates

There is no gas reaction since no difference is seen between quenching the gas to 0 °C or flow through a hot quartz tube [592].

2.7.4.1 Nitrogen Dissociation

An important question for the reaction mechanism for NH_3 synthesis is whether any H is added to N_2 before the fission of the N–N bond. Mechanisms involving the additions of H before the fission of the N–N bond are referred to as associative while mechanisms involving N* are called dissociative. Even if the kinetic expression by Temkin and Pyzhev was derived from considerations for an associative mechanism, the evidence for the associative mechanism is rather meager.

Attempts have been made to observe the intermediates under conditions where the formation of NH_3 could be hoped for. By field ion mass spectroscopy of a Fe tip at room temperature the observation of the ion N_2^+ [510, 511] and the observation of N_2* by laser Raman spectroscopy of the catalyst [647] has been interpreted as evidence for the associative mechanism [647]. Interestingly, other studies using field ion mass spectroscopy have shown data incompatible with the associative mechanism [648] or no reaction [649]. Presumably the reaction conditions in these experiments are too far removed from conditions where NH_3 may be formed in appreciable amounts.

More direct experimental evidence, such as N-isotopic exchange over the catalyst, the study of chemisorption of N_2, and the synthesis of NH_3 over single crystals, leaves no doubt that the mechanism is dissociative.

While assumption that

$$N_2* + * \rightleftharpoons 2N* \tag{28}$$

is build into the model by Stoltze and Nørskov, the models by Bowker, Parker, and Waugh and by Trivino and Dumesic makes no *a priori* assumption on the nature of the rate limiting step and actually calculate that the dissociation of N_2* is rate limiting under most reaction conditions.

2.7.4.2 Stability of Intermediates

While the thermodynamic stability may be deduced from experimental determinations of the concentration of intermediates, these concentrations are hard to determine in situ. In the numerical models the thermodynamic stability of the intermediates may easily be calculated. Actually, accurate calculations of the thermodynamic stability of the intermediates is a necessity for approaches starting at the description of the intermediates at the atomic level.

The calculated enthalpy and Gibbs free energy for the intermediates [396] are illustrated in Fig. 2.4 and 2.5. One arrives at somewhat different conclusions on the relative stability of the intermediates depending whether-one bases the conclusions on the enthalpy or on the Gibbs free energy. As a consequence there is no simple relation between the heat of formation and the equilibrium concentration.

As the model by Stoltze and Nørskov is based on a description of the spectroscopic properties of the intermediates, it is straightforward to use data

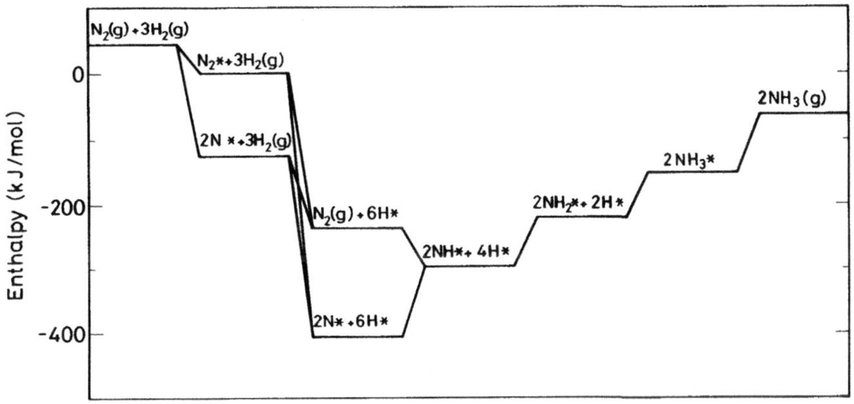

Fig. 2.4. Calculated enthalpy of the intermediates at 673 K for K-promoted Fe. Reproduced from [396]

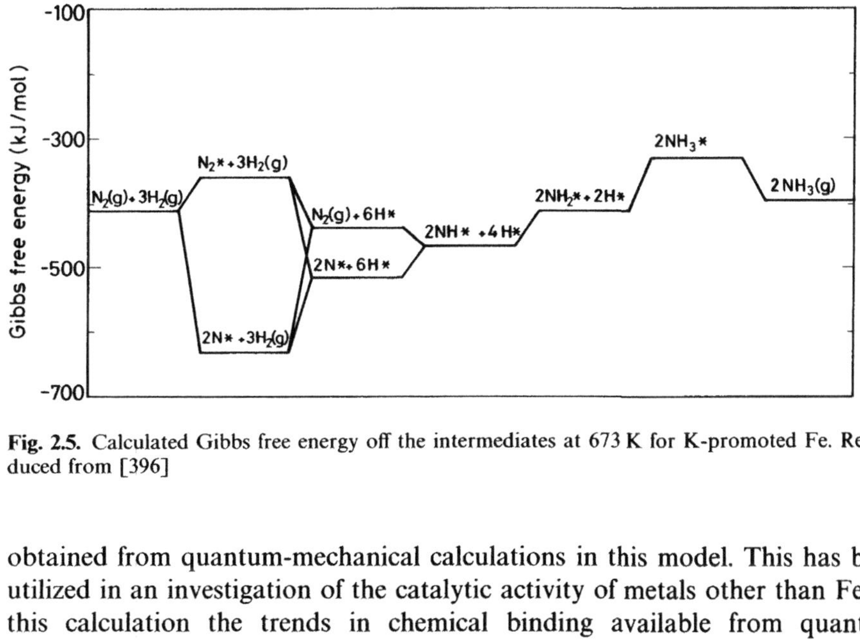

Fig. 2.5. Calculated Gibbs free energy off the intermediates at 673 K for K-promoted Fe. Reproduced from [396]

obtained from quantum-mechanical calculations in this model. This has been utilized in an investigation of the catalytic activity of metals other than Fe. In this calculation the trends in chemical binding available from quantum mechanical calculations were used in an extrapolation using Fe as a reference. The calculation show, in Fig. 2.6, that the reason why Fe is optimal is the correlation between the binding of N_2* and the binding of N* Cr and Mn bind both N_2* and N* more strongly than Fe, and the sticking coefficient for N_2 is close to unity. But the catalytic activity is negligible as N* adsorbs too strongly on the surface. Co, Ni, and Cu bind N_2* and N* more weekly than Fe, the coverage by N* is small but the catalytic activity is low as the sticking coefficient for N_2 is extremely low.

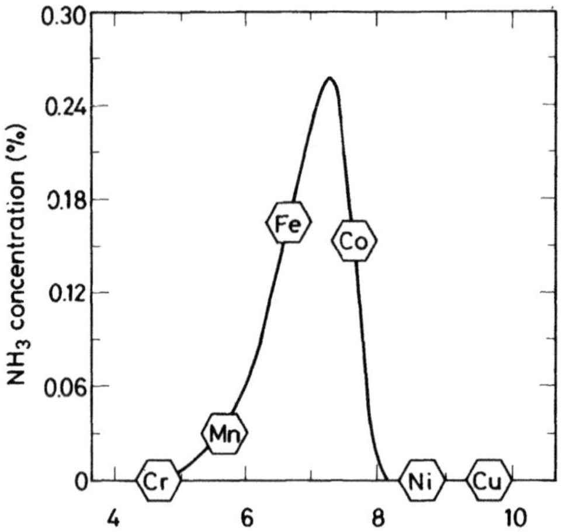

Fig. 2.6. Trends in the ammonia production from 1 m^2 of catalyst as a function of the number of d-electrons in the substrate. The reaction conditions are kept constant 1 atm, 400 °C, stoichiometric gas). Reproduced from [724]

2.7.4.3 Coverage by Intermediates

An interesting aspect of the study of the surface chemistry of the catalyst is the determination of the coverages by intermediates on the surface of the working catalyst. Attempts have been made to observe the intermediates on the surface of the working catalyst by various techniques. Laser Raman spectroscopy shows the following intermediates H* [647], N_2* [473, 647], N* [647], NH* [647], NH_2* [647], and NH_3* [647].

N_2* has been observed using infra-red spectroscopy [512] and secondary-ion mass-spectroscopy [379] on Fe surfaces exposed to $N_2 + H_2$. Arguments in opposition to N_2* being the most abundant reaction intermediate the weakness of adsorption for this species [650] and the rapid decomposition of N_2H_4 to NH_3 over Fe [651]

The surface coverages may easily be calculated from numerical models of ammonia synthesis. From the models [330, 396, 535, 625, 645] it is calculated, see Fig. 2.7, that N* is the most abundant reaction intermediate at most reaction conditions. Only at the conditions of vanishing partial pressure of NH_3 may H* become the most abundant reaction intermediate [330, 396, 625, 695]. For a reactor operating at 10.1 MPa, 673 K with no NH_3 in the inlet and approximately 70% of the thermodynamic equilibrium concentration of NH_3 at the outlet, the coverage by * is quite small, approximately 10^{-3} and decreasing through the reactor. The coverage by N_2* is smaller that 10^{-5} throughout the reactor. The coverage by N*, NH* NH_2* and NH_3* is zero at the inlet and then quickly increases to approximately 0.80, 0.12, $2 \cdot 10^{-2}$ and $4 \cdot 10^{-4}$. The coverage by H* is high, about 0.80, at the inlet and drops to a level of about 0.01 as the coverage by N* builds up.

The calculations show that there are few free sites and that the surface may be viewed as a surface nitride with a few vacancies. As the rate of ammonia synthesis may be expressed as a turnover frequency multiplied by the coverage

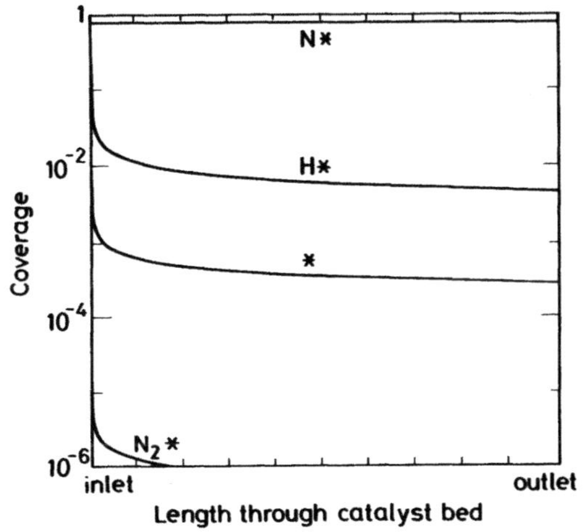

Fig. 2.7. Coverages of intermediates on the surface of a catalyst operating at 10.1 MPa, 673 K, The concentration of NH_3 is 0 at the inlet and about 70% of the thermodynamic equilibrium concentration at the outlet. Reproduced from [645]

by free sites, and since the calculated rate appears to be correct, it is unlikely that the calculated coverages should be significantly in error [396].

At conditions where the concentrations of NH_3 in the gas phase is significant, the coverages by NH_x* are high. At these conditions the sequence of coverages is $N* > NH* > NH_2* > NH_3*$. [396]. This sequence is determined by the difference in entropy between the intermediates [396].

Evidence for the existence of significant amounts of N* has been found from interpretation of reaction orders [607, 614, 617, 652–654], from a comparison of work function, electrical resistance and catalyst activity [655], from laser fluorescence [656], from the observation of N* by electron spectroscopy on the catalyst [52] or a Fe single crystal [414, 603, 604] after exposure to NH_3* synthesis conditions, and from thermodynamic estimates based on data measured for the intermediates on single crystal surfaces [550].

Gravimetric measurements of adsorbed N* during NH_3* synthesis have been performed [191, 494, 657]. For a (Fe, Al) catalyst model the coverage by N* is 0.52–0.69 [191] or 0.11–0.14 [378]. The coverage depends on the operating conditions [378].

Evidence for the existence of significant amounts of NH_x* has been deduced from interpretations of the reaction orders [512, 614, 658, 659], laser fluorescence [656, 660], and from the study of the dissociation of NH_3*/Fe(110) by electron spectroscopy [661]. It is a complication in the deduction of the surface coverages by intermediates that the coverages depend on the operating conditions of the catalyst [396]. These variations may be sufficient to change the nature of the most abundant intermediate [396]. Experimental evidence has been found for changes in the nature of the most abundant reaction intermediates with temperature or promoter concentration [614].

The numerical models of NH_3 synthesis predict that the coverages of H* is quite large if no NH_3 is present in the gas phase [396]. At low temperatures and low conversions a dramatic increase inactivation energy has been observed. The variations of the activation enthalpy with the partial pressure of NH_3 will be discussed in details in Sect. 7.6.2.

2.7.4.4 Lifetime of Intermediates

From the numerical models on the synthesis of ammonia the turnover frequency is readily available. The turnover frequency depends on the operating conditions [396]. The temperature is a particularly important parameter as the turnover frequency increases by 5 orders of magnitude from 500 to 1000 K at 10.1 MPa and a 28% approach to equilibrium [396].

From the turnover frequency the lifetime of N* may be determined. This lifetime is about 1 ms at 673 K and 10.1 MPa if no NH_3* is present in the gas phase and about 1 s at thermodynamic equilibrium [396]. The lifetime is a rapidly decreasing function of temperature and is almost independent of the pressure [396].

For intermediates other than N*, the model does not contain sufficient information to calculate the lifetime. However, the model does contain enough information to calculate an upper limit to the lifetime, see Fig. 8. At 673 K, 10.1 MPa, 10% approach to equilibrium this upper limit to the lifetime is 0.90 s for N*, 0.18 s for NH*, 24 ms for NH_2* and 0.4 ms for NH_3* [396].

The estimated lifetime for N_2* is extremely short, around 0.1 ps [396]. This is further evidence against the associative mechanism.

2.7.5 The Nature of the Rate Limiting Step

One important question about the mechanism of ammonia synthesis is the nature of the rate limiting step. The question about the nature of the rate

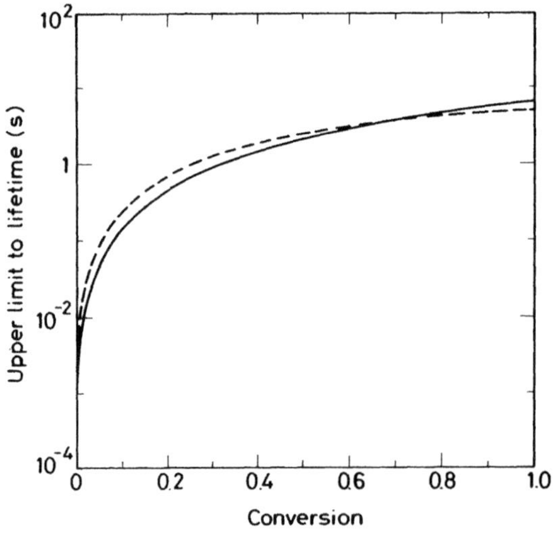

Fig. 2.8. Calculated upper limit to the lifetime of N-species on the surface at varying conversion. Pressure 10.1 MPa (solid curve) respective 101 kPa (dashed curve), N:H ratio 1:3, temperature 673 K. Reproduced from [396]

limiting step is closely related to the question whether the mechanism is associative or dissociative. The single crystal studies leave little doubt that the mechanism is dissociative that the dissociation of N_2* is rate limiting.

2.7.5.1 N_2 *Adsorption as the Rate Limiting Step*

For the associative mechanism with N_2, adsorption as the rate limiting steps one finds a kinetic expression of the same form as for the dissociative mechanism with N_2* dissociation being rate limiting. However, this has not prevented the use of experimental reaction orders for NH_3 synthesis [607, 662] and for NH_3 decomposition [663–666] as arguments for N_2 adsorption being the rate limiting step.

2.7.5.2 N_2 *Dissociation as the Rate Limiting Step*

Evidence for N_2 dissociation as the rate limiting step has been derived from a large number of experimental studies, such as measurements of isotopic exchange rates [356, 482, 521, 526, 532, 667], measurements of the isotopic exchange between NH_3 and N* [507, 528, 668], measurements of the stochiometric number [664] and measurements of the rate of dissolution of N in bulk Fe from N_2 and from NH_3 [423]. Further evidence has been found by the agreement between the rates of N_2 chemisorption and NH_3 synthesis [669], the agreement between adsorption rates and surface coverages [508], the agreement between the rate of synthesis for NH_3 and the coverage by N* [191], by interpretation of reaction orders for NH_3 synthesis [670, 664] and NH_3 decomposition [620], by the value of the activation energy [664] and activation entropy [597, 668] for NH_3 synthesis, and by the agreement between the calculated density of sites and crystallography [664]. Ultra-high-vacuum single crystal studies are consistent with $N_2(g) + 2* = 2N*$ being the rate limiting step for NH_3 synthesis [396, 414, 644].

Evidence against the N_2 dissociation being rate limiting has been postulated from the higher value of the activation energy for N_2 dissociation than for NH_3 synthesis [671], from inconsistent rates of N_2 chemisorption and NH_3 synthesis [518], from K promoting N_2-isotropic exchange but not NH_3 synthesis [611], from the lack of correlation between the rate of N_2 isotopic exchange and the catalytic activity [531], and from measurements of the stochiometric number [672–676]

2.7.5.3. N_2 *Hydrogenation as the Rate Limiting Step*

As for the case of N_2 adsorption as the rate limiting step, the arguments for hydrogenation as rate limiting step are weak.

Evidence for the hydrogenation of N_2 being rate limiting has been found from interpretations of reaction orders for NH_3 synthesis [652, 664, 677] and for NH_3 decomposition [678].

Evidence for the hydrogenation of NH_x* being rate limiting has been found from theoretical considerations [633], from interpretation of reaction orders for NH_3 synthesis [423] and for NH_3 decomposition [620, 664, 678–681].

2.7.5.4 Changes in the Rate Limiting Step

Evidence for a shift in the rate limiting step at low H_2 partial pressure [508, 652], at high temperatures [664], at low temperatures [663, 666] and far from equilibrium [652] has been reported. Transients in the rate during NH_3 cracking suggest that the rate limiting step for NH_3 decomposition is different on iron surfaces than on iron nitride surfaces [595, 665].

The synthesis of ammonia under transient [682, 683] or cyclic [684] operation has been studied. The rate of the catalytic reaction has been found to exhibit hysteresis during changes in the H_2:NH_3 ratio [665] and with changes in temperature [685]. The occurrence of hysteresis has been assigned to a recrystallisation of the surface [685].

Transients in the rate of ammonia synthesis have been observed when the flow is changed [418, 536, 686, 687], when the reaction mixture is changed from $N_2 + 3H_2$ to pure H_2 or to pure N_2 [536, 687, 688], and when the temperature is changed [686]. The behavior under transient operation has been taken as evidence for auto catalysis [686, 536, 689] or for the existence of parallel reaction pathways [682, 683]. By modelling the situation of a reactor under transient operation it has been concluded [690] that limited information can be derived from such experiments.

A change in reaction order at an extremely high space velocity has been found [652, 677]. This change was assigned to a change in the rate limiting step [652, 677]. However, using a numerical model for the kinetics of ammonia synthesis, Stoltze and Nørskov have found that the measured rates may be reproduced even with N_2* dissociation as the rate limiting step and that the unusual reaction orders are found because H* and not N* is the most abundant reaction intermediate.

2.7.6 The Kinetics of NH_3 Synthesis

The kinetics of ammonia synthesis is treated in detail in chapter □. The experimental results on the kinetics of ammonia synthesis and decomposition will be included in the present chapter to the extent that these results illustrate aspects of the mechanism.

Calculations using a numerical model for the kinetics and mechanism of ammonia synthesis show that the reaction orders and activation energy in this model are not constant but show some dependence on the reaction conditions for the catalyst [396].

2.7.6.1 Reaction Orders

In the model by Stoltze and Nørskov [396, 645] the reaction orders have a simple reaction to the surface coverages

$$\alpha N_2 = 1 - 2\theta_{N_2*} \tag{29}$$

$$\alpha H_2 = 3\theta_{N*} + 2\theta_{NH*} + \theta_{NH_2*} - \theta_{H*} \tag{30}$$

$$\alpha NH_3 = -2\theta_{N*} - 2\theta_{NH*} - 2\theta_{NH_2*} - \theta_{NH_3*} \tag{31}$$

The reaction order for N_2 is always very close to 1. Under vanishing partial pressure of NH_3 the reaction order for H_2 is $\alpha_{H_2} \sim -1$ and the reaction order for NH_3 is $\alpha NH_3 \sim 0$. At these conditions the relation is inhibited by H_2. At high partial pressures of NH_3 the reaction orders may approach their limiting values $\alpha H_2 \sim 3$ and $\alpha NH_3 \sim -2$ [396, 645]. The calculated reaction orders, Fig. 2.9, are fairly constant at typical reaction conditions and are in good agreement with experiments [396, 645].

The reaction orders may easily be calculated if the coverages are known. However, only in extreme cases may the coverages be calculated from the reaction orders [396].

At low pressures the synthesis [679] may be treated using pseudo-first-order kinetics, and the decomposition [675] may be treated using pseudo-second-order kinetics.

At low pressures the kinetics may be treated using Langmuir-Hinschelwood kinetics [691–694]. The data for NH_3 synthesis indicate that N_2 chemisorption is rate limiting [694] and that N* is the most abundant reaction intermediate rather than NH* [693]. Deviations from Langmuir-Hinschelwood kinetics have been explained as the dissolution of N in the bulk phase [691, 692].

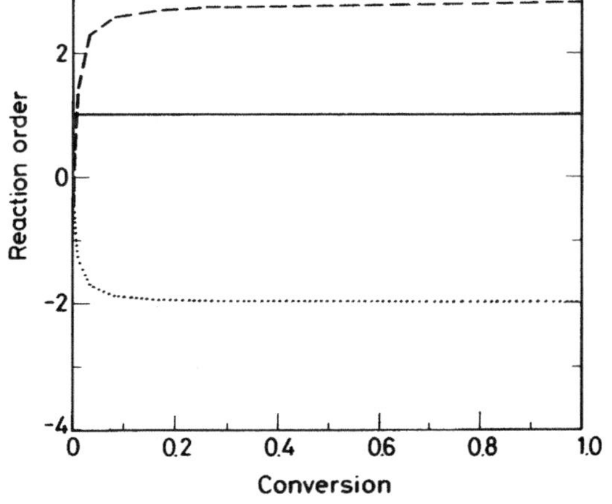

Fig. 2.9. Calculated reaction orders for N_2 (solid curve), H_2 (dashed curve), and NH_3 (dotted curve) for NH_3 synthesis at 10.1 MPa, N:H ratio 1:3, 673 K. Reproduced from [396]

However, the model of ammonia synthesis of Stoltze nd Nørskov shows that even if N* is the most abundant reaction intermediate at typical reaction condition, N*, NH*, NH_2*, H*, and NH_3* are all more abundant than *. This explains why a Langmuir-Hinschelwood expression with only one surface intermediate is not particularly successful [396, 645].

The Temkin-Pyzhev kinetics [671, 695] have been found for NH_3 synthesis [95, 617, 618, 623, 624, 696–698] and for NH_3 decomposition [678, 691, 692]. The Temkin-Pyzhev kinetics have been interpreted in terms of the existence of a continous distribution of binding energies on the surface [671, 699–703]. The reaction orders for NH_3 synthesis in the Temkin-Pyzhev kinetics are 1.0 [704, 670] for N_2; 0.98 [670], 1.0 [176], 1.8 [704], or 2.1 [704] for H_2 and − 1.2 [704], − 1.3 [670], or − 1.4 [704] for NH_3. The reaction orders for NH_3 decomposition are 0 [236] for N_2 − 0.85 [236, 662], − 0.64 [678], 1.2 [662] for H_2 and − 0.82 [662], 0.06–0.09 [662], 0.48 [678], 0.6 [236, 662] for NH_3

Experimental deviations from the Temkin-Pyzhev kinetics have been assigned to pore diffusion [705], to a shift of the rate limiting step at very high space velocity [677], or to a dependence of binding energy on the dispersion of Fe [623, 624]

The original derivation of the Temkin-Pyzhev kinetics was later generalized by Ozaki, Taylor, and Boudart [706]. The Ozaki-Taylor-Boudart kinetics has been found for NH_3 synthesis [613, 614, 617, 618, 620] and for NH_3 decomposition [620, 707]. The Ozaki-Taylor-Boudart kinetics are superior to the Temkin-Pyzhev kinetics [613, 617, 646] for the reproduction of experimental reaction rates over a range of operating conditions. The parameters in the Ozaki-Taylor-Boudart kinetics has been found to depend somewhat on the temperature [615] and on the concentration of K [614, 615]

2.7.6.2 Activation Energy

From the model by Stoltze and Nørskov an expression for the activation enthalpy for ammonia synthesis may be derived [396, 625, 645].

$$H^{\ddagger} = H_1 + H_2^{\ddagger} + 2H_1\theta_{N_2*} + 2H_3\theta_{N*} + 2H_4\theta_{NH*} + 2H_5\theta_{NH_2*} + 2H_6\theta_{NH_3*} + H_7\theta_{H*} \quad (32)$$

where the enthalpies are interpreted as

reaction	enthalpy
$N_2(g) + * \rightleftharpoons N_2*$	H_1
$NH_3 + * \rightleftharpoons N* + \frac{3}{2}H_2$	H_3
$NH_3 + * \rightleftharpoons NH* + H_2$	H_4
$NH_3 + * \rightleftharpoons NH_2* + \frac{1}{2}H_2$	H_5
$NH_3 + * \rightleftharpoons NH_3* +$	H_6
$H_2 + 2* \rightleftharpoons 2H*$	H_7

and $H_2^{\ddagger}$ is the activation enthalpy for the rate limiting step.

$$N_2* + * \rightleftharpoons 2N* \tag{33}$$

This expression may be interpreted by noting that $H_1 + H_2^{\ddagger}$ is the activation enthalpy for the dissociative adsorption of N*. The combinations of coverages and enthalpies in the expression are all enthalpies for desorption reactions. The equation thus says that the activation energy for ammonia synthesis equals the activation enthalpy for the rate limiting step plus the averaged cost of creating more free sites [396, 625, 645].

The contribution from the rate limiting step is small [396, 625, 645]. At typical conditions the dominating term originates from creation of free sites by desorption of N*

$$N* + \tfrac{3}{2}H_2 \rightleftharpoons NH_3 \tag{34}$$

The activation energy is not quite constant, see Fig. 2.10. The calculated values are in agreement with experimental values, provided the experimental values used do not span too large a range of operating conditions [396, 625, 645].

At conditions of a vanishing partial pressure of NH_3, θ_{N*}, θ_{NH*} θ_{NH_2*}, and θ_{NH_3*} will all vanish and the contribution from the reaction

$$2H* \rightleftharpoons H_2 + 2* \tag{35}$$

becomes detectable [396, 625, 645]. This results in a higher activation energy than under more usual conditions [396, 625, 645].

The reported values for the activation energy for NH_3 synthesis are 14 kcal/mole [671], 48 kcal/mole [676], 43.3 kcal/mole [617], 23 kcal/mole [670], 11.5 kcal/mole [677], 27 kcal/mole [708], 16 kcal/mole [524, 525], 50 kcal/mol [520] and 200 kJ/mole [696].

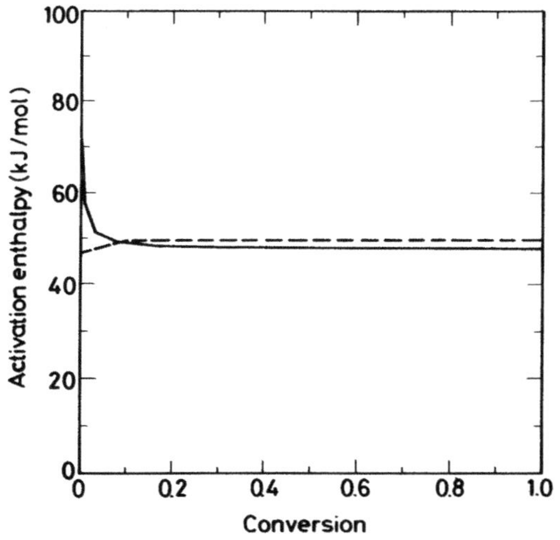

Fig. 2.10. Calculated activation enthalpy for NH_3 synthesis at 10.1 MPa (solid curve) and 1021 kPa (dashed curve), N:H ratio 1:3, temperature 673 K. Reproduced from [396]

The rate of NH_3 synthesis has been measured at 20 atm on single crystals of Fe [603, 604] in a high pressure microreactor. The activation enthalpy for NH_3 synthesis was 19.4 kcal/mole in the absence of K [603, 604] and 18.8 ± 0.5 kcal/mole for K/Fe(100) [232].

The reported values for the activation energy for NH_3 decomposition are 0–16.6 kcal/mole [662], 18.7–26.1 kcal/mol [662], 31.9 kcal/mole [679], 44 kcal/mole [236], and 46 kcal/mole [670]. The value depends on the temperature [236, 662] and on the bulk composition of the catalyst.

2.7.7 Deuterium Isotope Effect on NH_3 Synthesis

The NH_3 synthesis is 2.5 [709], or 3.45 [710] times faster in D_2 than in H_2.

The D-isotope effect is identical to the thermodynamic isotope effect if the most abundant surface intermediate is N* or NH* [693]. This is consistent with the dissociative mechanisms where H* is not involved in the rate-limiting step.

Based on Laser Raman spectroscopy failing to detect N* or NH*, the interpretation of the inverse D_2 isotope effect as a thermodynamic isotope effect has been questioned [473].

2.7.8 The Stoichiometric Number for Ammonia Synthesis

The stochiometric number [657, 680, 699, 711] for NH_3 synthesis is defined as the number of turnovers for the rate limiting step, necessary for one turnover of the total reaction written as $N_2 + 3H_2 = 2NH_3$. If chain reactions are neglected, the stochiometric number is 1 if the rate limiting step is N_2 chemisorption and 2 if the rate limiting step is hydrogenation.

The stochiometric number can be determined experimentally from detailed rate measurements using a reaction mixture of N_2 H_2 and NH_3, which is not in nitrogen isotopic equilibrium.

The stochiometric number for NH_3 synthesis is 1 [526, 527, 615, 712, 713] or 2 [672–676]. For NH_3 decomposition, the stochiometric number is 1 [527, 528] or 2 [673].

The interpretation of the measurements has caused some polemic [672, 674, 677]. The adsorption of NH_3 has been found to be important [527, 528]. Neglecting the adsorption of NH_3 at high partial pressures could give an apparent value of 0 for the stochiometric number [527]; this may have been a problem in some measurements [526].

It has been concluded that some of the earlier measurements were made at low conversion, where the rate is not inhibited by NH_3 and the stochiometric number is undefined [677]. Other complications are the suggestions that the stochiometric number should depend on the presence of K [667] or that some measurements may have been made after incomplete reduction [714].

From their kinetic model of ammonia synthesis Stoltze and Nørskov have found [396] that even if the rate is proportional to the affinity at equilibrium, the rate is first order only over a very narrow range of compositions. The usefulness of assuming proportionality between affinity and rate may be questionable.

2.7.9 Poisoning

A number of compounds including H_2O, H_2S and the halogens are strong poisons for NH_3 synthesis. This is reviewed in detail in Sect. □. In the present section, the literature on poisoning will be reviewed to the extent it illustrates the mechanism of NH_3 synthesis. A common feature of the poisons is that they are adsorbed at least as strongly on the surface of the catalyst as some of the intermediates of the reaction.

H_2O is a strong poison for NH_3 synthesis [194, 715–717]. The loss of activity during poisoning by H_2O is reversible only with mild poisoning [14, 18, 718]. The decrease in catalytic activity during poisoning has been found equal to [719], or much larger [533] than the expected value due to blockage of sites by O*.

The poisonous effect of H_2O has been included by Stoltze and Nørskov in their numerical models of NH_3 synthesis [396, 625] by adding the equation:

$$H_2O + * = H_2 + O* \tag{36}$$

The poisonous effect of H_2O is thus simulated by a simple site blocking. Unfortunately, the binding energy for O* is not readily available from single crystal experiments and had to be obtained by fitting experimental data for ammonia synthesis during partial poisoning by water [396, 625]. The result is $\Delta H = -117$ kJ/mol for reaction 36 [396].

From the model it is found that there is no effect of H_2O at a sufficiently low concentration, see Fig. 2.11. Depending on the reaction conditions, there is a transition from vanishing coverage by O* to close to complete coverage.

The reaction orders for the modified model are

$$\alpha N_2 = 1 - 2\theta_{N_2*} \tag{37}$$

$$\alpha H_2 = 3\theta_{N*} + 2\theta_{NH*} + \theta_{NH_2*} - \theta_{H*} + 2\theta_{O*} \tag{38}$$

$$\alpha NH_3 = -2\theta_{N*} - 2\theta_{NH*} - 2\theta_{NH_2*} - \theta_{NH_3*} \tag{39}$$

$$\alpha H_2O = -2\theta_{O*} \tag{40}$$

At the transition from vanishing coverage of O* to a significant coverage by this species, the reaction orders become abnormal as the reaction order for NH_3 increases from ~ -1 to ~ 0, and the reaction order for H_2O decreases from

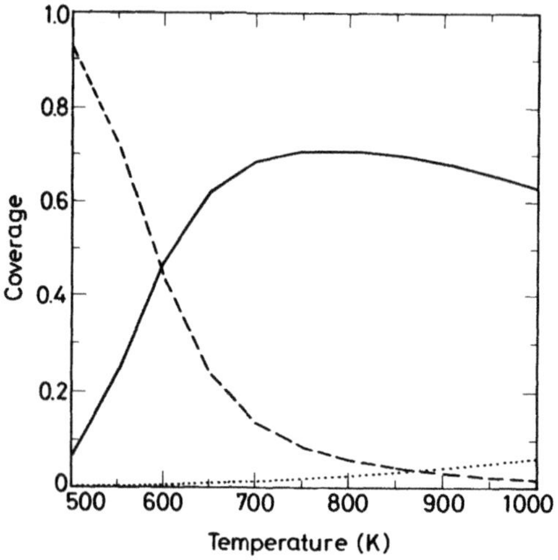

Fig. 2.11. Calculated coverages by N* (solid curve), O* (dashed curve) and H* (dotted curve) for a catalyst operating at 10.1 MPa, N:H ratio 1:3, 28% conversion, 10 ppm H_2O. Reproduced from [396]

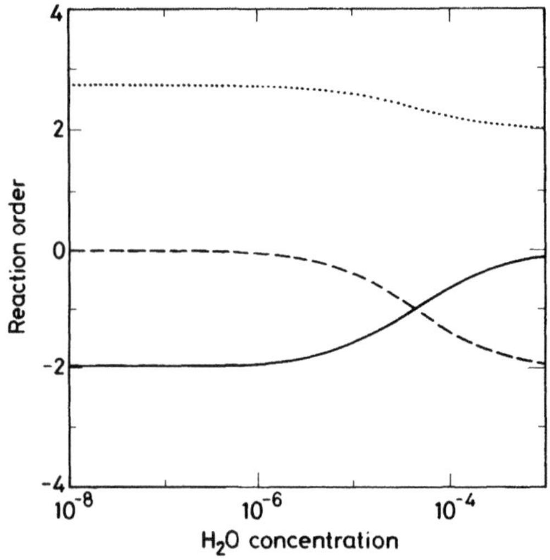

Fig. 2.12. Calculated reaction orders for NH_3 (solid curve), H_2O (dashed curve) and H_2 (dotted curve) for a catalyst operating at 10.1 MPa, N:H ratio 1:3, 673 K, 28% conversion. Reproduced from [396]

~ 0 toward ~ -2 [396], see Fig. 12. Experimentally, the reaction order for H_2O is -1.0 [670, 720].

The inclusion of Eq. 36 in the reaction scheme causes an additional term in the expressions for the activation energy [396, 625].

$$H^{\ddagger} = H_1 + H_2^{\ddagger} + 2H_1\theta_{N_2*} + 2H_3\theta_{NH*} + 2H_5\theta_{NH_2*} + 2H_6\theta_{NH_3*} + H_7\theta_{H*} + 2H_8\theta_{O*} \quad (41)$$

where the enthalpies are interpreted as

reaction	enthalpy
$N_2(g) + * \rightleftharpoons N_2*$	H_1
$NH_3 + * \rightleftharpoons N* + \frac{3}{2}H_2$	H_3
$NH_3 + * \rightleftharpoons NH* + H_2$	H_4
$NH_3 + * \rightleftharpoons NH_2* + \frac{1}{2}H_2$	H_5
$NH_3 + * \rightleftharpoons NH_3* +$	H_6
$H_2 + 2* \rightleftharpoons 2H*$	H_7
$H_2O + * \rightleftharpoons O* + H_2$	H_8

As before, $H_2^\ddagger$ is the activation enthalpy for the rate limiting step

$$N_2* + * \rightleftharpoons 2N* \qquad (42)$$

and the remaining combinations of coverages and enthalpies may be interpreted as the averaged cost of creating more free sites on the surface.

The activation energy is unchanged as long as the coverage by O∗ is negligible but increases rapidly with increasing coverage by this species, see Fig. 13. The limiting value under heavy poisoning is close to 200 kJ/mole [396]. Experimentally, the activation energy is unchanged or increased [718].

H_2S is very poisonous for NH_3 synthesis [14, 18, 721, 722] and N_2 adsorption [513]. The poisoning is irreversible [14, 590]. The poisonous effect is due to S∗ surface blockage [590, 723]. The activation energy of ammonia synthesis is unchanged during H_2S poisoning [723] After exposure to large partial pressures of H_2S, FeS may be observed by X-ray powder diffraction [590].

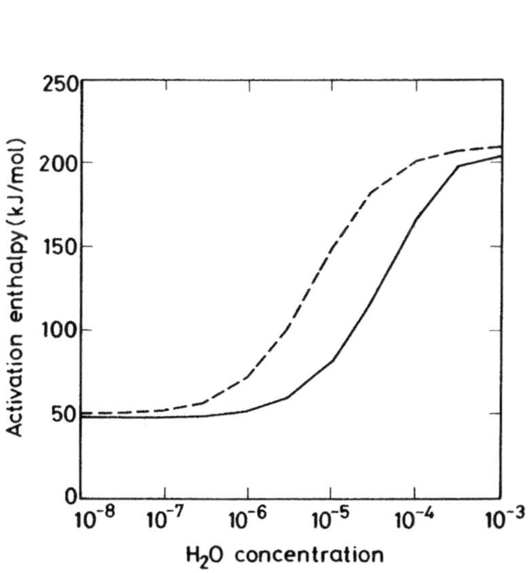

Fig. 2.13. Calculated activation enthalpy for NH_3 synthesis for a partially poisoned catalyst operating at 10.1 MPa (solid curve) resp. 101 kPa (dashed curve), N:H ratio 1:3, 673 K, 28% conversion. Reproduced from [396]

2.8 References

1. Mittasch A (1950) Adv Catal 2: 81
2. Mittasch A (1951) Geschichte der Ammoniak Synthese Verlag Chemie Weinheim
3. Nielsen A (1953) Adv Catal 5: 1
4. Bokhoven C, Van Herden C, Westrik R, Zwietering P (1955) In: Emmett P (ed) Catalysis, Vol 3 p 265
5. Frankenburg W (1955) In: Emmett PH (ed) Catalysis, Vol 3 p 171
6. Vancini CA (1971) Synthesis of Ammonia. McMillan
7. Emmett PH (1975) In: Dragulis E, Jaffee RI (eds) The Physical Basis for Heterogeneous catalysis. Plenum
8. Ozaki A, Aika K (1979) In: Hardy RWF, Bottomley F, Burns RC (eds) A Treatise on Dinitrogen Fixation, Wiley
9. Ozaki A, Aika K (1981) In: Anderson JR Boudart M (eds) Catalysis Science Technology, Vol 1 p 87
10. Boudart M (1981) Catal Rev Sci Eng 23: 1
11. Nielsen A (1981) Catal Rev Sci Eng 23: 17
12. Strel'tsov OA (1967) Kinet Katal 3: 140
13. Emmett PH (1940) 12th Report of the Committe on Catalysis. National Research Council
14. Nielsen A (1968) An Investigation on Promoted Iron Catalysts for the synthesis of Ammonia. Jul Gjelleup Copenhagen, 3rd edition
15. Nielsen A (1950) An Investigation on Promoted Iron Catalysts for the Synthesis of Ammonia. Jul Gjelleup Copenhagen
16. Nielsen A (1970) Catal Rev 4: 1
17. Nielsen A (1977) Fert Sci Technol Ser 2: 87
18. Nielson A (1986) Chem Age India 37: 267
19. Ertl G (1980) Catal Rev Sci Eng 21: 201
20. Ertl G (1982) CRC Crit Rev Solid State Mater Sci 10: 349
21. Ertl G (1983) In: Anderson JR, Boudart M (eds) Catalysis Science Technology, Vol 4 p 209
22. Strasser G, Grunze M, Golze M (1985) J Vac Sci Technol A 3: 1562
23. Ertl G (1980) NATO Adv Study Inst Ser Ser E 39: 271
24. Ertl G (1983) J Vac Sci Technol A 1:1247
25. Ertl G (1989) Stud Surf Sci Catal 44: 315
26. Chen H-C, Anderson RB (1973) J Catal 28: 161
27. Clausen BS, Morup S, Topsøe H, Candia R, Jensen EJ, Baranski A, Pattek A (1976) J Phys Colloq: 245
28. Malycheva TYa, Rabina PD, Kuznetsov LD (1968) Probl Kinet Katal Akad Nauk SSSR 12: 185
29. Yamaguchi S (1951) J Phys Colloid Chem 55: 1409
30. Aleksic BD, Terlecki-Baricevic A (1973) Rev Roum Chim 18: 575
31. Karibdazhanyan NA, Simulina IA, Lachinov SS, Vorotilina ZI, Mishchenko ShSh (1980) Khim Prom-st 294
32. Dimitrov M, Slavov S (1980) God Vissh Khim -Tekhnol Inst Sofia 26: 26
33. Slavov S, Dimitrov M (1980) God Vissh Khim -Tekhnol Inst Sofia 26: 19
34. Dimitrov M Slavov S (1980) God Vissh Khim -Tekhnol Inst Sofia 26: 11
35. Dimitrov IM, Slavov S (1980) Gd Vissh Khim -Tekhnol Inst Sofia 26: 3
36. Sasaki N, Osumi Y (1975) J Chem Soc Japan 73: 808
37. Uchida H, Todo N (1954) Bull Chem Soc Japan, 27: 585
38. Res Group Mossbauer Spectrosc Nanking Nan-ching (1978) Ta Hsueh Hsueh Pao Tzu Jan K'o Hsueh 48
39. Garbassi F, Fagherassi G, Calcaterra M (1972) J Catal 26: 338
40. Rabina PD, Malysheva TYa, Kuznetsov LD, Batyrev VA (1970) Kinet Katal 11: 1243
41. Saprykina TV, Rabina PD, Chudnilov MG, Alekseev AM, Kuznetsov LD (1976) Kinet Katal 17: 723
42. Norval GW, Phillips MJ (1986) J Phys Chem 90: 4743
43. Peev T, Visokov G, Czako-Nagy I, Vertes A (1985) Appl Catal 19: 301
44. Maksimov YuV, Dumesic DA, Suzdalev IP, Matveev AI (1977) Kinet Catal 18: 499

45. Peev T, Krylova AV, Bozhinova A (1981) Radiochem Radioanal Lett 47: 307
46. Bleskin OI, Lachinov SS, Mirkin AE (1984) Kinet Katal 25: 702
47. Topsøe H, Dumesic JA, Boudart M (1974) J Phys (Paris) 35: C6 411
48. Lachinov SS, Rubinshtein AM, Akimov VM, Klyachov-Gurvich AL, Konyukhova IN, Kuznetsov LD, Levitskaya TT, Pribytkova NA, Slinkin AA, Chesnokova RA (1964) Kinet Katal 5: 478
49. Lachinov SS, Torocheshnikov NS, Simulina IA, Lyudkovskaya BG, Rakmat-Zade AG, Ugnachev VI, Alipur G, Sushcheva AE (1975) Tr Mosk Khim-Tekhnol Inst 85: 20
50. Ludwiczek H, Preizinger A, Fischer A, Hosemann R, Schoenfeld A, Vogel W (1978) J Catal 51: 326
51. Michel A, Pouillard E (1948) Comptes Rend 227: 194
52. Ertl G, Thiele N (1979) Appl Surf Sci 3: 99
53. Peters C, Schafer K, Krabetz R (1960) Z Elektrochem 64: 1194
54. Wilchinsky ZW (1949) Anal Chem 21: 1188
55. Westrik R (1953) J Chem Phys 21: 2094
56. Dry ME, Ferreira (1967) J Catal 7: 352
57. Dmitrov M (1979) Geterog Katal 4: 349
58. Klisurski DG, Mitov IG, Petrov KP (1980) Thermochim Acta 41: 181
59. Tomov T, Klissurski D, Mitov I (1982) Phys Stat Sol (a) 73: 249
60. Tricker MJ, Vaishnava PP, Whan DA (1982) Appl Catal 3: 283
61. Glodeanu F, Spinzi A, Nicolaescu IV, Galatchi G, Spinzi M (1979) Rev Roum Phys 24: 161
62. Uchida H, Todo N (1955) Repts Govt Chem Ind Research Inst Tokyo 50: 23
63. Chen HC, Anderson RB (1972) J Coll Interfacial Sci 38: 535
64. Spinzi A, Galatchi G, Spinzi M (1979) Geterog Katal 4: 439
65. Rozin AT, Komarov VS, Efros MD, Lemeshonok GS (1982) Vestsi Akad Navuk BSSR Ser Khim Navuk 31
66. Uchida H, Todo N (1956) Bull Chem Soc Japan 29: 20
67. Krabetz R, Peters C (1965) Angew Chem 77: 333
68. Krabetz R, Peters C (1963) Ber Bunsen-Ges Phys Chem 67: 390
69. Egyhazi T, Scholtz J Beskov VS (1984) React Kinet Catal Lett 24: 1
70. Komarov VS, Efros MD, Rozin AT, Lemeshonok GS, Eremenko SI (1980) Dokl Akad Nauk BSSR 24: 1098
71. Rajaram RR, Sermon PA (1985) J Chem Soc Faraday Trans 1 81: 2577
72. Rajaram RR, Sermon PA (1985) J Chem Soc Faraday Trans 1 81: 2593
73. Brown R, Cooper ME, Whan DA (1982) Appl Catal 3: 177
74. Youssef AM, Ami NM (1978) Surface Technology 7: 469
75. Marakhovets LN, Simulina IA, Lachinov SS, Sobolevskii VS, Lytkin VP, Christozvonov DB (1972) Obl Neorg Tekhnol 3
76. Chernysheva LA, Tovbin MV, Zabuga VYa, Efimova NI (1980) Kinet Katal 18: 25
77. Timofeev VA, Lachinov SS, Torocheshnikov NS, Simulina IA, Lyudkovshaya BG, Rudnitskii LA, Chudinov MG (1973) Tr Mosk Khim-tekhnol Inst 738: 33
78. Baranski A, Bielanski A, Blasiak E, Dulski R, Rokosz A (1968) Chem Stosow Ser A 12: 45
79. Bielanski A, Baranski A, Musial U, Rokosz A (1967) Chem Stosow Ser A 11: 365
80. Jensen EJ, Topsøe H, Sørensen O, Kragh F, Candia R, Clausen BS, Morup S (1977) Sc J Metall 6
81. Lachinov SS, Torocheshnikov NS, Simulina IA, Rakhmat-Zade AG, Ugnachev VI, Alipur G Garanina EF (1975) Tr Mosk Khim- Tekhnol Inst 85: 22
82. Ertl G, Prigge D, Schlögl R, Weiss M (1983) J Catal 79: 359
83. Patyi L, Tsarev VI, Krylova AV, Oravets D, Farkas Z, Torocheshnikov NS (1979) React Kinet Catal Lett 12: 165
84. Maxwell LR, Smart JS, Brunauer S (1951) J Chem Phys 19: 30
85. Maxwell LR, Brunauer S (1949) Phys Rev 76: 175
86. Perel'man NI, Zubova IE, Rabina PD, Kuznetsov LD, Pavlova NZ (1978) Tr Mosk Khim-Tekhnol Inst im.D I Mendeleeva 99: 50
87. Pele L (1983) Rev Chim (Bucharest) 34: 168
88. Aleksic BD, Terelecki-Baricevic A (1973) Bull Chem Soc Beograd 38: 447
89. Medeleanu V, Doca N, Stefanescu M, Bibolaru A (1979) Rev Chim (Bucharest), 30: 751
90. Tovbin MV, Kuznetsov VA, Popovich ZP (1973) Kinet Katal 10: 55
91. Tovbin MV, Popovich ZP (1969) Ukr Khim Zh 35: 787

92. Pennock GM, Flower HM, Andrew SPS (1987) J Catal 103: 1
93. Bardik ZN, Loza AN, Rusov MT, Strel'tsov OA (1966) Khim Prom-st (Moscow) 42: 351
94. Tang Ren-Yuan, Zhang Su, Wang Chengyu, Liang Dongbai, Lin Liwu (1987) J Catal 106: 440
95. Strel'tsov OA, Antonyuk LF, Kuznetsova EP (1967) Kinet Katal 3: 155
96. Chistozvonov DB, Kirillov IP, Yagodkina GN, Dantsig GA, Lytkin VP, Marakhovets LN (1974) Izu Vyssh Ucheb Zaved Khim Khim Tekhnol 17: 870
97. Baranski A, Lagan JM, Pattek A, Reizer A (1982) Appl Catal 3: 207
98. Reizer A, Baranski A (1984) Appl Catal 9: 343
99. Baranski A, Lagan M, Pattek A, Reizer A (1980) Arch Hutn 25: 143
100. Strel'tsov OA, Rusov MA, Kokhar LA, Lonza AN (1960) Kinet Katal 1: 597
101. Dvornik OS, Strel'tsov OA, Artyukh YuN (1977) Ukr Khim Zh (Russ ed) 43: 375
102. Strel'tsov OA, Dvornik OS, Lytkin VP (1977) Izv Vyssh Uchebn Zaved Khim Khim Tekhnol, 20: 1662
103. Astakhov NN, Alekseev AM, Minaev DM, Rabina PD, Perel'man PD, Kuznetsov LD (1979) Ktim Prom-st Ser Azotn Prom-st, p 36
104. Pattek-Janczyk A, Hrynkiewicz AZ, Kraczka J, Kulgawczuk D (1983) Appl Catal 6: 35
105. Pattek-Janczyk A, Baranski A, Kotarba A, Kowalska A, Miczko B, Pyrczak E, Reizer A, Reubenbauer K, Sepiol B, Spiewak Z (1988) Appl Catal 39: 169
106. Visokov G, Ivanov D (1976) Zh Prikl Khim (Leningrad) 49: 1001
107. Minarev DM, Astakhov NN (1980) Deposited doc no SPSTL 575khp-D80
108. Kock AJHM, Fortuin HM, Geus JW (1985) J Catal 96: 261
109. Mosesman MA (1951) J Am Chem Soc, 73: 5635
110. Popovich ZP, Tovbin MV (1970) Ukr Khim Zh 36: 726
111. Klisurski D, Mitov I (1979) J Phys Colloq 353
112. Topsøoe H, Dumesic JA, Boudart M (1973) J Catal 28: 477
113. Hall WK, Tarn H, Anderson RB (1950) J Am Chem Soc 72: 5436
114. Mahapatra H, Ghorai DK, Ganguli NC, Sen SP (1978) Fert Technol 15: 226
115. Baranski A, Lagan M, Pattek A, Reizer A, Christiansen LJ, Topsøe H (1979) Stud Surf Sci Catal 3: 353
116. McCartney JT, Anderson RB (1951) J Appl Phys 22: 1441
117. Uvarova IV, Rusov MT, Samchenko NP (1969) Dopov Akad Nauk Ukr RSR Ser B, 31: 439
118. Uvarova IV, Rusov MT, Samchenko NP (1969) Kinet Katal 10: 558
119. Uvarova IV, Rusov MT, Samchenko NP (1969) Porosh Met 9: 51
120. Zhang H-B, Schrader GL (1985) J Catal 95: 325
121. Uvarova IV, Rusov MT, Samchenko NP (1969) Zh Fiz Khim 43: 1423
122. Baranski A, Pattek A, Reizer A (1978) Bull Acad Pol Sci Ser Chim 26: 353
123. Smith PJ, Taylor DW, Dowden DA, Kemball C, Whan DA (1982) Appl Catal 3: 303
124. Baranski A, Fulinski A, Pattek A, Reizer A (1976) Bull Acad Pol Sci Ser Sci Chim 24: 729
125. Baranski A, Bielanski A, Pattek A (1972) J Catal 26: 286
126. Baranski A, Hajduk J (1985) Chemik, 38: 228
127. Pattek-Janczyk A, Hrynkiewicz AZ (1983) Appl Catal 6: 27
128. Pattek-Janczyk A (1988) Wiss Z -Friderich-Schiller-Univ Jena: Naturwiss Reihe, 37: 835
129. Baranski A, Lagan JM, Pattek A, Reizer A (1980) React Kinet Catal Lett 15: 285
130. Yatsimirskii VK, Girenkova NI, Mel'nik PM, Zabyga VYa (1978) Ukr Khim Zh (Russ Ed.) 44: 239
131. Baranski A, Lagan JM, Pettek A, Reizer A (1982) Appl Catal 3: 201
132. Baranski A, Reizer A, Kotarba A, Pyrczak E (1988) Appl Catal 40: 67
133. Ivanov DG, Visokov GP (1972) Zh Prikl Khim (Leningrad) 45: 1456
134. Vink TJ, der Kinderen JM, Gijzeman OLJ, Geus JW (1986) Appl Surf Sci 26: 367
135. Baranski A, Reizer A, Kotarba A, Pyrczak E (1988) Appl Catal 40: 67
136. Baranski A, Reizer A, Kotarba A, Pyrczak E (1985) Appl Catal 19: 417
137. Perrichon V, Charcosset H, Barrault J, Forquy C (1983) Appl Catal 7: 21
138. Klisurski D, Mitov I, Tomov T (1979) Geterog Katal 4: 207
139. Klisurski D, Mitov I, Tomov T (1983) Stud Surf Sci Catal 16: 421
140. Zakieva KZ, Rabina PD, Zubova IE, Khaustova LE, Pavlova NZ, Kuznetsov LD (1973) Tr Mosk Khim-Tekhnol Inst 73: 120
141. Simulina IA, Lyudkovskaya BG (1969) Tr Mosk Khim-Tekhnol Inst 60: 86
142. Lachinov SS, Bleskin DI, Gandilyan SS (1980) Deposited Doc VINITI 3955
143. Tovbin MV, Zabuga VYa, Chernysheva LA (1977) 12th Tezisy Dokl.- UKr Resp Konf Fiz Khim 162

144. Lemeshko ND, Zabuga VYa, Chernysheva LA (1983) Ukr Khim Zh (Russ Ed) 49: 122
145. Tikhonova ON, Zubova IE, Pavlova NZ, Ivanova RF, Lyubchenko YuA, Dmitrenko LM (1973) Tr Mosk Khim -Tekhnol Inst 73: 126
146. Aleksic BD, Klisurski D, Mitov I (1981) Izu Khim 13: 660
147. Aleksic B, Bogdanov S (1985) Thermochim Acta 93
148. Mitov I, Klisurski D, Aleksic B, Gyurova L, Nikolov O (1984) Izu Khim 17: 305
149. Wachs IE, Dwyer DJ, Iglesia E (1984) Appl Catal 12: 201
150. Lachinov SS, Krylova AV, Pavlova NZ, Zubova IE (1978) Tr Mosk Khim Tekhnol Inst im D I Mendeleeva 99: 47
151. Lachinov SS, Simulina IA, Tikhonova ON,. Torocheshnikov NS, Alipur G, Strekalova NM (1975) Deposited Doc 3770
152. Yatsimirskii VK, Kovalenko VN, Ishchenko EV (1982) Ukr Khim Zh (Russ Ed.) 48: 614
153. Peev T (1985) Khim Ind (Sofia) 57: 250
154. Borghard WS, Boudart M (1983) J Catal 80: 194
155. Logan SR, Moss RL, Kemball C (1958) Trans Faraday Soc 54: 922
156. Pavlova NZ, Rogozhina SA, Kuznetsov DA, Zubova IE, Malysheva TYa (1968) Kinet Katal 9: 1390
157. Rogozhina SA, Lachinov SS, Pavlova NZ, Zubova IE, Kuznetsov DA (1966) Tr Mosk Khim-Tekhnol Inst 51: 173
158. Niemann W, Clausen BS, Topsøe H (1987) NATO ASI Ser B 158: 909
159. Niemann W, Clausen BS, Topsøe H (1987) Ber Bunsen-Ges Phys Chem 91: 1292
160. Fierro JLG, Homs N, Ramirez P del la Piscina, Sueiras JE (1983) Z Phys Chem (Wiesbaden) 135: 235
161. Rayment T, Schlögl R, Thomas JM, Ertl G (1985) Nature (London) 315: 311
162. Silverman DC, Boudart M (1982) J Catal 77: 208
163. Emmett PH, Brunauer S (1934) J Am Chem Soc 56: 35
164. Nielsen A, Bohlbro H (1952) J Am Chem Soc 74: 963
165. Fagherazzi G, Galante F, Garbassi F, Pernicone N (1972) J Catal 26: 344
166. Hosemann R, Hentschel MP (1986) Vak -Tech, 35: 3
167. Uchida H, Terao I, Ogawa K (1964) Bull Chem Soc (Japan) 37: 653
168. Rudnitskii LA, Demina GN, Makurina NA, Dmitrenko LM, Mazus EI, Alekseev AM (1984) React Kinet Catal Lett 25: 297
169. Westrik R, Zwietering P (1953) Proc K Ned Akad Wet 56: 492
170. Schafer K (1960) Z Elektrochem 64: 1191
171. Kolbl H, Schottle E (1961) Z Elecktrochem 65: 91
172. Zwietering P, Koks HTL (1954) Nature, 173: 683
173. Fischer A, Hosemann R, Vogel W, Kouteckly J, Pohl J, Ralek M (1981) Stud Surf Sci Catal 7: 341
174. Herbstein FH, Smuts J (1963) J Catal 2: 69
175. Hosemann R, Preizinger A, Vogel W (1966) Ber Bunsen-Ges Phys Chem 70: 796
176. Tamaru K (1964) Bull Chem Soc Japan 39: 771
177. Brunauer S, Emmett PH (1935) J Am Chem Soc 57: 1754
178. Emmett PH, Brunauer S (1937) Trans Electrochem Soc 71: 383
179. Topsøe H, Topsoe N, Bohlbro H, Dumesic JA (1980) Proc 7th Intern Congr Catal (Tokyo) 247
180. Artyukh YuN, Fedun OS, Zyuzya LA (1980) Kinet Katal 18: 20
181. Dvornik OS, Kozub GM, Strel'tsov OA (1977) Teor Eksp Khim 13: 546
182. Dvornik OS, Strel'stov OA, Lyubchenko YuA (1975) Dokl Akad Nauk SSSR 221: 648
183. Rabina PD, Perel'man ND, Zubova IE, Kuznetsov LD, Maravskaya GK (1977) Ukr Khim Zh (Russ ed) 47: 83
184. Dvornik OS, Strel'tsov OA, Lytkin VP (1975) Ukr Khim Zh (Russ ed) 41: 428
185. Dvornik OS, Strel'tsov OA, Maiboroda VP (1975) Ukr Khim Zh (Russ ed) 41: 1329
186. Dvornik OS, Strel'tsov OA, Chernobrivets VL (1975) Ukr Khim Zh (Russ ed) 41: 544
187. Limin Li, Wenxiang Wang (1983) Gaodeng Xuexiao Huaxue Xuebao 4: 763
188. Strel'tsov OA, Fedun OS, Artyukh YuN, Lytkin VP (1977) 12th Tezisy Dokl Ukr Resp Konf Fiz Khim 159
189. Dvornik OS, Strel'tsov OA, Lytkin VP (1975) Dopov Akad Nauk Ukr RSR Ser B 623
190. Bridger, Pole, Beinlich, Tomson (1947) Chem Eng Progr 43: 291
191. Scholten JJF, Konvalinka JA, Zwietering P (1960) Trans Faraday Soc 56: 262
192. Yatsimirskii VK, Vyaz'mitina OM, Kozlova TP (1971) Teor Eksp Khim 7: 645
193. Podgurski HH, Emmett PH (1953) J Phys Chem 57: 159

194. Artyukh YuN, Yas'mo VI, Loza AN (1973) Kinet Katal 14: 1599
195. Solbakken V, Solbakken A, Emmett PH (1969) J Catal 15: 90
196. Kuznetsov LD, Rabina PD, Dmitrenko LM, Mischenko ShSh, Zozulya VYu (1979) Khim Prom-st Ser Azotn Prom-st 19
197. Schultz JM (1972) J Catal 27: 64
198. Emmett PH, Brunauer S (1937) J Am Chem Soc 59: 1553
199. Dry ME, du Pleissis JAK, Leuteritz GM (1966) J Catal 6: 194
200. Dimitrov M, Rusev R (1980) God Vissh Khim -Tekhnol Inst Sofia 26: 120
201. Rabina PD, Kuznetsov LD, Anisimova MI (1969) Khim Prom 45: 350
202. Rozin AT, Komarov VS, Efros MD, Lemeshonok GS (1980) Vestsi Akad Navuk BSSR Ser Khim Navuk 27
203. Lachinov SS, Simulina IA, Alipur GA, Nefedova NV (1978) Tr Mosk Khim-Tekhnol Inst im D I Mendeleeva 99: 36
204. Kharchenko EV, Tovbin MV (1970) Khim Prom Ukr 15
205. Yatsimirskii VK, Girenkova NI (1979) Kinet Katal 20: 168
206. Yatsimirskii VK, Tovbin MV, Girenkova NI (1977) 2nd Tezisy Dokl Vses Simp Akt Poverkhn Tverd Tel 22
207. Brunauer S, Emmett PH (1940) J Am Chem Soc 62: 1732
208. Chudinov MG, Perov VM, Ksenzenko VI, Alekseev AM (1986) Khim Prom-st (Moscow) 92
209. Weiss M, Ertl G (1982) Stud Surf Sci Catal 11: 277
210. Hanji K, Shimizu H, Shindo H, Onishi T, Tamaru K (1980) J Res Inst Catal Hokkaido Univ 28: 175
211. Hanji K, Shimizu H, Shindo H, Onishi T, Tamaru K (1980) J Res Inst Catal Hokkaido Univ 28: 175
212. Pernicone N, Fagherazzi G, Galante F, Garbassi F, Lazzerin F, Mattera A (1972) Proc 5th Intern Congr Catal 2: 1241
213. Artyukh YuN, Boldyreva NA, Rusov MT (1970) Kinet Katal 11: 1531
214. Sasa Y, Uda M, Toyoshima I (1986) J Mater Sci Lett 5: 470
215. Dmitrov M (1980) Khim Ind (Sofia) 449
216. Chinh Chuong Zang, Krylova AV, Klimova GN, Lachinov SS, Torocheshnikov NS (1975) Tr Mosk Khim-Tekhnol Inst 85: 24
217. Ivanov MM, Rudnitskii LA, Rabina PD, Kuznetsov LD (1968) Kinet Katal 9: 1239
218. Poniewierski Z, Tchorzewski T (1975) Chem Stosow 19: 161
219. Alekseeva MP, Lachinov SS, Predoiu C, Totocheshnikov NS (1968) Probl Kinet Katal Akad Nauk SSSR 12: 175
220. Emmett PH, Brunauer S (1937) J Am Chem Soc 59: 310
221. Strongin DR, Somorjai GA (1988) Catal Lett 1: 61
222. Rudnitskii LA, Ivanov MM (1970) Kinet Katal 11: 207
223. Artyukh YuN, Bondarenko RN, Golovatyio VG, Kozub GM (1975) Dopov Akad Nauk Urk RSR Ser B 121
224. Pennock GM, Flower HM (1984) Conf Ser - Inst Phys 68: 263
225. Van Ommen JG, Bolink WJ, Prasad J, Mars P (1975) J Catal 38: 120
226. Bonzel HP, Broden G, Krebs HJ (1983) Appl Surf Sci 16: 373
227. Bare SR, Strongin DR, Somorjai GA (1986) J Phys Chem 90: 4726
228. Broden G, Bonzel HP (1979) Surf Sci 84: 106
229. Ertl G, Weiss M, Lee SB (1979) Chem Phys Lett 60: 391
230. Seip U, Bassignana IC, Küppers J, Ertl G (1960) Surf Sci 160: 400
231. Whitman LJ, Bartosch CE, Ho W (1986) J Chem Phys 85: 3688
232. Strongin DR, Somorjai GA (1988) J Catal 109: 51
233. Strongin DR, Bare SR, Somorjai GA (1987) J Catal 103: 289
234. Connell G, Dumesic JA (1985) J Catal 92: 17
235. Berengarten MG, Abdukadyrova SA, Zuvoba IE, Rabina PD, Rudnitskii LA, Pavlova NZ (1973) Tr Mosk Khim-Tekhnol Inst 72: 14
236. Love K, Brunauer S (1942) J Am Chem Soc 64: 745
237. Rudnitskii LA, Ivanov MM (1969) Dokl Akad Nauk SSSR 104: 139
238. Rudnitskii LA, Ivanov MM (1969) Dokl Akad Nauk SSSR 184: 886
239. Abdukadyrova SA, Rabina PD, Zubova IE, Lyudkovskaya B (1970) Tr Mosk Khim- Teknol Inst 67: 148
240. Kuznetsov LD, Rabina PD, Lopukhov GA (1971) Tr Nauch-Issled Proekt Inst Azoth Prom Prod Org Sin 11: 92

241. Sasa Y, Uda M, Toyoshima I (1982) Chem Lett 2011
242. Enikeev EKh, Krylova AV, Kuznetsov LD, Lachinov SS, Roginskii SZ (1960) Dokl Akad Nauk SSSR 131: 1126
243. Rudnitskii LA, Ivanov MM (1969) Kinet Katal 10: 349
244. Krylova EV, Kuznetsov LD, Konyukhova IN (1964) Kinet Catal 5: 948
245. Broden G, Gafner G, Bonzel HP (1979) Surf Sci 84: 295
246. Rao CNR, Rao G, Ranga PK (1987) Chem Phys Lett 134: 47
247. Rozin AT, Komarov VS, Efros MD, Lemeshonok GS, Eremenko SI (1980) Vestsi Akad Navuk BSSR Ser Khim Navuk 35
248. Dimitrov M, Rusev R (1980) God Vissh Khim -Tekhnol Inst Sofia 26: 111
249. Dimitrov M, Rusev R (1980) God Vissh Khim -Tekhnol Inst Sofia 26: 126
250. Dimitrov M, Rusev R (1980) God Vissh Khim -Tekhnol Inst Sofia 26: 103
251. Tikhonova ON, Zubova IE, Ivanova RF, Lyubchenko YuA, Dmitrenko LM, Mishchenko ShSh (1973) Tr Mosk Khim- Tekhnol Inst 73: 123
252. Berengarten MG, Rudnitskii LA, Zubova IE, Alekseev AM, Dmitrenko LM, Ivanova RF, Mischenko ShSh, Lyubchenko YuA, Tikhonova ON (1974) Kinet Katal 15: 250
253. Zakieva KZ, Rabina PD, Zubova IE, Batanova AM, Kuznetsov LD (1973) Tr Mosk Khim-Tekhnol Inst 73: 117
254. Tikhonova ON, Zubova IE, Pavlova NZ, Ivanova RF, Lyubchenko YuA, Dmitrenko LM (1973) Tr Mosk Khim-Tekhnol Inst 73: 126
255. Takeshita T, Wallace WE, Craig RS (1976) J Catal 44: 236
256. Wallace WE, France J, Shamsi A (1982) Rare Earths Mod Sci Technol 3: 561
257. Komarov VS, Dmitrenko LM, Efros MD, Ivanova RF, Karpinchik EV, Rabina PD, Lemeshonok GS, Mantseva GM (1977) Vestsi Akad Navuk BSSR Ser Khim Navuk 9
258. Komarov VS, Rabina PS, Efros MD, Kuznetsov LD, Rozin AT, Dmitrenko LM, Lemeshonok GS, Solov'eva AD (1979) Vestsi Akad Navuk BSSR Ser Khim Navuk 26
259. Kiss G (1983) Petrochemia 23: 17
260. Komarov VS, Rozin AT (1977) Dokl Akad Nauk BSSR 21: 1102
261. Lachnikov SS, Bleskin OI, Smirnova SA (1980) Deposited doc no VINITI 1014
262. Zakieva KZ, Rabina PD, Abdukadyrova SA, Zubova IE, Khaustova LE, Pavlova NZ (1973) Tr Mosk Khim-tekhnol Inst 72: 20
263. Sultanbekov GK, Kondrat'eva NM, Lachinov SS, Torochesnikov NS, Saltanova VP (1973) Tr Mosk Khim-Tekhnol Inst 72: 12
264. Sergeeva AN, Dovgei VV, Pavlenko LT, Zubritskaya DI, Tkachenko ZhI, Okoskaya AP, Lyubchenko YuA (1983) Kinet Katal 24: 1473
265. Simulina IA, Lachinov SS, Torocheshnikov NS, Sobolovskii VS, Christozvonov DB, Lytkin VP (1969) Tr Mosk Khim- Tekhnol Inst 60: 91
266. Chernysheva LA, Zabuga VYa, Pivovarova NS, Tovbin MV (1981) Visn Kiiv Univ (Ser.): Khim 22: 59
267. Zabuga VYa, Tovbin MV, Chernysheva LA, Drel II, Efimova NI (1975) Ukr Khim Zh (Russ ed) 41: 818
268. Maksimov YuV, Arents R, Suzdalev IP, Yatsimirskii VK (1980) J Phys Colloq 337
269. Maksinov YuV, Arents RA, Suzdalev IP, Yatsimirskii VK, Chernysheva LA (1980) Kinet Katal 21: 1555
270. Komarov VS, Rozin AT, Lemeshonok GS, Eremenko SI (1985) Vestsi Akad Navuk BSSR Ser Khim Navuk 30
271. Rudnitskii LA, Rabina PD, Melik'yan AA, Korneev VP, Kuznetsov LD, Saprykina TV, Alekseev AM (1976) Kinet Katal 17: 989
272. Rudnitskii LA, Rabina PD, Kuznetsov LD, Alekseev AM, Saprykina TV (1976) Dokl Akad Nauk SSSR 227: 919
273. Aleksic B, Jovanovic N, Terlecki-Baricevic A (1978) 3rd Tr Mezhdunar Simp Geterogennyi Katal p 147
274. Badik VS, Lyubchenko YuA, Sergeeva AN, Dmitrenko LM (1974) Zhur Prikl Khim 47: 2180
275. Badik VS, Lyubchenko YuA (1976) Zhur Prikl Khim 49: 876
276. Grisnaeva MP, Krylova AV, Marakhovets LN, Kondrat'eva NM, Nazarova IG, Lachinov SS, Torocheshnikov NS (1972) Tr Mosk Khim- Tekhnol Inst 72: 6
277. Karaslavova K, Anastasov M (1978) 3rd Tr Mezhdunar Simp Geterogenyi Katal 297
278. Krylova AV, Dimitrov M, Karaslavova K, Klimova GN, Lachinov SS, Torocheshnikov NS (1976) Deposited Doc VINITI 1117
279. Kagan TYu, Krylova AV (1968) Kinet Katal 9: 440

280. Dimitrov M, Karaslavova K (1975) Khim Ind (Sofia) 47: 19
281. Komarov VS, Efros MD, Lemeshonok GS, Rozin AT (1978) Vesti Akad, Navuk BSSR Ser Khim Navuk 15
282. Artyukh YuN, Zyuzya LA (1980) Kinet Katal 18: 23
283. Artyukh YuN (1977) 12th Tzeisy Dokl - Ukr Resp Konf Fiz Khim p 101
284. Artyukh YuN, Chernobrivets VL, Kuznetasova EP, Lakoza EL, Zyuzya LA (1977) Kinet Katal 18: 1614
285. Tovbin MV, Vyaz'mitina OM, Silkina SS (1967) Silkina SS Khim Prom-st Ukr 12
286. Huang KH (1985) Appl Catal 15: 175
287. Uebing C, Viefhaus H, Grabke HJ (1988) Appl Surf Sci 32: 363
288. Ozaki A, Aika K, Morikawa Y (1972) Proc 5th Intern Congr Catal 2: 1251
289. Bleskin OI, Lachinov SS (1980) Deposited Doc VINITI 1015
290. Bosch H, van Ommen JG, Gellings PJ (1985) Appl Catal 18: 405
291. Berengarten MG, Rudnitskii LA, Rabina PD, Kuznetsov LD, Zubova IE, Alekseev AM, Zakieva KZ (1974) Dokl Akad Nauk SSSR 214: 601
292. Lee SH, Ruckenstein E (1987) J Catal 107: 23
293. Ruckenstein E, Xu SD (1986) J Catal 100: 1
294. Ruckenstein E, Sushumna I (1986) J Catal 97: 1
295. Sushumna I, Ruckenstein E (1984) J Catal 90: 142
296. Sushumna I, Ruckenstein E (1985) J Catal 94: 239
297. Xin-Quan Xin, Xue-Qin Zhang, Long-Gen Zhu, Qing-Jin Meng, Zhao-Xian Wang, An-Bang Dai (1980) Ts'ui Hua Hsueh Pao 1: 98
298. Sudo M, Ichikawa M, Soma M, Onishi T, Tamaru K (1969) J Phys Chem 73: 1174
299. Mei-Zhi, Bei, Xiu-Zheng Zhang, Shu-Fen Liu, Chang-Ping Shao, Juan Li (1981) Ts'ui Hua Hsueh Pao, 2: 8
300. Pei-Qun Zhang, Gui-Quan Ji, Shi Li, Wei-Fang Wu, Jian-Xia Shen, Zong-Ce Wang (1981) Hua Hsueh Tung Pao, 26: 16
301. Jianxia Shen, Zongce Wang, Xianrong Huang (1983) Sci Sin Ser B (Engl ed) 26: 1
302. Guerrero A, Lopez-Gonzales J de D, Moreno-Castilla C, Rodriguez-Reinoso F (1983) Ext Abst Program 116th Bienn Conf Carbon 349
303. Kalucki K, Morawski W, Arabczyk W (1981) Stud Surf Sci Catal 7 B: 1496
304. Xinquan Xin, Lomggen Zhu, Qingjin Meng, Xuequin Zhang, Peicheng Wu, Anbang Dai (1982) Gaodeng Xuexiao Hauxue Xuebao, 3: 162
305. Nefed'ev AV, Stukan RA, Alekseev VP, Postnikov VA, Shur VB, Novikov YuN, Vol'pin ME (1977) Proc Int Conf Mössbauer Spectrosc 1: 259
306. Nitrogen Fixation Group, Chem Dep Nanking Univ (1977) Hua Hsueh Hsueh Pao 35: 141
307. Hsin-Chuan Hsin, Hsueh-Chin Chang, Chin-Chin Meng, Chao-Hsien Wang, Au-Pang Tai, Yuan-Fu Hsia,, Shen-Hao (1981) Yeh K'o Hsueh T'ung Pao 26: 93
308. Tovbin MV, Kharchenko EV, Yatsimirskii VK, Bogatova NF, Velichanskaya LA, Zhidkova TG (1975) Ukr Khim Zh (Russ ed) 41: 22
309. An-Bang Dai, Xin-Quan Xin, Long-Gen Zhu, Xue-Quin Zhang, Qing-Jin Meng (1979) Nanching Ta Hsueh Hsueh Pao 40
310. Bewer G, Wichmann N, Boehm HP (1977) Mater Sci Eng 31, 73:6
311. Nitrogen Fixation Group Chem Dep Nanking Univ (1977) Hsueh Hsueh Hua Pao 35: 153
312. Postnikov VA, Dmitrienko LM, Ivanova RF, Dobrolyubova NL, Golubeva MA, Gapeeva TI, Novikov YuN, Shur VB, Volpin ME (1975) Izu Akad Nauk SSSR Ser Khim p 2642
313. Rodriguez-Reinoso R, Guerrero-Ruiz A, Moreno-Castilla C, Rodriguez-Ramos I, Lopez-Gonzales JD (1986) Appl Catal 23: 299
314. Arkhipov IL, Stukan RA, Yunusov SM, Lokshin VB, Ezernitskaya MB, Shur VB, Vol'pin ME (1988) Metalloorg Khim 1: 314
315. Hegenberger E, Wu NL, Phillips J (1987) J Phys Chem 91: 5067
316. Chen AA, Vannice MA, Phillips J (1987) J Phys Chem 91: 6257
317. Homs N, Rimirez de la Piscina P, Fierro JLG, Sueiras JE (1984) Z Anorg Allg Chem 518: 227
318. Sueiras JE, Homs N, Ramirez de la Piscina P, Gracia M, Fierro JLG (1986) J Catal 98: 264
319. Marchetti SG, Alvarez AM, Mercader RC, Yeramian AA (1987) Appl Surf Sci 29: 443
320. Connell G, Dumesic JA (1986) J Catal 102: 216
321. Stockwell DM, Bertucco A, Coulston GW, Bennett CO (1988) J Catal 113: 317
322. Boudart M, Delbouille A, Dumesic JA, Khammouma S, Topsøe H (1975) J Catal 37: 486
323. Boudart M, Topsøe H, Dumesic JA (1975) Batelle Inst Mater Sci Colloq 9: 337
324. Dumesic JA, Topsøe H, Khammouma S, Boudart M (1975) J Catal 37: 503

325. Dumesic JA, Topsøe H, Boudart M (1975) J Catal 37: 513
326. Boudart M, Dumesic JA, Topsøe H (1977) Proc Natl Acad Sci USA 74: 806
327. Topsøe H, Dumesic JA, Derouane EG, Clausen BS, Mørup S, Villadsen J, Topsøe N (1979) In: Preparation of Catalysts II Elsevier Scientific Publishing Co (Amsterdam) p 365
328. Topsøe H, Topsøe N, Bohlbro H, Dumesic JA (1981) Stud Surf Sci Catal 7: 247
329. Ramirez de la Piscina P, Homs N, Fierro JLG, Sueiras JE (1985) Z Anorg Allg Chem 528: 195
330. Dumesic JA, Trevino AA (1989) J Catal 116: 119
331. Beckler RK, White MG (1988) J Catal 112: 157
332. Johnston C, Jorgensen N, Rochester CH (1988) J Chem Soc Faraday Trans 84: 2001
333. Johnston C, Jorgensen N, Rochester CH (1988) J Chem Soc Faraday Trans 1 84: 309
334. Rankin JL, Bartholomew CH (1986) J Catal 100: 526
335. Syrkov AG, Smirnov VM (1987) Kinet Katal 28: 1116
336. Lazar K, Reiff WM, Mörke W, Guczi L (1986) J Catal 100: 118
337. Rankin JL, Bartholomew CH (1986) J Catal 100: 533
338. Babb KH, White MG (1986) J Catal 98: 343
339. Conenell G, Dumesic JA (1986) J Catal 101: 103
340. Santos J, Dumesic JA (1982) Stud Surf Sci Catal 11: 43
341. Santos J, Phillips J, Dumesic JA (1983) J Catal 81: 147
342. Tatarchuk BJ, Dumesic JA (1981) J Catal 70: 308
343. Kirch G, Hempelmann R, Schwab E, Zuechner H (1981) Z Phys Chem (Wiesbaden) 126: 109
344. Kirch G, Zuechner H (1982) Springer Ser Chem Phys 19: 398
345. Schwab E, Wicke E (1980) Z Phys Chem (Wiesbaden) 122: 217
346. Kirch G, Schwab E, Wicke E, Zuechner H (1985) In: Proc 8th Inst Congr Catal 4: Verlag Chemie (Weinheim) 209
347. Biwer BM, Bernasek SL (1986) Appl Surf Sci 25: 41
348. Murell LL, Garten RL (1984) Appl Surf Sci 19: 218
349. Biwer BM, Bernasek SL (1986) Appl Surf Sci, 25: 41
350. Armbruster E, Baiker A, Baris H, Guentherodt HJ, Schlögl R, Walz B (1986) J Chem Soc Chem Commun 299
351. Walz B, Wiesendanger R, Rosenthaler L, Guentherodt HJ, Dueggelin M, Guggenheim R (1988) Mater Sci Eng 99: 501
352. Baiker A, Schlögl R, Armbruster E, Guntherodt HJ (1987) J Catal 107: 221
353. Schlögl R, Wiesendanger R, Baiker A (1987) J Catal 108: 452
354. Baiker A, Baris H, Schlögl R (1987) J Catal 108: 467
355. Kharchenko EV, Tovbin MV, Prisyazhnyuk KI, Maznitsa NF, Grion'ko AI (1969) Khim Prim Ukr 7
356. Taylor DW, Smith PJ, Dowden DA, Kemball C, Whan DA (1982) Appl Catal, 3: 161
357. Arthyukh YuN, Lavrentovich RF, Rusov MT (1968) Sci Selec Catal 214
358. Artyukh YuN, Rusov MT, Boldyreva NA (1967) Kinet Katal 8: 1319
359. Yatsimirskii VK, Ishchenko EV, Maksimov YuV, Arents RA, Suzdalev IP (1985) Teor Eksp Khim 21: 500
360. Yatsimirskii VK, Ischenko EV, Girenkova NI (1985) Ukr Khim Zh (Russ Ed.) 51: 599
361. Kalucki K, Narkiewicz U (1987) Pr Nauk Politech Szczecin, 354: 9
362. Yatsimirskii VK, Ishchenko EV, Nazarenko VA, Mischanchuk TB, Girenkova NI (1988) Teor Eksp Khim 24: 185
363. Nagorny K, Bubert S (1987) J Catal 112
364. Yatsimirskii VK, Girenkova NI, Panyuk EV (1987) Zh Fiz Khim 61: 1516
365. Paryjczak T, Zielinski P (1983) React Kinet Catal Lett 23: 165
366. Rabina PD, Kuznetsov LD, Enikeev EKh, Ivanov MM (1967) Kinet Katal 8: 167
367. Tovbin MV, Zabuga VYa, Perekhod NA, Stupak OP (1979) Kinet Katal 17: 49
368. Yatsimirskii VK, Girenkova NI, Ostapyuk VA, Maksimov YuV, Kushkov VD (1979) Kinet Katal 17: 53
369. Lachinov SS, Krylova AV, Vasserberg VE, Chinh Chuong Zang, Ratnikova LG, Torocheshnikov (1976) Deposited Doc VINITI 1241
370. Yatsimirskii VK, Ishchenko EV, Mischanchuk TB, Zo Zen Ho (1988) Teor Eksp Khim 24: 616
371. Matsuyama M, Ashida K, Takayasu O, Takeuchi T (1986) J Catal 102: 309
372. Tkachenko ZhI, Pavlenko LI, Lyubchenko YuA (1980) Zh Prikl Khim (Leningrad) 53: 2080
373. Lyubchenko YuA, Skolozdra OE, Sergeeva AN (1975) Zhur Prikl Khim 48: 958
374. Kuzora TV, Kozin LF, Sokol'skii DV (1975) Izv Akad Nauk Kaz SSR Ser Khim 25: 15
375. Egawa C, Sawabe K, Iwasawa Y (1988) J Chem Soc Faraday Trans 1 84: 321

376. Haase G, Asscher M, Linke U (1988) Appl Surf Sci 35:1
377. Zhou X-L, Yoon C, White JM (1988) Surf Sci 203: 53
378. Bozso F, Ertl G, Grunze M, Weiss M (1977) Appl Surf Sci 1: 103
379. Shvachko VI, Fogel YaM, Kolot VYa (1966) Kinet Katal 7: 834
380. Emmett PH, Takezawa N (1978) J Res Inst Catal Hokkaido Univ 26: 37
381. Emmett PH, Harkness RW (1935) J Am Chem Soc 57: 1631
382. Viswanathan B, Srinivasan V, Sastri MVC (1969) Indian J Chem 7: 363
383. Benzinger J, Madix R (1980) Surf Sci 94: 119
384. Kurz EA, Hudson JB (1988) Surf Sci 195: 15
385. Wedler G, Geuss, Colb, McElhiney (1978) Appl Surf Sci 1: 471
386. Wedler G, Borgmann D (1976) J Catal 44: 139
387. Sharnabanger (1985) Surf Sci 150: 451
388. Chornet E, Coughlin RW (1972) J Catal 27: 246
389. Bagg J, Tompkins F (1955) Trans Faraday Soc 51: 1071
390. Porter AAS, Tompkins FC (1953) Proc Roy Soc A217: 544
391. Kwan T (1949) J Res Inst Catal Hokkaido Univ 1: 100
392. Evdokimova ZhA, Valitov NKh, (1985) Zh Prikl Khim (Leningrad), 58: 672
393. Sastri MVC, Srikant H (1961) J Sci Ind Research (India), 20D: 321
394. Koebl H, Eschweiler R, Ralek M, (1973) Z Naturforsch A 28: 54
395. Ostrovskii VE, Igranova EG (1978) Kinet Katal 19: 681
396. Stoltze P (1987) Phys Scr 36: 824
397. Grunze M, Ertl G (1977) Proc 7th Int Vac Congr 2: 1137
398. Kurz EA, Hudson JB (1988) Surf Sci 195: 31
399. Nowicka E, Wolfram Z, Lisowski W, Dus R (1985) Surf Sci 151: L166
400. Vinals J, Gunton JD (1985) Surf Sci 157: 473
401. Arab A, Hudson JB (1987) Appl Surf Sci 29: 1
402. Porter AS, Tompkins FC (1953) Proc Roy Soc A217: 529
403. Artyukh, Yu N (1981) Kinet Katal 19: 14
404. Kai. Huei. Huang, Xiao.Ming.Zeng, Ji.Tao.Li. (1983) J Catal 81: 259
405. Krylova AV, Vokhmyanin NN, Lachinov SS, Torocheshnikov NS (1974) Tr Mosk Khim-Tekhnol Inst 79: 14
406. Amenomiya Y, Pleizier G (1973) J Catal 28: 442
407. Balint I, Segal E, Bucur I, Chirulescu T (1983) Thermochim Acta 67: 103
408. Kummer JT, Emmett PH (1951) J Phys Colloid Chem 55: 337
409. Enikeev EKh, Ivanov MM (1973) Kinet Katal 14: 1597
410. Artyukh YuN, Lunev NK, Rusov MT (1970) Dopov Akad Nauk RSR Ser B 32: 605
411. Ertl G, Lee SB, Weiss M (1981) Surf Sci 111: L711
412. Artyukh YuN, Il'yushonok VA (1976) Kinet Katal 17: 1078
413. Brunauer S, Emmett PH (1940) J Am Chem Soc 69: 1432
414. Ertl G, Huber M, Lee SB, Paal Z, Weiss M (1981) Appl Surf Sci 8: 373
415. Russo C, Kaplow R (1978) J Vac Sci Technol 15: 479
416. Martsenyuk MG, Rusov MT, Samchenko NP (1968) Kinet Katal 4: 129
417. Ustinov YuK, Urazaev RSh (1979) Kinet Katal 20: 1513
418. Rambeau G, Amariglio H (1978) J Chim Phys-Chim Biol 75: 333
419. Artyukh YuN, Batychko SV, Roev LM (1972) Katal 9: 72
420. Tret'yakov II, Shub BR, Sklyarov AV, Roginskii SZ (1967) Dokl Akad Nauk SSSR 175: 1332
421. Tret'yakov II, Shub BR, Shlyarov AV (1968) Dokl Akad Nauk SSSR 181: 648
422. Boreskova EG, Kuchaev VL, Temkin MI (1984) Kinet Katal 25: 112
423. Grabke HJ (1970) AGARD Conf Proc 52: 8
424. Kummer JT, Emmett PH (1952) J Phys Chem 56: 258
425. Emmett PH, Harkness RW (1935) J Am Chem Soc 57: 1624
426. Chavalier JC, Chornet E (1976) Surf Sci 60: 125
427. Brundle CR (1978) IBM J Res Develop 22: 235
428. Erley W (1981) J Vac Sci Technol 18: 472
429. Yoshida K, Somorjai GA (1978) Surf Sci 75: 46
430. Moon DW, Cameron S, Zaera F, Eberhardt W, Carr R, Bernazek SL, Gland JL, Dwyer DJ (1987) Surf Sci 180: L123
431. Rhodin TN, Rsai M-H, Kasowski RV (1985) Appl Surf Sci 22/23: 426
432. Watanabe M, Kadowaki T (1987) Appl Surf Sci 28: 147

433. Lu J-P, Albert MR, Bernasek SL, Dwyer DJ (1988) Surf Sci 199: L406
434. Kummer JT, Emmett PH (1951) J Am Soc 73: 2886
435. Eischens RP (1952) J Am Chem Soc 74: 6167
436. Wedler G, Ruhmann R (1982) Appl Surf Sci 14: 137
437. Seip U, Tsai C, Christmann, Kuppers J, Ertl G (1984) Surf Sci 139: 29
438. Wedler G, Colb KG, McElhiney G, Heinreich W (1978) Appl Surf Sci 2: 30
439. Beebe, Stevens (1940) J Am Chem Soc 62: 2134
440. Krylova AV, Chechulina GN, Koroleva TL, Lachinov SS, Torocheshnikov NS (1973) Deposited Doc. VINITI 7698
441. Ueda K, Enatsu M (1985) Surf Sci 159: L421
442. Ueda K, Enatsu M (1985) Surf Sci 159: L421
443. Benndorf C, Krüger B, Thieme F (1985) Surf Sci 163: L675
444. Cameron SD, Dwyer DJ (1988) Surf Sci 198: 315
445. Alshorachi G, Wedler G (1985) Appl Surf Sci 20: 279
446. Moon DW, Dwyer DJ, Bernasek SL (1985) Surf Sci 163: 215
447. Huang Y-Y, Emmett PH (1972) J Catal 24: 101
448. Gafner G (1979) S Afrl J Phys 2: 129
449. Rao CNR, Ranga G (1988) Chem Phys Lett 146: 557
450. Behner H, Spiess W, Wedler G, Borgmann D (1986) Surf Sci 175: 276
451. Pirner M, Bauer R, Borgmann D, Wedler G (1987) Surf Sci 189/190: 147
452. Peev T, Krylova A, Nefedova N, Stoilova T (1986) J Radioanal Nucl Chem 105: 157
453. Krylova AV, Nefedova NV, Peev TM, Stoilova TI (1986) Kinet Katal 27: 520
454. Krylova AV, Ustimenko GA, Nefedova NV, Peev TM, Torocheshnikov NS (1986) Appl Catal 20: 205
455. Krylova AV, Nefedova NV, Torocheshnikov NS (1988) Appl Catal 39: 325
456. Krylova AV, Nefedova NV, Peev TM (1987) Stud Surf Sci Catal 34: 625
457. Peev T, Krilova A, Nefedova N, Stoilova T (1986) Khim Ind (Sofia) 58: 2
458. Ponec V, Knor Z (1968) J Catal 10: 73
459. Hall PG, Hope CJ (1970) J Chem Soc A 1970: 2003
460. Tomanek D, Bennemann KH (1985) Phys Rev B 31: 2488
461. Freund HJ, Bartos B, Messmer RP, Grunze M, Kuhlenbeck H, Neumann M (1987) Surf Sci 185: 187
462. Grunze M, Golze M, Fuhler J, Neumann M, Schwarz E (1985) In Proc 8th Int Congr Catal 4: 133
463. Grunze M, Golze M, Hirschwald W, Freund H.-J, Pulm H, Seip U, Tsai MC, Ertl G, Küppers J (1984) Phys Rev Lett 53: 850
464. Grunze M, Strasser G, Golze M (1987) Appl Phys A A44: 19
465. Borgmann D, Kroninger F, Steidl P, Wedler G (1987) Surf Sci 189-190: 485
466. Ertl G, Lee SB, Weiss M (1982) Surf Sci 114: 527
467. Ertl G, Lee SB, Weiss M (1982) Surf Sci 114: 515
468. Bozso F, Ertl G, Weiss M (1977) J Catal 50: 519
469. Wedler G, Borgmann D, Geuss J (1975) Surf Sci 47: 592
470. Tsai MC, Seip U, Bassignana IC, Küppers J, Ertl G (1985) Surf Sci 155: 387
471. Kishi K, Roberts MW (1977) Surf Sci 62: 252
472. Baro AM, Eley W (1980) Surf Sci 112: L759
473. Liao D, Zhang H, Wang A, Cai Q (1987) Sci Sin Ser B (Engl. Ed.) 30: 246
474. Bozso F, Ertl G, Grunze M, Weiss M (1977) J Catal 49: 18
475. Whitman LJ, Bartosch CE, Ho W, Strasser G, Grunze M (1986) Phys Rev Lett 56: 1984
476. Toyoshima I (1981) Shinku, 24: 216
477. Ertl G, Huber M (1980) Z Phys Chem (Wiesbaden), 119: 97
478. Grunze M, Kind In D, Woodruff DP (1982) eds, The Chemistry and Physics of Solid Surfaces and Heterogeneous Catalysis, Vol 4, p 413
479. Wedler G, Steidl G, Borgmann D (1980) Surf Sci 100: 507
480. Wedler G, Borgmann D, Alter W, Witan K (1986) Ber Bunsen-Ges Phys Chem 90: 235
481. Kummer JT, Emmett PH (1951) J Chem Phys 19: 289
482. Magomedbekov EP, Kasatkina LA (1976) Zhur Fiz Khim 50: 2414
483. Morikawa Y, Ozaki A (1968) J Catal 12: 145
484. Ertl G, Grunze M, Weiss M (1976) J Vac Sci Technol 13: 314
485. Imbihl R, Behm RJ, Ertl G, Moritz W (1982) Surf Sci 123: 129
486. Arabczyk W, Muessig HJ (1987) Vacuum, 37: 137

487. Dowben PA, Grunze M, Jones RG (1981) Surf Sci 109: L519
488. Wedler G, Borgmann D, Forchhungsber Wehrtech (Bundesmisist. Verteidigung)., No. BMVg-FBWT 76-20 Luft-Raumfahrt Teil. 2: 151
489. Evdokimova ZhA, Valitov NKh (1985) Zh Prikl Khim (Leningrad) 58: 2121
490. Vitol EN, Orlova KB (1973) Kinet Katal 14: 1514
491. Kwan T (1953) J Res Inst Catal Hokkaido Univ 3: 16
492. Kwan T (1955) J Res Inst Catal Hokkaido Univ 3: 109
493. Brito V, Ralek M (1978) React Kinet Catal Lett 9: 15
494. Zwietering EP, Roukens JJ (1954) Trans Faraday Soc 50: 178
495. Scholten JJF, Zwietering P, Konvalinka JA, de. Boer HJ (1959) Trans Faraday Soc 55: 2166
496. Boreskova EG, Kuchaev VL, Temkin MI (1979) Kinet Katal 20: 147
497. Minachev KhM, Evdokimova ZhA, Valitov NKh (1982) Dokl Akad Nauk SSSR 266: 296
498. Grabke HJ (1976) Z Phys Chem (Frankfurt am Main), 100: 185
499. Ostrovskii VE, Igranova EG (1978) Kinet Katal 19: 538
500. Paal Z, Ertl G, Lee SB (1981) Appl Surf Sci 8: 231
501. Boheim J, Brenig W, Engel T, Leuthauser U (1983) Surf Sci 131: 298
502. Rettner CT, Stein H (1987) J Chem Phys 87: 770
503. Rettner CT, Stein H (1987) Phys Rev Lett 59: 2768
504. Rettner CT, Pfnur HE, Stein H, Auerbach DJ (1988) J Vac Sci Technol A 6: 899
505. Dumesic JA, Boudart M (1975) Proc Symp Catal Chem Nitrogen Oxides p 95
506. Egawa C, Iwasawa Y Chem Lett 1987: 959
507. Nakata T, Matsushita S (1982) J Chem Phys 76: 6335
508. Tamaru K (1965) In: Sachtler WMH, Schuit GCA, Zwietering P (eds), Proc 3rd Intern Conf Catal 1: 664
509. Takezawa N (1972) J Catal 24: 417
510. Schmidt WA (1968) Angew Chem Int Ed Engl 7: 139
511. Artyukh YuN, Golovatyi VG, Korol EN (1980) Teor Eksp Khim 16: 714
512. Okawa T, Onishi T, Tamaru K (1977) Z Phys Chem (Wiesbaden), 107: 239
513. Karpinski W, Palczewska W (1974) Bull l'Acad Polon Sci Chim 22: 159
514. Kozub GM, Strel'tsov OA, Rusov MT (1973) Kinet Katal 10: 52
515. Paal Z, Ertl G (1982) Kem Kozl 57: 51
516. Arthyukh YuN, Yas'mo VI, Rusov MT (1971) Kinet Katal 8: 51
517. Gay ID, Textor M, Mason R, Iwasawa Y (1977) Proc R Soc London A356: 25
518. Tamarau K Prep. 2nd Intern Congr Catal (Paris), 1: 325
519. Tamaru GK (1963) Trans Faraday Soc 59: 979
520. Joris GG, Taylor HS (1939) J Chem Phys 7: 893
521. Boreskov GK, Kolchanova VM, Rachkovskii EE, Filimonova SN, Khasin AV (1975) Kinet Katal 16: 1218
522. Takezawa N, Toyoshima I (1970) J Catal 19: 271
523. Kazusaka A, Toyoshima I (1981) Z Phys Chem (Wiesbaden) 128: 111
524. Yatsimirskii VK, Girenkova NI, Maksimov YuV (1976) Teor Eksp Khim 12: 263
525. Urabe K, Ozaki A (1978) J Catal 52: 542
526. Tanaka K, Matsuyama A (1971) J Res Inst Catal Hokkaido Univ 19: 63
527. Tanaka K (1966) J Res Inst Catal Hokkaido Univ 14: 153
528. Tanaka K (1965) J Res Inst Catal Hokkaido Univ 13: 119
529. Napol'skikh GA, Kasatkina LA (1973) Kinet Katal 14: 653
530. Miyahara K (1956) J Res Inst Catal Hokkaido Univ 4: 193
531. Vol'pin ME, Novikov YuN, Postnikov VA, Schur VB, Bayerl B, Kaden L, Wahren M, Dmitrienko LM. Stukan RA, Nefed'ev AV (1977) Z Anorg Allg Chem 428: 231
532. Morikawa Y, Ozaki A (1971) J Catal 23: 97
533. Boreskova EG, Kuchaev VL, Temkin MI (1984) Kinet Katal 25: 116
534. Schulz G, Schaefer H (1969) Z Phys Chem (Frankfurt am Main) 64: 333
535. Bowker M, Parker IB, Waugh KC (1985) Appl Catal 14: 101
536. Amariglio H, Rambeau G (1976) Proc 6th Int Congr Catal 2: 1113
537. Bowker M, Parker I, Waugh KC (1988) Surf Sci 197: L223
538. Hinda F, Hirokawa K (1977) J Electron Spectrosc Related Phenom 12: 313
539. Benndorf C, Madey TE, Johnson AL (1987) Surf Sci 187: 434
540. Erley W, Ibach H (1983) J Electron Spectrosc Related Phenom 31: 61

541. Weiss M. Ertl G, Nietsche F (1979) Appl Surf Sci 2: 614
542. Grunze M, Bozso F, Ertl G, Weiss M (1978) Appl Surf Sci 1: 241
543. Yoshida K (1980) J Res Inst Catal Hokkaido Univ 28: 15
544. Okawa T, Onishi T, Tamaru K (1977) Chem Lett 1977: 1077
545. Drechsler M, Hoinkes H, Kaarmann H, Wilsch H, Ertl G, Weiss M (1979) Appl Surf Sci 3: 217
546. Erley W, Ibach H (1981) Surf Sci 119: L357
547. Veselkova AA, Krylova AV, Torocheshnikov NS (1980) Deposited doc no VINITI 3109-80
548. Block J, Schulz-Ekloff G (1973) J Catal 30: 327
549. Al-Haydari YK, Saleh JM, Matloob MH (1985) J Phys Chem 89: 3286
550. Grunze M (1978) Surf Sci 81: 603
551. Ustimenko GA, Krylova AV, Torocheshnikov NS (1978) Deposited doc no VINITI 3611-78
552. Bruker CF, Rhodin TN (1976) Surf Sci 57: 523
553. Pignocco AJ, Pellisier GE (1967) Surf Sci 7: 261
554. Pirug G, Broden G, Bonzel HP (1980) Surf Sci 94: 323
555. Miyano T, Saklisaka Y, Komeda T, Onichi M (1986) Surf Sci 169: 197
556. Vink TJ, Kinderen JM, Gijzeman OLJ, Geus JW, van Zoest JM (1986) Appl Surf Sci 26: 357
557. van Zoest JM, Fluit JM, Vink TJ, van Hassel BA (1987) Surf Sci 182: 179
558. Horgan AM, King DA (1970) Surf Sci 23: 259
559. Vasilevich AA, Blokhina LN, Chesnokova RV, Minaev DM (1988) React Kinet Catal Lett 36: 467
560. Gimzewski JK, Padalia BD, Affrosman S, Watson LM, Fabian DJ (1977) Surf Sci 62: 386
561. Simons GW, Dwyer DJ (1975) Surf Sci 48: 373
562. Papagno L, Caputi LS, Chiarello G, Delogu P (1986) Surf Sci 175: L767
563. Aleksic BD, Mitov IG, Klissurski DG, Petranovic NA, Jovanovic NN, Boganov SS (1984) Bull Soc Chim Beograd 49: 477
564. Chudnikov MG, Minaev DM, Zaichko GN, Alekseev AM (1984) Kinet Katal 25: 1205
565. Arents RA, Maksimov Yuv, Baldokhin YuV, Suzdalev IP, Chesnokova RV, Minarev DM (1983) Poverkhnost no 12: 52
566. Maksimov YuV, Arents RA, Baldokhin YuV, Sudalev IP, Minarev DM, Chesnokova RV (1982) React Kinet Catal Lett 21: 81
567. Enikeev EZh, Loskutov AI, Rozenfel'd IL (1977) 2nd Tezisy Dokl Vses Simp Akt Poverkhn Tverd Tel 1977: 23
568. Tsarev VI, Aptekar EL, Krylova AV, Torocheshnikov NS (1980) Izv Akad Nauk SSSR Ser Khim p 1404
569. Morozov VV, Vokhmaynin NN, Krylova AV, Pograbennyi YuV, Kasatkina LA, Torocheshnikov NS (1979) Tr Mosk Khim-Tekhnol Inst im D I Mendeleeva 107: 144
570. Vasilevich AA, Chesnokova RV, Minaev DM, KUznetsov LD (1979) Kinet Katal 20: 984
571. Burnett JA, Allgood HY, Hall JR (1953) Ind Eng Chem 45: 1678
572. Krylova AV, Morozov VV, Lachinov SS, Torocheshnikov NS (1978) React Kinet Catal Lett 9: 125
573. Tsarev VI, Aptekar EL, Krylova AV, Torocheshnikov NS (1980) React Kinet Catal Lett 14: 279
574. Tsarev VI, Aptekar EL, Krylova AV (1981) Izu Akad Nauk SSSR Ser Khim p 1658
575. Krylova AV, Tsarev VI, Peev T, Kushnarenko TI, Torocheshnikov NS (1986) React Kinet Catal Lett 30: 229
576. Krylova AV, Spirina LN, Rakhmat-Zade AG, Mokhova VN, Lachinov SS, Torocheshnikov NS (1974) Tr Mosk Khim-Tekhnol Inst 79: 24
577. Dorfeld WG, Hudson JB, Stuhr R (1976) Surf Sci 57: 460
578. Brundle CR (1979) Surf Sci 66: 581
579. Krylova AV, Vokhmanin NN, Koroleva TL, Lachinov SS (1973) Deposited Doc no VINITI 7699
580. Sakisaka Y, Komeda T, Miyano T, Onichi M, Masuda S, Harada Y, Yagi K, Kato H (1985) Surf Sci 164: 220
581. Hall GK, Mee CHB (1971) Surf Sci 28: 598
582. Alshorachi G, Wed;er G (1985) Appl Surf Sci 20: 279
583. Kasatkina LA, Krylova AV, Morozov VV, Holu A (1977) Kinet Katal 18: 1603
584. Krylova AV, Morozov VV, Kasatkina LA (1979) Kinet Katal 20: 806
585. Patyi L, Tsarev VI, Krylova AV, Orovac D, Farkas Z (1978) Deposited doc no VINITI 3096
586. Heras JM, Albano EV (1983) Appl Surf Sci 17: 220
587. Heras JM, Albano EV (1983) Appl Surf Sci 17: 207

588. Baer DR, Thomas MT (1986) Appl Surf Sci 26: 150
589. Nakanishi S, Sasaki K (1988) Surf Sci 194: 245
590. Brill R, Schafer H, Zimmermann G (1968) Ber Bunsen Ges Phys Chem 72: 1218
591. Zubova IE, Rabina PD, Pavlova NZ, Kuznetsov LD, Chudinov MG, Le. Cong. Shang (1974) Kinet Katal 15: 1261
592. Tovbin MV, Psheichnaya OV (1966) Kinet Katal 2: 149
593. Nwalor JU, Goodwin JG Jr, Biloen P (1989) J Catal 117: 121
594. Ertl G, Huber M, Thiele N (1979) Z Naturforsch 34A: 30
595. Loeffler DG, Schmidt LD (1976) J Catal 44: 244
596. Brill R (1970) J Catal 19: 236
597. Yatsimirskii VK (1978) Zh Fiz Khim 52: 584
598. Kuznetsova EP, Samchenko NP, Rusov MT (1974) Kinet Katal 11: 92
599. Mahapatra H, Chhabra DS, Puri VK, Sen SP (1981) Fert Technol 18: 160
600. Samchenko NP, Rusov MT, Strel'tsov OA (1966) Kinet Katal 2: 96
601. Vasilevich AA, Rabina PD, Alekseev AM, Dmitrenko LM, Mosolova E, Lyubchenko YuA, Muravskaya GK, Kuznetsov LD (1973) Izu Otd Khim Mauki Bulg Akad Nauk 6: 225
602. Brill R, Kurzidim J (1970) Colloq Int Cent Nat Rech Sci 187: 99
603. Spencer ND, Schoonmaker RC, Somorjai GA (1982) J Catal 74: 129
604. Spencer ND, Schoonmaker RC, Somarjai GA (1981) Nature (London) 294: 643
605. Strongin DR, Carrazza J, Bare SR, Somorjai GA (1987) J Catal 103: 213
606. Yatsimirskii VK (1982) Izv Sib Otd Akad Nauk SSSR Ser Khim Nauk 1982: 131
607. Rambeau G (1987) Bull Soc Chim Fr 1987: 815
608. Artyukh YuN, Zyuzya LA (1981) Kinet Katal 19: 17
609. Törnqvist E, Chen AA (1991) Catal Lett 8: 359
610. Fierro JLG, Homs N, Ramirez P, Sueiras J (1984) React Kinet Catal Lett 24: 179
611. Magomedbekov EP, Kasatkina LA (1978) Tr Mosk Khim-Tekhnol Inst im DI Mendeleeva 99: 60
612. Chirstozvonov DB, Lytkin VP, Sobolevskii VS, Kozlov LI, Markovets LN, Kirillov IP, Korbutova ZV (1969) Izv Vyssh Ucheb Zaved Khim Khim Teknol 12: 1388
613. Miki Y, Terao I, Uchida H (1967) Tokyo Kogyo Shikensho Hokoku 62: 64
614. Altenburg K, Bosch H, Van. Ommen JG, Gellings PJ (1980) J Catal 66: 326
615. Scholten JJF, Zwietering P (1957) Trans Faraday Soc 10: 1363
616. Parker IB, Waugh KC, Bowker M (1988) J Catal 114: 457
617. Nielsen A, Kjaer J, Hansen B (1964) J Catal 3: 68
618. Krabetz R, Peters C (1963) Ber Bunsen-Ges Phys Chem 67: 381
619. Bosch H, van Ommen JG, Gellings PJ (1985) Appl Catal 18: 405
620. Takezawa N (1966) Shokubai, 8: 390
621. Rudnitskii LA, Berengarten MG, Alekseev AM (1973) J Catal 30: 444
622. Rudnitskii LA, Berengarten MG (1972) Kinet Katal 13: 115
623. Samchenko NP, Golodets GI (1986) Kinet Katal 27: 378 (russ) 324 (eng)
624. Samchenko NP, Golodets GI (1986) Kinet Katal 27: 384 (russ) 329 (eng)
625. Stoltze P, Nørskov JK (1987) J Vac Sci Technol A 5: 581
626. Nørskov JK (1981) J Vac Sci Technol 18: 420
627. Holloway S, Lundqvist BI, Nørskov JK (1984) Proc 8th Intern Congr Catal (Berlin) 4: 85
628. Nørskov J, Holloway S, Lang ND (1984) Surf Sci 137: 65
629. Rudnitskii LA, Berengarten MG (1971) Dokl Akad Nauk SSSR 201: 396
630. Markert K, Wandelt K (1985) Surf Sci 159: 24
631. Schlögl R, Schoonmaker RC, Muhler M, Ertl G (1988) Catal Lett 1: 237
632. Konoplya MM, Gorlov YuI, Yatsimirskii VK (1982) Teor Eksp Khim 18: 398
633. Kai-Hui.Huang. (1981) Sci Sin (Engl. ed) 24: 800
634. Kai Huei Huang (1981) Stud Surf Sci Catal 7 A: 554
635. Tomanek D, Kreuzer HJ, Block JH (1986) J Phys Colloq 1986: 139
636. Tomanek D, Kreuzer HJ, Block JH (1985) Surf Sci 157: L315
637. Ortoleva E, Simonetta M (1985) Croat Chem Acta 57: 1387
638. Gagarin SG, Chuvylkin ND (1981) Zh Fiz Khim 55: 3094
639. Guzikevich AG, Chuiko AA, Yatsimirskii VK (1988) Teor Eksp Khim 24: 263
640. Tomanek D, Kreuzer HJ, Block JH (1985) Surf Sci 157: L315
641. Holloway S, Hodgson A, Halstead D (1988) Chem Phys Lett 147: 425
642. Boudart M (1988) Catal Lett 1: 21
643. Boudart M, Löffler DG (1984) J Phys Chem 88: 5763

644. Stoltze P, Nørskov JK (1985) Phys Rev Lett 55: 2502
645. Stoltze P, Nørskov JK (1988) J Catal 110: 1
646. Kjaer J (1958) Measurement and calculation of temperature and conversion in fixed bed catalytic reactors. Gjellerup
647. Zhang HB, Schrader GL (1986) J Catal 99: 461
648. Lui W, Tsong TT (1986) Surf Sci 165: L26
649. Li Y, Drachsel W, Block JH, Okuyama F (1986) J Phys Colloq 1986: 413
650. Wedler G, Borgmann D (1971) Angew Chem Internat Ed 10: 562
651. Aika KI, Ohhata T, Ozaki A (1970) J Catal 19: 140
652. Temkin MI, Morozov NM, Shapatina EN (1963) Kinet Katal 4: 565
653. de Bruijn H (1950) Disc Faraday Soc 1950: 69
654. Igranova EG, Ostrovskii VE, Temkin MI (1976) Kinet Katal 17: 1257
655. Kozub GM, Turchaninov AM, Zyuzya LA, Voroshilov IG (1977) Katal 15: 64
656. Melton FCE, Emmett PH (1964) J Phys Chem 68: 3318
657. Mars P, Scholten JJF, Zwietering P (1960) In: de Boer HJ (ed) The Mechanism of Heterogeneous Catalysis p 66
658. Brill R, Jiru P, Schulz G (1969) Z Phys Chem (Frankfurt am Main), 64: 215
659. Nakata T, Matsushita S (1968) J Phys Chem 72: 458
660. Selwyn GS, Lin MC (1982) Chem Phys 67: 213
661. Bhattacharya AK, Chesters MA (1987) J Catal 108: 484
662. Love K, Emmett PH (1941) J Am Chem Soc 63: 3297
663. Takezawa N, Toyoshima I (1966) J Res Inst Catal Hokkaido Univ 14: 41
664. Takezawa N, Toyoskima I (1968) J Res Inst Catal Hokkaido Univ 15: 111
665. Toyoshima I, Horiuti J (1958) J Res Inst Catal Hokkaido Univ 6: 146
666. Ertl G, Huber M (1980) J Catal 61: 537
667. Tanaka K. Yakamoto O, Matusuyama A (1965) Proc 3rd Intern Congr Catal 1965: 676
668. Nakata T, (1982) J Chem Phys 76: 6328
669. Gauchman SS, Roiter WA (1938) Zh Fiz Khim 11: 569
670. Kwan T, (1956) J Phys Chem 60: 1033
671. Temkin M, Pyzhev V (1940) Acta Physiochem URSS 12: 327
672. Kodera T, Takezawa N (1960) J Res Inst Catal Hokkaido Univ 8: 157
673. Enomoto S, Horiuchi J (1953) J Res Inst Catal Hokkaido Univ 2: 87
674. Horiuti J, Takezawa N (1960) J Res Inst Catal Hokkaido Univ 8: 127
675. Horiuti J, Toyoshima I (1957) J Res Inst Catal Hokkaido Univ 5: 120
676. Enomoto S, Horiuti J, Kobayashi H (1955) J Res Inst Catal Hokkaido Univ 3: 185
677. Temkin MI, Morozov NM, Shapatina EN (1963) Kinet Katal 4: 260
678. Takezawa N, Toyoshima I (1966) J Phys Chem 70: 594
679. Horiuti J, Takezawa N (1960) J Res Inst Catal Hokkaido Univ 8: 170
680. Horiuti J, Toyoshima I (1958) J Res Inst Catal Hokkaido Univ 6: 68
681. Takezawa N, Toyoshima I (1966) J Catal 6: 145
682. Li C, Hudgins RR, Silveston PL (1985) Can J Chem Eng 63: 795
683. Li C, Hudgins RR, Silveston PL (1985) Can J Chem Eng 63: 803
684. Rambeau G (1988) Bull Soc Chim Fr 1988: 450
685. Richard MA, Vanderspurt TH (1985) J Catal 94: 563
686. Rambeau G, Amariglio H (1978) J Chim Phys Phys-Chim Biol 75: 110
687. Rambeau G, Amariglio H (1978) J Chim Phys Phys-Chim Biol 75: 397
688. Jain AK, Hudgins RR, Silverston PL (1982) Can J Chem Eng (1982) 60: 809
689. Rambeau R (1988) Bull Soc Chim Fr 1988: 941
690. Jain AK, Li C, Silveston PL, Hudgins RR (1985) Chem Eng Sci 40: 1029
691. Horiuti J, Kita H (1956) J Res Inst Catal Hokkaido Univ 4: 132
692. Schwab GM, Krabetz R (1956) Z Electrochem 60: 855
693. Aika K, Ozaki A (1970) J Catal 19: 350
694. Brill R (1970) J Catal 16: 16
695. Temkin M, Pyzev V (1939) Zh Fiz Khim 13: 851
696. Panov GI, Kharitonov AS (1985) React Kinet Catal Lett 29: 267
697. Shapatina EN, Kuchaev VL, Temkin MI (1988) Kinet Katal 29: 603 (russ) 520 (eng)
698. Kuchaev VL, Shapatina EN, Temkin MI (1988) Kinet Katal 29: 610 (russ) 526 (eng)
699. Boudart M, Djega-Mariadassou G (1984) Kinetics of Heterogeneous Catalytic Reactions. Princeton University Press
700. Brunauer S, Love KS, Keenan RG (1942) J Am Chem Soc 64: 751

701. Tamaru K (1964) Bull Chem Soc Japan, 37: 1087
702. Temkin MI (1967) Kinet Katal 8: 1005
703. Boreskova EG, Kuchaev VL, Pen'kovoi BE, Temkin MI (1972) Kinet Katal 13: 358
704. Brill R (1951) J Chem Phys 19: 1047
705. Peters C, Krabetz R (1957) Z Electrochem 60: 859
706. Ozaki A, Taylor H, Boudart M (1960) Proc Roy Soc 258: 47
707. Takezawa N, Toyoshima I, Kazusaka A (1972) J Catal 25: 118
708. Anokhin VN, Menshov VN, Zuev AA (1975) Zhur Prikl Khim 48: 489
709. Shapatina EN, Kuchaev VL, Temkin MI (1972) Kinet Catal 12: 1307
710. Kita K, Ozaki A (1969) J Catal 13: 232
711. Horiuti J. (1957) J Res Inst Catal Hokkaido Univ 5: 1
712. Bokhoven C, Gorgels MJ, Mars P (1959) Trans Faraday Soc 55: 315
713. Yamamoto O, Tanaka S, Tanaka K (1974) Sci Pap Inst Phys Chem Res (Jap) 68: 8
714. Enomoto S, Horiuti J (1952) Proc Jap Acad 28: 493
715. Baranski A, Reizer A, Kotarba A, Pyrczak E (1985) Appl Catal 19: 417
716. Brill R, Hensel J, Schaefer H (1969) Ber Bunsen-Ges Phys Chem 73: 1003
717. Smirnov IA, Morozov, NM, Temkin MI (1965) Kinet Katal 6: 351
718. Brill R, Hensel J, Schaefer H (1969) Ber Bunsen-Ges Phys Chem 73: 1003
719. Royen P, Langhans GH (1962) Z Anorg Allgem Chem 315: 1
720. Kiperman SL (1954) Zjur Fiz Khim 28: 389
721. Le Cong Shang, Kuznetsov DA, Rabina PD, Zubova IE, Semenov GM (1967) Tr Mosk Khim-Tekhnol Inst 56: 174
722. Sarkisyan AA, Kuznetsov LD, Rabina PD, Vavilov NS (1980) Khim Prom 1980: 92
723. Brill R, Tauster S (1963) Ber Bunsen-Ges Phys Chem 67: 390
724. Nørskov JK, Stoltze P (1987) Surf Sci 189/190: 91

Chapter 3

Ammonia Synthesis over Non-Iron Catalysts and Related Phenomena

Ken-ichi Aika[1] and Kenzi Tamaru[2]
[1]Department of Environmental Chemistry and Engineering, Interdisciplinary Graduate School of Science and Engineering, Tokyo Institute of Technology, Yokohama, Japan
[2]Department of Chemistry, Faculty of Science, Science University of Tokyo, Tokyo, Japan

Contents

Nielsen (Ed.), Ammonia

3.1 Introduction

Industrial ammonia synthesis, now known as the Haber-Bosch process, began in 1913. For this process, the doubly promoted iron catalyst ($Fe–Al_2O_3–K_2O$) was synthesized in 1909 and the preparation concepts are still applied today [1]. During the research of that time, most elements other than iron were also examined. Osmium was found to have the excellent activity, giving as high as 8% NH_3 at 550 °C and 19 MPa for a long time. Continuous production of NH_3 (80 g/h) was first demonstrated in July 1909 using Os [2]. However, osmium was too expensive and has not produced in quantities sufficient for commercial use. Uranium as its carbide was also very active, but it was irreversibly poisoned by traces of O_2 of water vapor. These examples suggested to us that we might be able to develop catalyst containing active elements other than iron. In this chapter, various catalyst systems other than iron, which have not been applied commercially, will be reviewed. Several elements, for example ruthenium and rhenium, are quite interesting as active elements which might lead to second generation ammonia catalysts.

In the second section, attention is generally focused on catalyst materials such as the elements, alloys, supports, promoters and precursor materials. How these materials are related to nitrogen activation and ammonia synthesis will be shown through the reaction mechanism or creation of active centers. The developments of material science and preparation techniques lead to new catalysts for ammonia synthesis, and will be reviewed in this section. The promoter effect and structure sensitivity are important concepts for catalyst preparation.

In this third section, classical kinetic studies will be surveyed first. Here, attention is again focused on the active elements with respect to the important step, dinitrogen chemisorption. Every element has different characteristics for ammonia synthesis, because N_2 adsorption, NH_3 retardation, H_2 retardation and nitridation are each different among the elements. Thus activation conditions and proper reaction conditions may differ. Catalyst materials can be characterized through instrumental techniques such as TEM, SEM, XPS, AES, LEED and EXAFS. These techniques, which have been developed by the advances in surface science, contributed to the understanding of the catalysis and the detailed mechanism of N_2 activation.

Ammonia is used as fertilizer, chemical reagents, on both large and small scales, and recently as a reagent for NO_x removal. The production conditions can differ depending on the final use. This may demand the availability of various catalysts, each with different characteristics. Related phenomena with respect to ammonia synthesis are not discussed in detail in this monograph. Although the kinetics of ammonia decomposition is described in Sect. 3.3.1, the surface science study is not reviewed in detail. Ammonia decomposition is easily studied and is often used as a model reaction in the field of surface science [3, 4]. N_2 coordinated metal complexes were studied recently in great detail [5, 6].

Photosynthesis and plasmasynthesis of ammonia could be important in the future [7–13]. There are many good reviews on ammonia synthesis, mainly based on an iron catalyst [1, 14–24], as well as other various catalysts [22, 24–26].

3.2 Ammonia Synthesis Activity of Elements and Promoter Effects

3.2.1 Properties of the Elements in the Activation of Dinitrogen

The ammonia synthesis reaction from N_2 and H_2 is composed of several steps, namely: 1) dissociation of the $N \equiv N$ bond, 2) dissociation of the H–H bond and 3) formation of a N–H bond. Generally speaking, step 1 (dissociation of the $N \equiv N$ bond) is the most difficult step (rate-determining step) because the bond energy of N_2 is the highest among the diatomic molecules (941 kJ/mol). Thus, one of the important roles of the catalyst is to dissociate the $N \equiv N$ bond on the surface (dissociative adsorption or surface nitride formation). A reaction path in which N_2 is hydrogenated prior to $N \equiv N$ splitting is not known in the field of metal catalysis. Thus, it is important to study the nature of dissociative adsorption of nitrogen (surface nitride formation) and its reactivity with hydrogen, which must be related to the catalytic properties of ammonia synthesis. Before discussing surface nitrides the properties of bulk nitrides are described. Table 3.1 is a periodic presentation of nitrides and their properties. Most elements form nitrides. Elements of groups IA and IIA form ionic nitrides. Lithium and group IA elements readily react with N_2 to form stable nitrides, but the heavier alkali metals do not react directly with N_2. Group IIIA elements also react directly with N_2, forming stable covalent nitrides. Nitrides of Group IVA to VIII are interstitial compounds. The affinity toward N_2 is still large for group IVA, but it decreases toward group VIII, where only Fe, Co and Ni form nitrides, but only from NH_3. Both Fe_2N and Fe_4N are prepared by reaction with NH_3 at 673–773 K [27]. Although Mo_2N can be formed from N_2, nitride formation is enhanced by H_2 [28]. Elements of groups IB and IIB are less reactive toward N_2 and their nitrides, prepared only by indirect methods, are unstable. Group IIIB elements form stable covalent nitrides.

Those elements active as NH_3 synthesis catalysts are found in groups IVA through VIII, which form interstitial nitrides. Interstitial nitrides have expanded metal lattices in which nitrogen occupies interstital positions. They are referred to as "metallic nitrides" because of their resemblance to the metals [29].

Chemisorption is generally easier than the formation of a nitride from N_2, because the clean surface atoms are more active (having an available bond) than the bulk metal atoms. Chemisorption of N_2 takes place on some metals even at room temperature if it is a vapor deposited film. Those metals that can

Table 3.1. Reactivity of the elements with N_2 and properties of their nitrides

Element	Reactivity with N_2[a]	Nitride	Standard heat of formatiion (kJ/N-atom 25 °C)	Decomposition temperature (°C)
IA				
Li	+ +	Li_3N	− 197	(S)
Na	−	Na_3N	− 151	150
K	−	K_3N	+ 84	L
Rb	−	Rb_3N	+ 180	L
Cs	−	Cs_3N	+ 314	L
IIA				
Be	+	Be_3N_2	− 285	> 220
Mg	+	Mg_3N_2	− 230	700
Ca	+ +	Ca_3N_2	− 213	H
Sr	+ +	Sr_3N_2	− 197	H
Ba	+ +	Ba_3N_2	− 184	H
IIIA				
Sc	+	ScN	− 285	H
Y	+	YN	− 301	H
La	+	LaN	− 301	H
IVA				
Ti	+	TiN	− 305	H
Zr	+	ZrN	− 343	> 3000
Hf	+	HfN	− 326	H
VA				
V	+	VN	− 172	> 2300
Nb	+	NbN	− 247	> 2300
Ta	+	TaN	− 243	> 3000
VIA				
Cr	+	CrN	− 121	H
Mo	+	Mo_2N	− 71	H
W	+	WN	− 71	H
U	+	UN	− 335	H
VIIA				
Mn	+	Mn_5N_2	− 117	> 1200
Tc		TcN		
Re		Re_2N		
VIII				
Fe	−	Fe_4N	− 12	440
Co		Co_3N		
Ni	−	Ni_3N	+ 0	
Ru	− −			
Rh	− −			
Pd	− −			
Os	− −			
Ir	− −			
Pt	− −			
IB				
Cu	−	Cu_3N	+ 75	450
Ag	−	Ag_3N	+ 285	Ex
Au	−	Au_3N		Ex
IIB				
Zn	−	Zn_3N_2	− 12	H
Cd	−	Cd_3N_2	+ 79	
Hg	−	Hg_3N_2	+ 8	

Table 3.1. (Continued)

Element	Reactivity with N_2[a]	Nitride	Standard heat of formatiion (kJ/N-atom 25 °C)	Decomposition temperature (°C)
IIIB				
B	+	BN	− 134	> 3000
Al	+	Al	− 243	2000
Ga	−	GaN	− 105	H
In	−	InN	− 21	H
Tl	−	Ti_3N	+ 84	
IVB				
C	+	$(CN)_2$	+ 155	
Si	+	Si_3N_4	− 188	H
Ge	−	Ge_3N_4	− 17	450
Sn	−	Sn_3N_4		< 360
Pb	− −			
VB				
P	+	PN	− 84	750
As	− −			
Sb	− −			
Bi	−	BiN		
VIB				
O	+	NO	+ 92	
S	−	S_4N_4	+ 134	178 Ex
Se	−	Se_4N_4	+ 176	U
Te	−	Te_3N_4		
VIIB				
F	−	NF_3	− 109	S
Cl	−	NCl_3	+ 230	Ex
Br	−	NBr_3	+ 335	U
I	−	NI_3	+ 272	Ex

[a] + + : Reacts with N_2 directly below 300 °C, + : Reacts with N_2 directly with N_2 above 300 °C, − : Made from nitrogen compounds, − − : nitride unknown
[b] L: Low temperature, H: High temperature, S: Stable, U: Unstable, Ex: Explosive

chemisorb N_2 at room temperature are: IIA (Ca, Sr and Ba), IVA (Ti, Zr and Hf), VA (V, Nb and Ta), VIA (Cr, Mo and W), VIIA (Re), and VIII (Fe) [30, 31]. Note that they are mostly found in Group IVA through VIII, which are known to form interstitial nitrides. It is interesting to note that IIA metals chemisorb N_2 and these metal "nitrides" are quite effective catalysts for the isotopic equilibration of N_2. However, they are inactive for ammonia synthesis because they react to form hydrides in an N_2–H_2 mixture [4, 32, 33], which will be shown in Sects. 3.3.2 and 3.3.3.4.

A similar chemisorption can take place on other metals which do not form a nitride from N_2. The much lower ability of other metals to chemisorb N_2 seems to come primarily from the difficulty in activating the N_2 molecule. Even a copper surface can chemisorb N_2 when the copper surface is activated by ion bombardment [34], even though copper nitride, Cu_3N, is unstable. Chemisorption of N_2 was found on reduced cobalt oxide with a potassium oxide promoter at room temperature [35, 36] and even on noble metals (Ru, Rh,

Os and Ir) promoted with alkali metals. Promoter action will be shown in Sect. 3.2.4.

The second reason for an inability for N_2 chemisorption may come from an insufficient metal-nitrogen bond energy. The heat of chemisorption of N_2 has been measured over metals and is shown in Table 3.2. The differential heat of chemisorption generally decreases with an increase in the surface coverage [37]. Although the initial heat of chemisorption is much larger on vapor deposited films than on powder or supported metal, the value on the vapor deposited films decreases more rapidly, presumably because of a larger degree of disorder in crystallinity [38]. The data shown in Table 3.2 are the initial heat of adsorption and the desorption energy.

Table 3.2. Observed and calculated heats of dissociative adsorption of N_2

Metals	Heat of adsorption kJ/mol Calc.[a]	Obs,[b, c, d, e]	$-2\times$(Heat of nitride formation)[f] kJ/N-atom
IVA			
Ti	481		610
Zr	657		686
Hf	816		652
VA			
V	469		344
Nb	582	> 502[c] [39]	494
Ta	732	590[b] [40]	486
VIA			
Cr	410	439[b] [38]	242
Mo	335	263[c] [41], 289[c] [42], 259[c] [43]	142
W	536	397[b] [38], 385[c] [44], 314[c] [45], 389[c] [46], 334–372[c] [47]	142
VIIA			
Mn	465		234
Tc	126		
Re	167	284–313[c] [48]	
VIII			
Fe	205	293[b] [49]	24
Co	134		
Ni	138		0
Ru	−117	92–167[g] [50]	
Rh	−146		
Pd	−209	−159[e] [51]	
Os	−67		
Ir	−109	242[d] [52]	
Pt	−142	25[c] [53], 92[d] [53]	

[a] Ref. [54]
[b] Initial heat of adsorption on film at room temperature
[c] Initial heat of adsorption on filament at room temperature.
[d] Desorption energy on filament
[e] Estimation from the heat of dissociative adsorption of NO and O_2
[f] From Table 3.1
[g] Ru or Ru/Al_2O_3 with K

The initial heat of chemisorption of N_2 can be evaluated by a semiempirical approach [54]. This approach is based on correlations between the initial heat of chemisorption and the heat of formation of the corresponding compound as demonstrated for hydrides, nitrides, and oxides by Sachtler and van Reijen [55] or others [56–58]. The calculated results are shown in Table 3.2 together with the observed data shown in a review [26, 59]. The calculated values roughly correspond with the observed value. The negative value of the heat of nitride formation per N_2 is also shown in Table 3.2. These values are always lower than the heat of chemisorption. However, the two values are parallel [38, 40, 60]. It is noteworthy that the initial heat of chemisorption of N_2 on Fe is 293 kJ/mol, whereas the negative heat of formation of Fe_4N is only 12.5 kJ/mol. This indicates that chemisorption evolves a large amount of excess energy due to the the bonding ability of surface available bonds.

A recent adsorption study on single crystal planes disclosed that there can be many binding states of adsorbed species on a given surface and that the heat of adsorption many differ by as much as 80 kJ/mol. Thus, it is not possible to identify one value for the heat of chemisorption on a given transition metal unless the binding state is specified or unless it is certain only one binding state exists [59].

It is notable that the heats of N_2 chemisorption on noble metals are negative as shown in the Table. The negative values are caused by the large value of D_{N2} (bond energy) as compared with the metal-nitrogen bond energy. If the heat of chemisorption is really negative, the adsorption state of nitrogen atoms on noble metals should be unstable. Mimeault and Hansen showed that nitrogen atoms (desorded from a hot tungsten filament (2000 K)) can be adsorbed on an iridium or rhodium filament at 300 K, as demonstrated by the desorption of dinitrogen upon flashing [52]. Since contamination of the filament with tungsten vapor was carefully avoided, the nitrogen atoms must be held by the iridium surface. It seèms that the nitrogen atoms stay on the surface because of a very slow rate of the second order desorption at lower temperatures (with its high activation energy). Stimulated nitrogen adsorption by electron bombardment on polycrystalline Pd [61], Pd (331) [62] and Pd (110) [63] has been reported. Nitrogen accumulation on polycrystalline Pd by a reaction of NO with H_2 or CO has been observed [51]. Use of the same reaction also enabled the adsorption of nitrogen on Pd (100), where a C(2 × 2)–N surface structure was confirmed [64]. Active elements for N_2 activation are summarized in Fig. 3.1 in the form of the periodic table.

3.2.2 Properties of the Elements in Ammonia Synthesis

Systematic studies of the catalytic activity of single metals of NH_3 synthesis were first made by Haber as described in the preceeding paragraphs. In these studies not only readily reducible metals were tested but also less reducible ones as well,

IIa	IIIa	VIa	VIIa	VIII		
			Mn	Fe	Co	Ni
Sr		Mo	Tc	Ru	Rh	
Ba	Ce	W	Re	Os	Ir	
		U				

Fig. 3.1. Active elements for N_2 activation

such as cerium which was reduced with magnesium. The results of these early studies are summarized in Fig. 3.2 [65]. In addition to the metals shown, Re [66], Cr [67], V [68], Rh [69], Ir [69] and Tc [70] act as NH_3 catalysts. Platinum was also used [69, 71], but has a poor activity. Some of the above metals, such as Mo, V and U are transformed into nitrides during the reaction. The reverse reaction, NH_3 decomposition, has been studied over a series of vapor deposited metal films [72]. The catalytic activities are also shown in Fig. 3.2.

It is obvious that osmium and iron are the most effective elements under the conditions studied by Haber. On the other hand, ruthenium is the most active metal in ammonia decomposition. Since the two reactions are forward and backward steps of the same reaction, the most active metals should be the same, at least near equilibrium. The difference disclosed above is probably caused by some discrepancy in the reaction conditions. It is obvious that the H_2/NH_3 ratio is much larger in the synthesis reaction than in the decomposition reaction. Adsorbed nitrogen and hydrogen often become a retarding species, which depends upon the H_2/NH_3 ratio. The effect of such variables will be discussed later.

In the 1970s, a catalyst system promoted by metallic potassium [73, 74] was studied. The ammonia synthesis rates at 80 kPa and 588 K over transition metals supported on active carbon and promoted by metallic potassium are given in Fig. 3.2 [69]. The activity of isotopic equilibration of N_2 over the same series of catalysts at 30 kPa of N_2 and at 588 K are shown in Fig. 3.3 [75]. The same reaction over Raney metals are also shown in this figure [76]. In these cases ruthenium is the most active metal. There is a common belief that Fe, Ru and Os are the most active elements in ammonia synthesis, ammonia decompo-

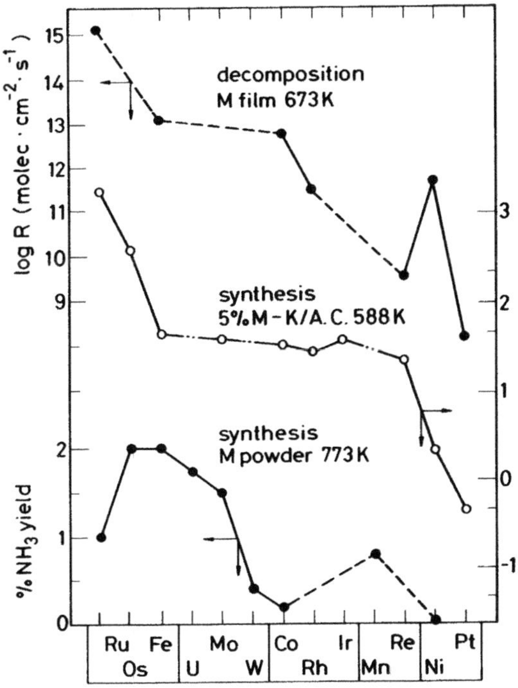

Fig. 3.2. Rate of ammonia synthesis and decomposition over various metal catalysts: Decomposition at 0.2–0.8 kPa [72], synthesis at 80 kPa (5% M–K/AC) [69, 74] and 5 M Pa (M powder) [65]

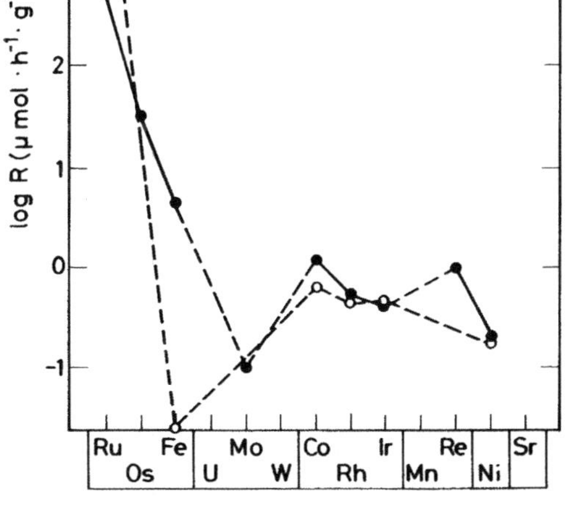

Fig. 3.3. Rate of isotopic equilibration of dinitrogen ($^{28}N_2 + {}^{30}N_2 = 2{}^{29}N_2$) over various metal catalysts at 588 K and a pressure of 20 kPa (150 Torr) [75, 76, 87]

sition and isotopic equilibration of N_2. These elements are known to have a medium metal-nitrogen bond energy among the elements listed in Table 3.1.

A radioactive element, technetium (Tc, fission product of U), was also proved to be one of the most active ammonia catalysts. Per-weight activities of Tc powder, 5.9%Tc/$BaTiO_3$, 5.3%Tc-4.1%BaO/Al_2O_3 were comparable with that of Fe–K_2O–Al_2O_3–CaO–SiO_2. The results are shown in Table 3.3 [70]. The supported Tc catalyst was prepared by impregnating with NH_4TcO_4 and reducing at 593 to 773 K for 5 h by a N_2–H_2 mixture. It was reduced almost completely to the metal with only 0.01% of the oxide of the technetium remaining. The catalytic activity is stable over several months. A constant radiation from ^{99}Tc (E (max) = 0.29 MeV, half life = 2.12×10^5 year) emits counts of 4–8 × 10^3 rad/day. The radiation is thought to create a defect on the support which may stabilize the technetium metal cluster or create an active center on technetium which activates dinitrogen [70].

3.2.3 Alloying Effect

As will be described in the section on kinetics, the rate of ammonia synthesis is a function of the rate of N_2 chemisorption, the amount of adsorbed nitrogen (retardation), and the amount of adsorbed hydrogen (retardation). These factors in turn depend on the reaction conditions (temperature, pressure and flow rate) and the nature of the element. Thus, we might change the reaction rate or kinetics on a new active center which is composed of two elements (ensemble effect). A support and a promoter also influence such characteristics and will be described in the next section. However, if an alloy is used as a starting material, and the two elements are separated and turned into an active metal and an inert oxide, then the resulting activity should be classified as a support effect.

Table 3.3. Ammonia synthesis at 673 K with $N_2 + 3H_2 = 101$ kPa on technetium catalysts[a] [70]

Catalyst	Surf. area (m^2/g)	k_{400}[b] ($atm^{0.5}$/h)	x[c]
Tc powder	0.9	6050	0.54
5.9%Tc/$BaTiO_3$	9.0	4868	0.50
13.8%Tc/γ-Al_2O_3	109	2182	0.36
5.3%Tc-4.1%BaO/$\gamma - Al_2O_3$	110	6772	0.56
CA-1[d]	10.2	5124	0.51

[a] Space velocity 15000 h^{-1}, catalyst volume 0.5 ml, catalyst weight 0.5–1 g, Tc; radioactive element

[b] Rate constant from Temkin equation ($\alpha = 0.5$, $k_{400} = 15000x^2(1 - x^2)^{-1}$). Activation energy of k_{400} = 40–50 kcal/mol

[c] Relative concentration of ammonia compared to the value at equilibrium

[d] 31.5%FeO-61.0%Fe_2O_3 – 0.81%K_2O – 3.92%Al_2O_3 – 2.4%CaO – 0.33%SiO_2

A number of bimetallic catalysts have been examined for their activity in ammonia synthesis. An Fe–Mo (1/1) catalyst gives a high activity, although it decreases remarkably during a prolonged synthesis run unless the Mo content is a higher than 80 per cent [77]. Ni–Mo, and Co–Mo systems behave similarly. The catalysts are prepared by calcination of metal nitrate-ammonium molybdate mixtures, followed by a reaction with N_2–H_2 gas [77]. These catalysts absorb N_2 at the beginning of the operation to an extent that increases with Mo content, whereas the absorbed N_2 is gradually desorded during the operation resulting in the decrease in activity. X-ray analysis of the catalyst indicates that the desorption of N_2 is accompanied by the formation of Mo-metal (Ni, Co or Fe) mixed crystals. It appears that the real active component in this system is molybdenum. The catalytic activity of either iron or molybdenum seems to be lost upon the formation of a mixed crystal. The results of other metals mixed with Mo show that Cu behaves as a simple diluent. Mn gives a favorable effect, and both Cr and W are unfavorable [67].

The activity of Fe is decreased by addition of Ni [78, 79], and a Mn-Fe alloy of 30–40 percent Mn gives a high activity [80]. The conventional promoted iron catalyst is further promoted by the addition of Co. The catalysts are prepared by burning a Fe-Co alloy in O_2 followed by the addition of promoters [81]. The alloying effect of Fe-Co, and Fe-Ni was studied in detail [82]. The addition of 3–7 wt% RuO_2 to a catalyst composed of Co ferrite (25–35%), Mg ferrite (20–25%), K_2O (0.5–2%), and Fe oxide (rest%) increases the ammonia activity and heat resistance of the catalyst [83].

Transition metals are also activated by alloying with electropositive metals. Raney Ru prepared from Ru-Al alloy is more active than pure Ru in terms of specific activity per surface of Ru, suggesting promotion by residual Al remaining after leaching. It is further activated by the addition of metallic potassium to give a highly active catalyst which works even at 373 K [84]. $CsNO_3$ promoted Raney Ru is also very active [85, 86].

Rare earth-transition metal intermetallics such as $CeCo_3$, $CeCo_5$, $CeRu_2$ and $CeFe_2$ are claimed to be more active than doubly promoted iron catalyst on a unit (BET) area basis for ammonia synthesis under elevated pressures (5 MPa) and at temperatures up to 873 K, although the actual activity is lower because of the relatively small surface area of the intermetallics. The rates at 588 K under a total pressure of 5 MPa were re-calculated and are shown in Table 3.4. Cerium intermetallics with Fe, Co and Ru were active. Praseodymium intermetallics with Co and $ErFe_3$ were also active. $GdFe_3$, Gd_2Fe_{17}, Th_2Fe_{17}, Er_2Fe_{17}, $HoFe_2$, $TbMn_2$, $HoMn_2$, $CeCu_5$, $LaCu_5$, Y_6Mn_{23}, Tb_6Mn_{23}, CeNi, $CeNi_5$, $CeCo_2$, CeIn, $CeOs_2$, $CeRe_2$, $HoCo_2$ and $DyCo_3$ had lower activities than those shown in Table 3.4. The intermetallics are decomposed into finely dispersed transition metals and cerium nitrides during the reaction [88, 89]. The high specific activities observed suggest an electron donation from the rare earths to the transition metals which are the active materials [88]. Recently, $CeRu_2$, $CeCo_2$ and $CeFe_2$ were studied by an in situ powder X-ray diffraction technique. These intermetallic alloys were found to be converted to cerium hydride

Table 3.4. Rage of ammonia synthesis on rare earth intermetallics at $N_2 + 3H_2 = 5$ MPa and Space velocity of 120000 h^{-1} [88]

Catalyst[a]	Surf area (m^2/g)	Rate at 588 K (mmol NH_3/g/h)	Act. energy (kcal/mol)
$Fe\text{-}Al_2O_3\text{-}K_2O$[b]	5.21	3.50	20.5
$CeFe_2$	0.45	0.41	13.2
Ce_2Fe_{17}	–	1.01	8.0
$TbFe_3$	–	0.04	23.0
$DyFe_3$	–	0.10	7.4
$HoFe_3$	–	0.02	18.6
$ErFe_3$	–	1.24	(4.3)
$ThFe_3$	0.60	0.36	10.6
$CeRu_2$	0.33	1.41	10.3
$Ce_{24}Co_{11}$	–	1.06	5.6
$CeCo_2$	0.61	1.45	9.2
$CeCo_3$	0.12	1.33	8.7
Ce_2Co_7	0.11	0.30	13.6
$CeCo_5$	0.21	0.73	9.3
$PrCo_2$	0.38	0.13	14.4
$PrCo_3$	0.12	0.17	11.9
$PrCo_5$	0.20	0.51	11.9

[a] Catalyst weight ca. 3.3 g,volume ca. 1 cm^3
[b] Catalyst 416 (Fe – 0.97% Al_2O_3 – 0.65% K_2O)

(CeH_{2+x}) and transition metal phases during the synthesis condition ($N_2 + 3H_2 = 50$ bar, 450 to 550 °C). These phases are considered to be the active component. However, these states are quite sensitive to oxygen containing compounds (air, H_2O and CO) forming an inactive phase CeO_2/transition metal [272].

TiFe alloy turned out to be an active ammonia catalyst which is composed of the mixture of Fe, TiN and TiO_x surface phases mounted on a TiFe bulk phase [30]. However, TiRu has no activity because TiN and TiO_2 cover the surface of TiRu and Ru bulk phase [90]. $Fe_{91}Zr_9$ alloy is found to be an active catalyst, Fe–ZrO_x, under the reaction conditions [91]. A Re-Pt bimetallic cluster is thought to be formed on an Al_2O_3 support [92]. It does not seem that alloying induces drastic effects.

3.2.4 Support and Promoter Effects

Precious metals which can be easily reduced are usually supported on non-reducible oxides. Starting metal compounds may be reduced, forming metal cluster particles. The activity is a function of the number of exposed metal atoms. The particles have various crystal planes on the surface, each of which may have different activities. The chemical and physical properties of the support, concentration of metal compounds, and the reduction temperature may influence the cluster forming process, which in turn influences the size of metal particles (structural effect of the support). Supports and the third additives (promoters) may have some electronic interaction with the metal clusters, which may influence the activity (chemical effect).

Common metals often form mixed oxides with the support compounds. For that reason common metals are usually used as massive metal catalysts. In the case of massive metal catalysts, a few weight percent of a promoter is added. Some promoters make mixed oxides with the active element and influence the reduction process and the surface area (structural promoter). Others are deposited on the metal surface and have an electronic interaction with the surface (chemical effect). Ammonia activity on Fe is known to be enhanced by adding Al_2O_3 and K_2O. It is believed that Al_2O_3 stabilizes the high surface area of Fe (structural effect) and K_2O promotes the ammonia activity per Fe surface area (chemical effect). The structural effect is well studied on Fe single crystal surfaces, where Fe(111) is the most active plane and Fe(110) is the next and Fe(100) is the least active plane [93]. Such studies have been expanded to other catalysts such as Re, and will be reviewed in Section 3.2.4.2.

Various alloy catalysts have been studied as was shown in the previous section. Raney metal catalyst which has small amounts of aluminum, is classified as a massive catalyst. However, some intermetallic are transformed to the supported metal catalysts when one of the components is separated and oxidized or nitrided.

3.2.4.1 Electron Donation to the Active Center

A chemical promoter such as K_2O in an Fe catalyst is thought to be an electron donor to Fe. In the case of Ru, the electron donating properties of the support and the promoter are quite important for the activity. Table 3.5 and Fig. 3.4 summarize the effect of the electron donor. The compounds' electronegativities, which are represented as the geometric average of the elements electronegativities, are 0.8 (K), 1.20 (Cs_2O), 1.31 (K_2O), 1.73 (CsOH), 1.80 (KOH), 1.87 (CaO), 2.05 (MgO), 2.2 (Ru), 2.29 (BeO), 2.49 (Al_2O_3), and 2.5 (C), respectively. Either the support or the promoter which interacts more strongly with Ru particles is arranged based on the order of the compounds' electronegativity. The turnover frequency (TOF) of ammonia synthesis and that of isotopic equilibration of N_2 obeys this order as is shown in Table 3.5. Although the promoter interaction should be discussed on the atomic scale, a rough criteria of the promoter can be estimated from the intrinsic chemical properties. It should be noted that ruthenium has quite a low activity when the metal is supported on Al_2O_3 and has almost no activity when it is on AC (active carbon). This is ascribed to the acidic nature of Al_2O_3 or the electron-accepting properties of AC.

Since the alkali "metal" is a superior promoter, the effect has been studied in detail. Fig. 3.5 shows the effect of the kind and amount of alkali metal promoter. Unpromoted Ru/AC is not active, however, the activity increases with the amount of added alkali metal. The maximum activity obtained at 4 mmol-alkali metal/g-catalyst was 3.30 for Cs, 1.28 for K, and 0.12 mmol NH_3/g/h. for Na

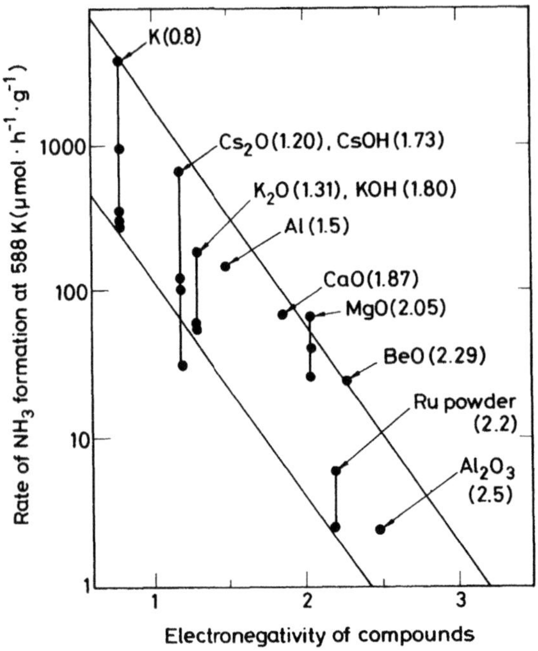

Fig. 3.4. Rate of ammonia synthesis at 588 K and $N_2 + 3H_2 = 80$ kPa on Ru (2 or 5 wt%) catalysts with various supports and promoters as a function of average electronegativities of compounds [94, 96]

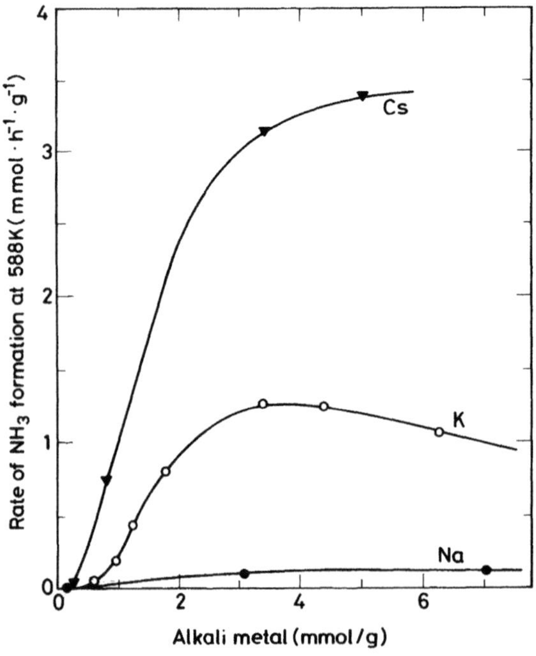

Fig. 3.5. Effect of alkali metal addition on the rate of ammonia synthesis at $N_2 + 3H_2 = 80$ kPa on 2.5% Ru–AC (1.0 g) at 588 K [73]

Table 3.5. Support and promoter effect in ammonia synthesis ($N_2 + 3H_2 = 80$ kPa) and N_2 isotopic equilibration ($N_2 = 20$ kPa) on Ru catalysts [94, 96]

Main promoter	Catalyst	Ru dispersion (%)	NH_3 Rate at 588 K (TOF × 10^4)	NH_3 Rate at 588 K (mmol/h/g)[b]	N_2 equilibration at 588 K (TOF × 10^4)	N_2 equilibration at 588 K (mmol/h/g)[b]
K(metal)	4.7% Ru-K/AC	10.3	205	3.3	–	–
	5% Ru-K/AC	12.7	–	–	24	0.51
	5% Ru-K/BeO	8.6	204	2.9	175	2.5
	4.6% Ru-K/MgO	2.9	197	0.87	147	0.66
	2.0% Ru-K/Al_2O_3	5.0	100	0.34	–	–
	0.5% Ru-K/Al_2O_3	22.1	–	–	24	0.094
	4.2% Ru-K/CaO	2.5	75	0.26	139	0.49
	Ru powder-K	0.30	29	0.31	–	–
	Ru powder-K	0.25	–	–	6.9	0.061
	Raney Ru-K	3.1	17.5	1.76	–	–
Cs^+(Cs_2O)[a]	1.96% Ru-Cs^+/MgO	8.5	19	0.083	–	–
	1.96% Ru-Cs^+/MgO	4.1	11	0.022	–	–
	1.96% Ru-Cs^+/MgO	35	27	0.47	–	–
	2% Ru-Cs^+/Al_2O_3	20	7	0.073	–	–
K^+(K_2O)	2% Ru-K^+/Al_2O_3	21	3.8	0.046	–	–
	2% Ru-K^+/Al_2O_3	34	7.5	0.146	–	–
	2% Ru-K^+/Al_2O_3	18	4.2	0.052	–	–
CaO	4.2% Ru/CaO	2.5	18.7	0.066	–	–
MgO	4.6% Ru/MgO	2.9	14.3	0.063	–	–
	1.96% Ru/MgO	8.4	7.2	0.041	–	–
	1.96% Ru/MgO	12.5	3.0	0.026	–	–
Al and/or Al^{3+}	Raney Ru	3.1	1.40	0.144	–	–
None	Ru powder	0.30	0.54	0.006	–	–
	Ru powder	0.25	–	–	0.020	0.0002
	Ru powder	0.18	0.41	0.0025	–	–
BeO	5% Ru/BeO	8.6	1.63	0.023	–	–
Al_2O_3	2% Ru/Al_2O_3	5.0	0.67	0.0023	–	–
	0.5% Ru/Al_2O_3	22.1	–	–	0.12	0.0005
AC	4.7% Ru/AC	10.3	Neg	Neg	–	–

[a] State of Cs^+ is estimated as CsOH and/or Cs_2O
[b] Weight of nontreated catalyet ($RuCl_3 \cdot 3H_2O$ + support + promoter) is the base of activity except for Ru-K/support for which weight of metallic K is ignored

promoted catalysts, respectively. This follows the order of electronegativity (0.7 for Cs, 0.8 for K, 0.9 for Na) or ionization potential of the elements (90 for Cs, 100 for K, 118 kcal/mol for Na). XPS chemical shifts suggest that there is an electron transfer from the promoter to the Ru surface [94]. Ammonia synthesis is observed at a steady rate even at room temperature and atmospheric pressure over 4.5%Ru-K/AC. Since the activity (0.45 μmol/h/g) was higher than the extrapolated value from the kinetic data above 473 K, a different reaction mechanism was suggested to occur at room temperature [95].

The effect of added metallic potassium is similarly observed with respect to the isotopic equilibration of N_2 on Ru/AC [97] as well as on pure ruthenium [50]. The addition of metallic potassium results in about a 500 fold increase in the activity of Ru at 673 K, while the increase in the ammonia synthesis rate is less extensive (25 fold), due to the inhibition by the presence of hydrogen (cf. Sections 3.3.2, 3.3.3.1) [98]. The inhibition by hydrogen is accounted for by the competitive adsorption of hydrogen as well as by a decreased adsorption constant for N_2 [86, 98].

An analogous catalyst system including graphite has been reported [99]. It is known that transition metal chlorides are intercalated into the layered lattice of graphite [100]. On reduction of these compounds, transition metals remain intercalated as metal atoms giving no activity, although they can become activated for ammonia synthesis by the addition of metallic potassium [101]. These studies originated from the finding that ammonia was catalytically produced over electron donor-acceptor complexes of graphite or metal phthalocyanines with alkali metals [102].

The detailed mechanism of the electronic promoter action is not well understood. Since the apparent activation energy is not changed by applying the promoter, several mechanisms are suggested; 1) molecularly adsorbed N_2, which is the intermediate species for N(a), is stabilized (increases of the sticking probability); or 2) generation of active sites by an electronic promoter. The first mechanism is shown in Fig. 3.6.

Various transition metal/active carbon (AC) catalysts are also promoted by the addition of an alkali metal [73, 74]. The synthesis activities of transition metals in this system are shown in Fig. 3.2.

3.2.4.2 Structure Sensitivity

Since Re is not easily reduced completely, the activity has not been reported in detail. This element becomes active when promoted with metallic potassium. Rhenium also become quite active when activated in an ultrahigh vacuum chamber [103], the activity promotion of platinum by a vacuum treatment is known with respect to the hydrogenation reaction, where chlorine evaporation may play a role [103, 104]. Polycrystalline Re is reported to be nearly 10 times more active than Fe(111) at low conversion under 2 MPa and at a temperature between 603 and 713 K. The activation energy was 16.2 kcal/mol. Although Re becomes reversibly poisoned by product ammonia (1–2 Torr), exposure to water

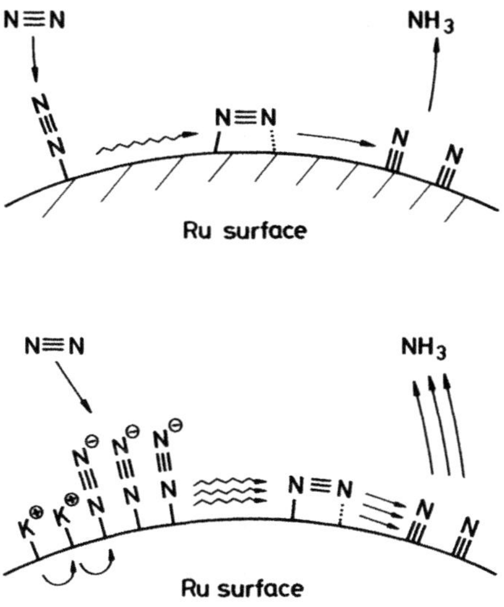

Fig. 3.6. A model of the alkali promoter effect in the activation of N_2 on Ru surface

Table 3.6. Structure sensitivity of ammonia synthesis on Re single crystals

Face	Rate[a] (NH_3 molec./cm^2 s)	TOF[a]
$(11\bar{2}1)$	–	314
$(11\bar{2}0)$	8.5×10^{16}	98
$(10\bar{1}0)$	2.5×10^{15}	6
(0001)	5×10^{13}	0.03

[a] Activity at 870 K at 20 atm (2 MPa) of $N_2 + 3H_2$ [105, 106]

vapor or presulfidation has little effect on its activity. This is contrast to the behavior of iron. Ammonia synthesis rates were measured on Re single crystals at 870 K under 2 MPa of $N_2 + 3H_2$. The results are shown on Table 3.6. The basal plane (0001) was inactive. The open and rough planes (1010), and especially (1120) or (1121) were quite active for this reaction. The activity on (1121) was ca. 10^4 times greater than that on (0001). The activation energy was 19.4 kcal/mol for all of the surfaces. The structure sensitivity of Re surfaces was verified for ammonia synthesis [105, 106].

3.2.5 Preparation, Activation and Nitridation

As shown in the previous section, ammonia catalysts are classified as two types: The supported metal catalyst and the massive metal catalyst. For the supported catalyst, metal salts such as chlorides or nitrates are impregnated on the support materials and usually reduced with H_2 or N_2/H_2 mixture. Promoter salts, usually nitrates, are also impregnated on the supports. Mixed metal compound salts, which are composed of metal elements and promoter elements, can also be used as the starting materials. If the chloride ion may possibly remain after the reduction, and its poisoning is not negligible, compounds free of chlorine should be used. The starting materials and preparation methods may influence the activity through the degree of metal dispersion or metal-support interaction as was shown in the previous section.

Massive metal catalysts are usually made from a mixture of both metal and promoter oxides by reduction with H_2 or N_2/H_2 gas. Many studies of activation processes have been carried out on Fe catalysts which are written elsewhere [107]. The commercial iron catalyst needs to be reduced for a long period because it is in a massive form and the oxide is not easily reduced completely. The iron catalyst is activated under the ammonia synthesis condition [108], during which the surface structure probably changes to form the active sites [109]. On the other hand, supported precious metal catalysts such as Ru catalysts need a short reduction time because their starting compounds are easily reduced. Generally, they have no induction period with respect to the activity.

Alkali metals, especially potassium, were proved to be excellent promoters for Ru catalysts. However, since metallic potassium is very reactive to water vapor or oxygen-containing gases, the catalyst system has several problems, such as a handling problem or deactivation. Several other preparation methods were developed. $K_4Ru(CN)_6$ was prepared by alkali fusion of Ru metal and KNO_3 or by adding KCN to $RuCl_3$ [110]. Although the hydrogenation of $K_4Ru(CN)_6$ alone was very slow even at 773 K, alumina supported $K_4Ru(CN)_6$ was readily hydrogenated at 653 K. The reaction stoichiometry was suggested as follows:

$$K_4Ru(CN)_6 + 14H_2 = Ru + 2K + 2KCN + 4CH_4 + 4NH_3$$

It should be noted that $K_4Ru(CN)_6$ gives metallic potassium and ruthenium on an alumina support, although KCN cannot be reduced to K by hydrogen alone. Activated $K_4Ru(CN)_6/Al_2O_3$ catalysts gave catalytic activities for ammonia synthesis higher than $RuCl_3 \cdot 3H_2O/Al_2O_3$, and the activities correspond to 1/6 to 1/2 of that of $Ru\text{-}K/Al_2O_3$, one of the most active catalysts [110]. This high activity of $K_4Ru(CN)_6/Al_2O_3$ was irreversibly poisoned by water vapor, which suggested the existence of metallic potassium [110]. Various alkali metal or alkaline earth metal salts of hexacyanometal (metal: Fe, Os, Co) were activated on SiO_2 or other supports and showed ammonia activities [111, 112]. Alkali salts such as Na_2CO_3 were effectively added to such complex salts [113].

$CsNO_3$ was also found to be an excellent promoter [110, 114]. Hydrogenolysis of $CsNO_3$ only occurs in the presence of Ru metal as hydrogen activator.

$$CsNO_3 + 4H_2 = 0.5Cs_2O + NH_3 + 2.5H_2O$$

Thermal gravimetric analysis suggested that Cs_2O was partly converted to Cs_2O_2 and Cs, and finally to CsOH. The activity of Ru-Cs^+/MgO (Cs^+ means Cs_2O and/or CsOH), which is shown in Table 3.5, is quite high. It is second to a Ru-K/Support catalyst. This promoter is stable in the presence of water vapor. The activity can be controlled by the amount of nitrate addition, which depends on the surface area of the support. In the case of $Ru_3(CO)_{12}$ supported on a graphite containing carbon support, $RbNO_3$ was effective as a promoter [115, 116]. $CsNO_3$ promoted $Ru_3(CO)_{12}$/MgO was also quite active [117]. Careful studies showed that a small amount of $CsNO_3$ was enough to give the highest activity for Ru/MgO as is shown in Fig. 3.7. 5% Ru-Cs^+/MgO was as active as 4.7%Ru-K/AC as is shown in Fig. 3.8 [117]. Since Ru-Cs^+/MgO is not poisoned irreversibly by oxygen containing compounds, it is hoped to be one of the second generation catalysts after the iron catalyst. Barium compounds as well as an alkali salt were also useful promoters in a $RuCl_3$/carbon catalyst [118]. $RbNO_3$ and $Ba(NO_3)_2$ are effective for Ru/AC which is prepared from $RuCl_3$ [119]. Lanthanide nitrates were found to be effective promoters for $Ru_3(CO)_{12}/Al_2O_3$ [120, 121].

$CsNO_3$ was also proved to be an excellent promoter for Raney Ru catalyst. The activity was compared with other alkali metal nitrate-promoted catalysts and is shown in Fig. 3.8. $CsNO_3$ was suggested to remain mainly as Cs^+ (Cs_2O, CsOH) and partly as Cs metal because water vapor decreased the activity. The activity sequence is in the order of electropositivity (Cs > Rb > K > Na). Since Raney Ru contains aluminium (mainly Al_2O_3, some Al), $CsNO_3$-Raney Ru quite resembles the $Fe–Al_2O_3–K_2O$ catalyst with respect to the composition.

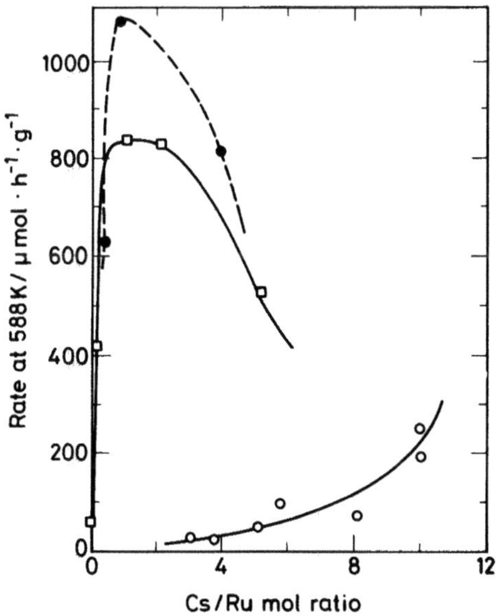

Fig. 3.7. Effect of $CsNO_3$ addition on ammonia synthesis at 588 K on 2 wt% Ru catalysts; $Ru_3(CO)_{12}$/support is treated with H_2 at 623 K for 4 h. Ru-CsOH/MgO (□ and ●); Ru–CsOH/Al_2O_3 (○)

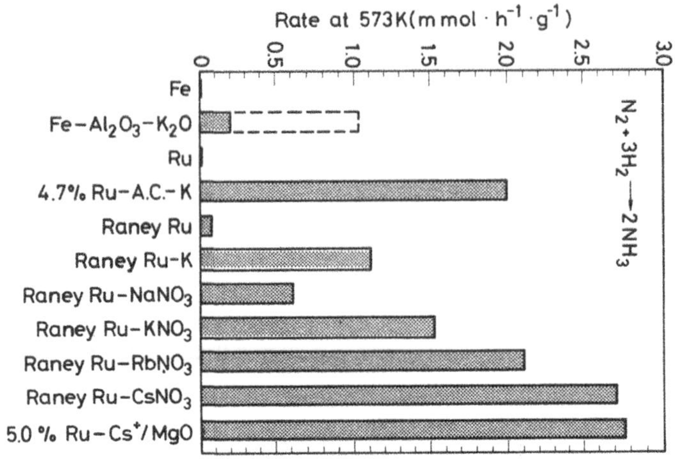

Fig. 3.8. Rate of ammonia synthesis at 573 K and 80 kPa on various Fe and Ru catalysts. Fe [123], Fe–Al_2O_3–K_2O [108, 124], Ru and 4.7% Ru–A.C.–K [73], Raney Ru and Raney Ru–K [84], Raney Ru-alkali nitrate [85, 86], 5.0% Ru–Cs^+/MgO [117]

$CsNO_3$-Raney Ru is more active than the commercial iron catalyst and is one of the most active ammonia catalysts under atmospheric pressure conditions [85, 86] (see Fig. 3.8). However, the Ru catalysts might not be highly active at high pressure because of hydrogen inhibition. Nevertheless, a recent work suggests that hydrogen promotes the activation of N_2 on Ru catalysts if a lanthanide promoter is present [122].

Elements between groups IIIA and VIIA usually become nitrides under ammonia synthesis conditions. Nitrides of uranium and molybdenum are known to be active catalysts. When MoO_3 is reduced with H_2–CH_4 mixture, a high surface area Mo_2C is formed. The oxide is first heated to a temperature 75 K below the temperature at which reduction of the oxide and reaction with CH_4 occur. The temperature was then raised at a rate of 0.1–5 K/min until the reaction ceased (1041 to 1273 K). The quenched sample (Mo_2C) had a surface area of 51 m^2/g, which was higher than the sample (13 m^2/g) made by the usual method. The activity of the Mo_2C was higher than 5% Ru/Al_2O_3, but was lower than Fe–Al_2O_3–K_2O catalyst [125, 126]. Oxycarbonitride of molybdenum ($Mo_xC_yN_z$) was also active for ammonia synthesis [127]. The rate of ammonia synthesis on reduced molybdenum (Mo), its carbide (Mo_2C), and the oxycarbide (MoO_xC_y) was measured with simultaneous gravimetric determination of the nitrogen uptake of the catalysts. On prereduced Mo_2C and MoO_xC_y the activity increased with time as reduction proceeded and reached a constant value when the catalyst had taken up about one monolayer of nitrogen, corresponding to a surface stoichiometry of MoN. On Mo the steady state was reached after taking up the same monolayer of MoN plus two layers of Mo_2N. In all three cases the steady-state rates were found to be very similar. The small amounts of nitrogen taken up to reach similar rates suggested that the bulk plays no role in determining the catalytic activity [128].

With respect to ammonia synthesis at 773 K, tungsten carbide (WC, $2 m^2/g$) also seems to behave like tungsten metal (W), although their physical properties differ considerably. 390 Pa of ammonia was produced at a contact time of 0.10 s under $N_2 + 3H_2 = 101$ kPa and at 773 K. H_2S adsorption on only a fraction of the surface of WC inhibited the ammonia synthesis. This suggested that the three side planes $(10\bar{1}0)$, $(\bar{1}100)$ and $(0\bar{1}10)$ of hexagonal WC crystal exposed tungsten atoms in the uppermost layer. These faces would not hold carbon atmos strongly and the "metallic" face was considered to be the active sites for ammonia synthesis [129].

Atomically dispersed low-valency Mo, W, V and Zr ions can be deposited on surface oxides with the use of π-allyl (Mo, W, and Zr) and σ-benzyl (V) compounds and subsequent hydrogen treatment. $Mo(C_3H_5)_4$ reacted with surface hydroxyl ions and formed Mo(II). Such heterogenized metal ion catalysts showed some ammonia activity, although the levels were low. Some examples are listed in Table 3.7. Among four ions, Mo(II) ions were the most active. TOF (NH_3 molec./Mo ions/s) did not change when the Mo ion concentration was increased from 10^{17} to 10^{18} ions/m^2-SiO_2. The activation energy (25 kcal/mol) was also constant. The activity of the deposited Mo ion was shown to depend on the nature of the oxide carrier as is shown in Table 3.7 [130]. The reaction mechanism on transition metal ions might be different from that on reduced metals [5].

Table 3.7. Activity of atomically dispersed transition metal ions (Me) in ammonia synthesis at $N_2 + 3H_2 = 5$ MPa [130]

Me/wt.%	Support/$m^2 g^{-1}$	Reduction temp. (K)	Reaction temp. (K)	Contact time (10^4 h)	Conversion (%)	TOF $\times 10^4$
Zr/4.3	Al_2O_3/200	823	823	1	0.002	0.46
V/2.0	SiO_2/250	823	723	1	0.007	0.16
W/1.5	SiO_2/250	873	723	1	0.015	2.2
Mo/4.3	SiO_2/250	823	723	1	0.7	320
Mo/4.3	SiO_2/250	823	673	1	0.135	–
Mo/7.1	Al_2O_3/200	823	723	1	0.012	–
Mo/2	Cr_2O_3/19	673	673	0.6	0.024	–
Mo/0.5	Fe_2O_3/21	483	483	0.6	0.0040	–
– /0	Fe_2O_3/21	483	483	0.6	0.0	–
Mo/1	Co_3O_4/11	473	473	1.9	0.0013	–

TOF: NH_3 molec./Me ions

3.3 Mechanism of Ammonia Synthesis over Metals

3.3.1 Kinetics of Ammonia Synthesis and Decomposition

3.3.1.1 Synthesis

According to Temkin and Pyzhev [131], the rate of ammonia synthesis is controlled by the rate of N_2 adsorption (step 1), whereas the H_2 adsorption (step 2) and the hydrogenation of adsorbed nitrogen (step 3) are fast enough to be in equilibrium.

$$N_2 \rightarrow 2N(a) \tag{1}$$

$$H_2 \leftrightarrows 2H(a) \tag{2}$$

$$N(a) + 3H(a) \leftrightarrows NH_3 \tag{3}$$

The rate of adsorption of $N_2(V_s)$ is expressed as follows:

$$V_s = k[N_2]\exp(-g\Theta_N) \tag{4}$$

where $[N_2]$ is the N_2 pressure, Θ_N is the coverage of adsorbed nitrogen, and k and g are constants, Θ_N is a function of the fictitious N_2 pressure $[N_2]^*$ which is in equilibrium with H_2 and NH_3.

$$\begin{aligned}\Theta_N &= (1/f)\ln(A[N_2]^*)\\ &= (1/f)\ln(AK[NH_3]^2/[H_2]^3)\end{aligned} \tag{5}$$

where A is a constant and K is an equilibrium constant of ammonia decomposition. Thus the synthesis rate is expressed as follows:

$$\begin{aligned}V_s &= k[N_2](AK[NH_3]^2/[H_2]^3)^{-g/f}\\ &= k_s[N_2]([H_2]^3/[NH_3]^2)^{g/f}\end{aligned} \tag{6}$$

The corresponding rate of N_2 desorption which is assumed to be rate-determining in ammonia decomposition is similarly given by

$$V_d = k_d(NH_3]^2/[H_2]^3)^{(1-g/f)} \tag{7}$$

The Temkin and Pyzhev equation expressed above was in agreement with a number of kinetic measurement made on various catalysts as summarized in Table 3.8 for the synthesis and in Table 3.9 for the decomposition. One characteristic feature of the ammonia synthesis rate is the retardation by the product ammonia, and this is reasonably explained by the Temkin theory. The basic assumption of the rate-determining step, N_2 chemisorption was also supported.

Although Eq. (4) was successful in described a number of experimental data, the experimental values of g/f are not always the same. Sometimes it ranges from

Table 3.8. Kinetic parameters of NH_3 synthesis according to the temkin-phzhev equation

Metal	Temperature (K)	Pressure (MPa)	g/f	E_{act} [a] (kJ/mol)	Ref.
VIA					
Mo	720–823	0.1	0.5	178	[132]
W	855–951	0.1	0.5	190	[133]
W	673–773	0.1	0.5	(203)	[129]
VII					
Tc	623–698	0.1	0.5	190 ± 20	[70]
VIII					
Fe(– Al_2O_3-K_2O)	673–723	0.1	0.5	167	[131]
	723–773	30	0.5	176	[134]
Fe(– Al_2O_3)	645–651	0.1	0.5–0.6		[135]
Fe(– MgO)	570–680	0.1	0.5	(156)	[136]
Ru	881–1013	0.1	0.5	249	[137]
Os	673–723	0.1	0	164	[138]
	823–873	0.1	0.5	174	[138]

[a] Activation energy of NH_3 decomposition (k_d)

0.5 to 0.8 depending on temperature. This is partly natural because Eq. (4) is valid only for an intermediate range of coverage ($\Theta_N = 0.2$–0.8). Thus, when $[NH_3]$ in the gas mixture is very low or very high, the results tend to give lower or higher value of g/f [124, 163]. Even with moderate coverage, the rate constants given by the Temkin equation are dependent on the H_2/N_2 ratio, and/or total pressure [21, 135, 164, 165]. Thus, the value of g/f obtained from rate data can be an average value under the experimental conditions, and is not a constant as postulated in the original theory. However, this rate expression is widely used in commercial operations at high pressure with an Fe catalyst.

If the nitrogen adsorption is described by the Langmuir equation: $V = k[N_2](1 - \Theta_N)^2$, the rate of ammonia synthesis is described as follows:

$$V_s = k_s[N_2]/(1 + K[NH_3]/[H_2]^{1.5})^2 \tag{8}$$

This equation can be applied even under low NH_3 pressure. When various catalysts are examined below a total pressure of 1 atm, the activity (or NH_3 pressure on the catalyst) is sometimes quite low. Under these conditions, the Langmuir type equation is quite often used.

Equation (8) is based on the mechanism (1 to 3), where N(a) is the main retarding species. If NH(a) is the main retarding species, the equation can be written as follows:

$$V_s = k_s[N_2]/(1 + K[NH_3]/[H_2])^2 \tag{9}$$

In the case of Temkin expression, the equation is written as follows:

$$V_s = k_s[N_2]([H_2]^2/[NH_3]^2)^{g/f} \tag{10}$$

Table 3.9. Kinetic parameters of NH_3 decomposition according to the equation $V = k(P_{NH3})^m/(P_{H2})^n$

Metal	Temperature (K)	Pressure (kPa)	m	n	n/m	E_{act} (kJ/mol)	Ref.
VA							
V	673–753	ca. 100 (Low P_{H2})	0.5	0	0	138	[139]
	673–753	High P_{H2}	1.0	1.5	1.5	138	[139]
VIA							
Mo	1027–1273	13.3	0	+		134–180	[140]
W	548–873	2–20	0–1	–		205	[141]
	903–1023	4.7–20	0	–		146	[142]
	953–1153	0.1	0	0		176	[143]
	904–1214	7–26	0	0		163	[144]
	1073 – 1523	2–35	0	+		146–197	[140]
	950–1150	1–5	0	0		113–130	[145]
	1073	10^{-9}–10^{-6}	0.8				[146]
VIIA							
Re	653–713	ca. 100	0.53	0.89	1.68	134	[147]
	728–843	1.3–7	0.7	1.4	2.0	205	[72]
VIII							
Fe	608–743	1–4.4	0.5	0.7	1.4	163	[148]
	653–773	1.3–10.6	0.9–1.0·	1.4–1.5	1.5	176–209	[149]
	765–797	ca. 100	0.9	1.5	1.67	226	[150]

Fe($-Al_2O_3$	608–703	100	0.6	0.85	1.42	192	[151]
$-K_2O$)	693	13	0.48	0.72	1.5	–	[152]
	752	13	0.75	0.38	0.5	–	[152]
Co	643–753	1.3–7	0.85	1.42	1.67	188	[72]
Ni	573	0.2–5.3	1.0	1.5	1.5	–	[153]
	663–773	1.3–7	0.96	1.53	1.59	180	[72]
	623–723	4×10^{-3}–0.17	1	–			[154]
Ru	543–738	1.3–7	1.2	2.0	1.67	188	[72]
	623–673	80–106	0.6	0.9	1.5	130	[155]
	825–1009	–	1.0	1.75	1.75	247	[137]
Rh	693–773	1.3–7	1.35	2.45	1.81	239	[72]
Pt	763–833	0.001	1.5–2.0	2.2–3.0	1.5	205	[156]
	793–883	1.3–7	0.96	1.63	1.7	247	[72]
	978	10^{-5}	0.63–0.84	0.25	0.34	17	[157]
	1045–1131	Low P_{H2}	1	1	1	167	[158]
	1045–1131	High P_{H2}	1	0	0	167	[158]
	1100–1485	0.03–0.5	1	1	1	184	[159]
	1100–1485	1.3–39	1.4	2.3	1.64	419	[159]
	1206–1488	13–26	1	1	1	586	[144]
	1173–1623	ca. 10^{-3}	1	0	0	21	[160]
IB							
Cu	768–893	5–67	1	1	1	192	[161]
	993–1183	100	1	1.5	1.5	243–259	[162]

Practically, the power rate expression is convenient. This is evident when the mechanism is complicated.

Ru catalysts for ammonia synthesis were found to be retarded by hydrogen adsorption. In this case both N(a) (or NH(a)) and H(a) may retard the reaction.

$$V_s = k[N_2](1 - \Theta_N - \Theta_H)^2$$
$$= k[N_2](1 + K_1[NH_3][H_2]^{-1.5} + K_2[H_2]^{0.5})^{-2} \quad (11)$$

Since this expression is complicated to use in analysis, it can be changed to the power rate form although it is not correct algebraically.

$$V_s = k[N_2](K_1[NH_3][H_2]^{-1.5})^{-2y}(K_2[H_2]^{0.5})^{-2z}$$
$$= k_s[N_2][H_2]^{3y-z}[NH_3]^{-2y} \quad (12)$$

Table 3.10 shows the kinetic parameters of ammonia synthesis over Ru catalysts. On Ru powder and Ru-alkali/support, hydrogen has a negative order and the ammonia order is close to zero. This suggests strong hydrogen inhibition and weak nitrogen inhibition. On the other hand, it is suggested that the nitrogen inhibition is extensive and hydrogen inhibition is weak on the support Ru catalysts [122, 166]. The extent of ammonia inhibition also depends on the kind of promoter as is shown in Table 3.10 [92, 167, 170].

When the synthesis is carried out far from equilibrium as happens by increasing the space velocity or by decreasing the reaction temperature, two steps can be the rate-determining. Under such a condition where the reverse rate is negligible, the overall rate (V_s) in the steady state is equal to the rate of

Table 3.10. Kinetic parameters* in ammonia synthesis on various Ru catalysts

Form of Catalyst	Temp. (K)	Press. (kPa)	n	h	a	Ref.
Ru Powder	598	80	1	− 0.72	− 0.15	[166]
	553–581	101	1	− 1	0**	[192]
Ru/SiO_2	825–1009	101	1	1.25	− 1	[137]
	574–671	101	–	–	− 0.56	[167]
Ru/Celite	619–683	80	1	0	0	[168]
Ru/Al_2O_3	610–691	101	–	–	− 1.0	[167]
	638	80	1	0.30	− 0.74	[166]
	500–725	150	1.1	0.6	–	[169]
Ru/MgO	593	80	1	0.66	− 0.78	[166]
	500–725	150	0.9	0.2	–	[169]
$Ru\text{-}Cs^+/Al_2O_3$	595	80	1	− 0.55	0	[166]
$Ru\text{-}K^+/Al_2O_3$	500–725	150	1	− 0.3	–	[169]
$Ru\text{-}Cs^+/MgO$	548	80	1	− 0.43	− 0.12	[166]
$Ru\text{-}K^+/MgO$	500 – 725	150	1	− 0.5	–	[169]
$Ru\text{-powder-}Cs^+$	598	80	1	− 0.90	− 0.07	[166]
$Ru\text{-}K/Al_2O_3$	500–725	150	0.9	− 1.1	–	[169]
Ru-K/MgO	500–725	150	1.2	− 0.5	–	[169]

* $V = k[N_2]^n[H_2]^h[NH_3]^a$

** High flow rate (100–1000 ml/min) condition

nitrogen chemisorption (V_n), and possibly to the rate of hydrogenation of adsorbed nitrogen, step 3, (V_h).

If V_n and V_h are given by:

$$V_n = k_n[N_2](1 - \Theta_N)^2 = k_n[N_2](1 - 2\Theta_N)$$

$$V_h = k_h[H_2]\Theta_N \qquad (13)$$

and if $V_s = V_n = V_h$
V_s is given by eliminating Θ_N

$$V_s = k_n k_h[N_2][H_2](2k_n[N_2] + k_h[H_2])^{-1}$$

$$= k'[N_2]^m[H_2]^n \qquad (15)$$

where m and n are apparent orders of reaction with $m < 1$, $n < 1$, and $m + n = 1$. Temkin et al. found kinetics corresponding to Eq. (15) with $m = n = 0.5$ under reaction conditions far from equilibrium with very high space velocities of about $10^6\,h^{-1}$ at 627–723 K on an iron catalyst [171]. The rates of each elemental step can be better elucidated by a dynamic approach, measuring the reaction rates as a function of adsorption amounts during catalysis. At a sufficiently low pressure of H_2 and high temperature, the rate of hydrogenation was found to be lower than the rate of N_2 chemisorption on doubly promoted iron catalyst by measuring the adsorption during the reaction [172].

3.3.1.2 Decomposition

The rate equation of ammonia decomposition was expressed in Eq. (7) according to the Temkin-Pyzhev mechanism (Eq. 1–3). The observed parameters in Table 3.9 show that n/m values are mostly 3/2, which indicates that the Temkin-Pyzhev mechanism is applicable on most catalysts (Re, Fe, Co, Ni, Ru and Rh) under normal ammonia decomposition conditions.

However, the hydrogen order (n) is often zero for the reaction on W and Pt catalysts, for which a different mechanism should be applied [173].

Ammonia decomposition on a clean tungsten surface was studied by direct and simultaneous measurements of the reaction at 773–1473 K under an ammonia pressure of 10^{-6}–10^{-3} Pa [174]. The order of the reaction rate with respect to ammonia pressure changed with temperature from first order at 1478 to 2/3 order at 773 K. No hydrogen was adsorbed on the surface in any form above 973 K and an increase in the hydrogen partial pressure (P_{H2}) during the reaction had no effect on either the reaction rate or on the amount of surface nitrogen (Θ_N).

Under higher ammonia pressures, up to 100 Pa, thick surface nitride layers were formed during the decomposition, which were decomposed at the same rate in vacuo as in the steady state NH_3 decomposition, provided that the uptake of nitrogen was the same.

The rate of nitrogen desorption from the surface depended only on Θ_N, being faster with increasing Θ_N.

$$V_d = k_d \exp(h\Theta_N) \tag{16}$$

whereas the rate of nitrogen uptake decreased with increasing Θ_N and increased with ammonia pressure (P_{NH3}).

$$V_a = k_a P_{NH3} \exp(-g\Theta_N) \tag{17}$$

On the basis of these data, it was concluded that the overall reaction proceeds through a reaction mechanism of "dynamic balance" between two consecutive rate-determining steps: $2NH_3 \rightarrow 2N(a) + 3H_2(V_a)$ and $2N(a) \rightarrow N_2(V_d)$. The rate of the overall reaction (V_s) is obtained as follows:

$$V_s = V_a = V_d = k_d(KP_{NH3})^{h/(h+g)} \tag{18}$$

Here no hydrogen order is reasonably explained [174, 175]. The isotope effect observed with NH_3/ND_3 is consistent with the above theory [146]. Such kinetics are obtained when the rate of hydrogenation of N(a) ($2N(a) + 3H_2 \rightarrow 2NH_3$; V_h) is negligibly lower than V_d. This mechanism (tungsten-type mechanism) is seen at high temperatures and low pressures on W, Pt, and even on Fe [176, 177].

The kinetics of ammonia decomposition on platinum wire under pressures below 1.3 Pa are found to conform to that of the rate-determining dissociation of NH_3 at around 1273 K, while to that of desorption of N_2 at around 773 K [156, 157]. Details of ammonia decomposition were studied over a polycrystalline Pt wire in a continuous flow microreactor [53, 176, 177]. Kinetics were analyzed by the steady-state intermediate method. For the high temperature and/or the low pressure (for example, 550–1200 K at 10^{-3} Torr or 550–1400 K at 10^{-6} Torr), a surface reaction involving the dissociation of the first N–H bond controls the rate of ammonia decomposition ($V_h \ll V_d$, tungsten-type mechanism). The activation energy in this regime is 4.2 kcal/mol, and the rate of decomposition exhibits a primary isotope effect with a first order dependence of ammonia pressure. For the low temperature and/or the high pressure (for example, 480–550 K, at 10^{-3} Torr or 550–800 K at 0.5 Torr), the surface is nearly saturated with nitrogen, and the rate of nitrogen desorption controls the rate of reaction ($V_h \gg V_d$).

Consequently, the reaction rate is independent of ammonia pressure, and the observed activation energy, 22 kcal/mol, is equal to the activation energy for desorption of nitrogen. Including several other works on Pt, the activation energies (E_a) and desorption energies (E_d) of the elementary steps are estimated as follows [53]:

$$NH_3(g) \leftrightarrows NH_3(a) \qquad \Delta H = -E_d = -12\ \text{kcal/mol} \tag{19}$$

$$NH_3(a) \rightarrow NH_2(a) + H(a) \qquad E_a = 16\ \text{kcal/mol} \tag{20}$$

$$ND_3(a) \rightarrow ND_2(a) + D(a) \qquad E_a = 17.2\ \text{kcal/mol} \tag{21}$$

$$2N(a) \rightarrow N_2(g) \qquad E_a = 22 \text{ kcal/mol} \tag{22}$$

$$N_2 \rightarrow 2N(a) \qquad E_a = 16 \text{ kcal/mol} \tag{23}$$

$$2H(a) \leftrightarrows H_2(g) \qquad E_d = 19 \text{ kcal/mol} \tag{24}$$

This value can provide a one-dimensional potential energy diagram of the reaction. Ammonia is assumed to be adsorbed with a sticking probability of unity [53].

Over the Ru (1, 1, 10) surface, the tungsten-type mechanism is operative on its terrace sites, whereas the Temkin and Pyzhev type is operative on the step site under similar condition [175]. Sebba and coworkers investigated ammonia decomposition over the metal nitrides TiN, VN and Cr_3N_2, finding VN the most active [118]. The kinetics of ammonia synthesis [68] and decomposition [139] on VN at around 773 K follow the Temkin-Pyzhev equation, while the VN catalyst was contaminated with oxygen. At lower pressures of hydrogen the decomposition rate on VN has a zero order for hydrogen [139], suggesting a tungsten-type mechanism. The kinetics of ammonia synthesis over uranium nitride (U_2N_3–UN_2) at 648–823 K and 3 MPa [179], as well as over molybdenum nitride (Mo_2N) at 720–828 K [132], also obey the Temkin-Pyzhev equation.

3.3.2 Isotopic Equilibration of Dinitrogen on Various Metals

If the chemisorption or dissociation of dinitrogen is rate-determining in the ammonia synthesis as has been demonstrated, the isotopic equilibration of dinitrogen

$$^{15}N_2 + {}^{14}N_2 \leftrightarrows 2\,{}^{15}N^{14}N \tag{25}$$

must be a slow process. The two reaction rates must not be exactly the same because the surface in isotopic equilibration is more strongly covered with N, and the surface in ammonia synthesis must be partly covered with hydrogen. Kummer and Emmett confirmed that the rate of equilibration is comparable with the rate of nitrogen desorption from the same Fe catalyst at 773 K [180]. Catalysts which are active for nitrogen equilibration are also active for ammonia synthesis. Known active metals include Fe [181–183] Os [184], Co [185], Raney Ni [185], and others [76]. The equilibration on Os is measurable at a temperature as low as 473 K, and is rapid at 573 K [184]. The equilibration requires higher temperatures over Re [147], W [186–187] and Mo [188–190]. The activities of transition metals on carbon promoted by potassium for ammonia synthesis [21] and for isotopic equilibration [75] are shown in Figs. 3.2 and 3.3. The relative activities run parallel in the two reactions. Generally, ammonia activities are higher than the isotopic equilibration. The reasons are suggested to be: a) the adsorption step is in a dynamic state under ammonia synthesis, whereas it is in the adsorption equilibrium under isotopic equilibration, and 2) possibly, incomplete reduction of the surface in isotopic equilibration. The rates of nitrogen isotopic equilibration on Raney metals are not high except for Raney Ru, probably due to the second reason. However, the rate is inversely related to the desorption energy of atomic nitrogen. Adsorbed states on Raney-type metals are studied by the TPD method. Atomically adsorbed species, which is related to isotopic exchange, is believed to be the active intermediate species for ammonia synthesis [76].

The rate of isotopic equilibration is the rate of forward and backward reaction of step (1). If the rate-determining step of ammonia synthesis is step (1), the rate of isotopic equilibration (backward reaction) reaches zero at ammonia synthesis conditions. In this way the rate-determining step was determined on Ru-K/AC [98] and on Mo_2N [191].

3.3.3 Dinitrogen Adsorption Study and Retarding Species

3.3.3.1 Ruthenium and Osmium

N_2 adsorption and hydrogenation of absorbed nitrogen was studied in detail on Ru powder [192]. One g of Ru powder (0.1 m^2/g) gives an ammonia activity of 0.011 mmol/h/g at 573 K under 1 atm of $N_2 + 3H_2$. Under this condition, 1 g of Fe powder (0.1 m^2/g) gives the activity of 0.040 mmol/h/g when the ammonia retardation is extrapolated to zero. Thus, Ru seems to be less active than Fe. However, when N_2/H_2 ratio is increased, the activity of Ru surpasses that of Fe because the rate of nitrogen adsorption (the rate-determining step) increased with the decrease of hydrogen retardation. The rate of nitrogen adsorption at $[N_2] = 0.027$ atm and zero coverage of nitrogen in the absence of H_2 was estimated to be 40 mmol/h/g at 573 K. This value is 10^3 times higher than the ammonia activity (N_2 adsorption rate) of Fe. Thus, the intrinsic activity of N_2 activation of Ru is 10^3 times higher than Fe. The former suffers from hydrogen inhibition, while the later does not. The rate of hydrogenation of adsorbed nitrogen is so fast on Ru powder above 431 K that it is controlled by the flow rate of hydrogen and the equilibrium ammonia pressure. The true rate is so fast that it is measurable only at low temperature (203 to 298 K). Both the rate of N_2 adsorption and the rate of hydrogenation of adsorbed nitrogen are much faster on Ru than those on Fe. However, the rate of ammonia synthesis on Ru is comparable with that on Fe because of the hydrogen inhibitation on Ru [192]. An ammonia process in which N_2 and H_2 is supplied alternatively was proposed [193, 194].

The kinetic analysis of this reaction was also studied on 1 g of Os powder (7.5 μmol of sites/g) [195], the rate is expressed as follows:

$$V_s = k_s[N_2][H_2]^{-1}(1 + K[NH_3])^{-2} \quad (26)$$

The reaction is therefore limited by N_2 adsorption and inhibited by H_2. It is also inhibited by NH_3, but to a lesser degree. The high rate of hydrogenation of adsorbed nitrogen even at 473 K exemplifies the rate-determining step of N_2 adsorption. The situation is quite similar to Ru powder. The rate of ammonia synthesis on a bare site of Os (rate of N_2 adsorption) is 100 times higher than that of Fe but 100 times lower than that of Ru at 473 K. However, under 1 atm of a stoichiometric mixture and at 673 K, Os is 100 times less active than Fe powder [195].

The rates of isotopic equilibration are retarded by hydrogen on Ru catalysts: Ru-K, Ru-K/Al_2O_3, Ru-K/active carbon [98] and Raney Ru-$CsNO_3$ [86]. This corresponds to the hydrogen retardation for N_2 adsorption on Ru powder [192] or to the negative order of hydrogen pressure in the ammonia synthesis rate expression (see 3.3.1). In the case of $CsNO_3$ promoted Raney Ru (one of the most active ammonia catalysts, the rate of isotopic equilibration is expressed as being proportional to $(1 + b[H_2]^{0.5})^{-2}$ [86]. The reaction order in the isotopic equilibration of N_2 on Ru catalysts increases with temperature as follows: from 0.26 at 553 K to 0.45 at 593 K on Ru-K/Al_2O_3, from 0.44 at 593 K to 0.79 at 653 K on Ru-K, and first order on Ru/Al_2O_3 at 653–713 K [50]. The kinetics on Ru-K and Ru-K/Al_2O_3 are represented by a Langmuir type equation for dissociative adsorption:

$$V = k[N_2](1 + K^{0.5}[N_2]^{0.5})^{-2} \tag{27}$$

The heat of chemisorption of nitrogen on Ru-K is estimated from the adsorption constant in the above equation to be 167 kJ/mol. It is to be noted, however, that both isotopic equilibration and N_2 chemisorption are enhanced by hydrogen on potash-promoted iron catalysts [26]. No enhancement by hydrogen was found on pure iron [181, 182]. Recently, N_2 isotopic equilibration was found to be enhanced by hydrogen on rare earth-promoted Ru catalysts [122]. The hydrogen effect on Ru seems to depend on the kind of support and promoter. The guiding principle of this phenomena has not been solved, yet, the study is quite important to construct an effective Ru catalyst under high pressure.

It is interesting to note that a clean single crystal of Ru (0001) does not adsorb nitrogen effectively at a high temperature. It has a very small sticking coefficient for dissociatively adsorbed nitrogen. Exposing the Ru crystal to be a does of 10 L of N_2 at temperatures between 400 and 760 K showed no N_2 peaks in the desorption spectrum [196, 197]. At lower temperatures (78 K), Ru (001) can adsorb molecular species with a high sticking coefficient (ca. 0.4). Two molecular forms are suggested from the unusual coverage dependence of a sticking coefficient at 78 K [198, 199].

3.3.3.2 Tungsten, Molybdenum and Rhenium

Adsorbed nitrogen on metal films is desorbed by increasing the temperature at a constant rate. This gives TPD spectra or thermal desorption spectra, and the adsorbed states were analyzed on W, Fe, Mo or Re. Three peaks of adsorbed nitrogen have been observed in TPD on W. The typical feature of the three types of nitrogen on a W surface are summarized in Table 3.11. The first order kinetics with respect to the nitrogen coverage, as well as the lower energy of desorption observed with γ and α types, suggest a molecular or dinitrogen nature of the adsorbed site. This was confirmed by the absence of isotopic mixing in dinitrogen desorbed from the surface on which a mixture of $^{28}N_2$ and $^{30}N_2$ was preadsorbed [200, 201]. On the other hand, the β type was regarded as an atomic state on the basis of the second order desorption kinetics, the higher energy of desorption, and the occurrence of isotopic mixing [42, 47, 201–204].

The TPD spectra were studied on single crystal surfaces of W, and the results are summarized on Table 3.11. As expected from the weak interaction, the γ state is found on any of three planes, W(100) [203], (110) [205, 206], and (111) [207]; and α state is only found on W(111) as confirmed by FEM [203] or probe-hole emission [207]; and the β state is predominant on (100) as demonstrated by the large differences in sticking probability of nitrogen between W(100) (0.3–0.6) [208–210] and W(110) (0.01–0.005) [203, 210, 211] or W(111) (0.08) [209, 210] (see also Table 3.12). The desorption energies from single crystalline planes have been measured and are shown in Table 3.11 [59]. Adams and Germer demonstrated by LEED techniques that the β nitrogen atoms on W(100) are located in an interstitial position surrounded by 4-square W atoms as is shown in Fig. 3.9 [208, 212].

The sticking probability s (rate of adsorption divided by the collision flux) was studied as well on the clean W surface as a function of the coverage Θ. The initial values of s (on a bare site) are as high as 0.3 to 0.4 on four-fold symmetry sites such as (100) (see Table 3.12 and Fig. 3.9). This situation is contrary to the case of Fe single crystals. Molecular beam experiments suggest that the dissociation preferably takes place at a vacant pair of (100) sites [73, 210]. On W(100) polycrystalline foil, s decreases gradually at lower coverage and then more steeply with increasing Θ showing a convex curve. A constant value of s, independent of coverage, indicates the presence of a precursor state of suffi-

Table 3.11. Adsorbed states of nitrogen on tungsten surfaces

		γ-state	α-state	β-state
Desorption temperature (K)		ca. 150	ca. 300	ca. 1,000
Desoprtion energy (kJ/mol)		8–40	80	300–320
Desorption kinetics		1st order	1st order	2nd order
Isotopic mixing of N_2		no	no	yes
Suggested form		molec.	molec.	atomic
Plane selectivity	(110)	formed	no	a little
Desorp. energy (kJ/mol)		< 38 [203]	–	330 [211]
	(111)	formed	formed	a little
		38 [203]	67 [203]	314 [203]
	(100)	formed	no	formed
		38–44 [213, 214]	–	314–334 [203, 213, 214]

Table 3.12. Sticking probability at zero converage (s_0) at 300 K on various W planes [215]

Plane	s_0	Reference
(110)	$< 10^{-3}$	[211]
(211)	$< 10^{-2}$	[203]
(321)	$< 10^{-2}$	[212]
(111)	ca. 10^{-2}	[208]
(310)	0.28	[208]
(411)	0.31	[212]
(210)	0.25	[208]
(311)	0.29	[212]
(100)	0.4	[203]

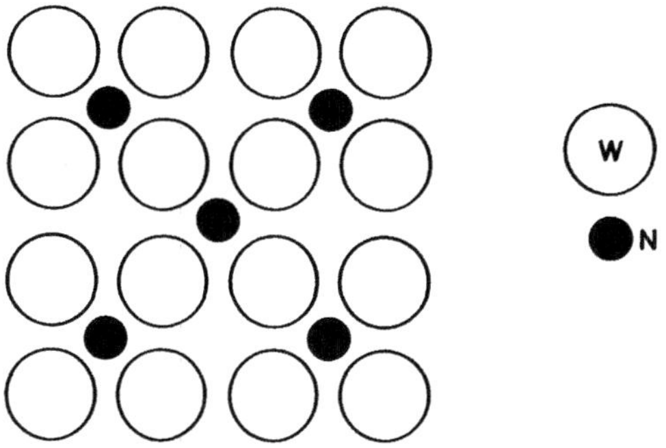

Fig. 3.9. Surface Structure of W(100) with nitrogen atoms

ciently long life such that there is a high probability of transformation to the chemisorbed state before desorption occurs. For the chemisorption of nitrogen on W(100), the problem precursor is the γ-state with a heat of adsorption of 43 kJ/mol (see Table 3.11). The value of s as a function of Θ, $s = \alpha Z/(1 + k_d/kf(\Theta))$, indicates the following scheme:

$$N_2(g) \underset{k_d}{\overset{\alpha Z}{\rightleftarrows}} N_2(a) \xrightarrow{kf(\theta)} 2N(a)$$

where α is a condensation coefficient, Z is the collision flux, k_d is a rate of desorption, k is a conversion rate constant. The experimental value of s is about 0.4 between 190 and 600 K, and the activation energy of k is 39 kJ/mol [213]. On the other hand, for N_2 on W(111), s decreases linearly with Θ, indicating the absence of a precursor state and direct molecular chemisorption (α-state) on specific sites [73]. The adsorption and decomposition of ammonia on a clean W(100) surface has been studied by photoemission spectroscopy (XPS and UPS), and the results have been compared with those on $c(2 \times 2)$ –N ordered W(100) [216].

Although the fresh surfaces of W have a strong affinity with N_2, the ammonia activity is not high. The W surface is partly turned to nitride (WN) under the ammonia synthesis condition, so that the rate is determined by a dynamic balance of nitride formation and its hydrogenation (see Sect. 3.3.1.2).

Analogous types of adsorbed nitrogen have been found or suggested by thermal desorption studies on Mo [188, 214, 217–219] and Re [48] in addition to W. Desorption energies were measured to be 41 kJ/mol for the γ-state on Mo(100), 364 kJ/mol for the β-state on Mo(100) [214], and 339 kJ/mol for the β-state on Mo(110). Ammonia activity on Re single crystals is reviewed in Section 3.2.4.2. β-type nitrogen is suggested as the intermediate in ammonia synthesis.

Adsorption of N_2 has also been studied on Mo single crystals. Mo(111) was changed to the (433) face by faceting at 850 K. The nitrogen atom is adsorbed on this surface up to 0.35 of the coverage. The sticking coefficient at this state is about 0.05, which is less than that on Mo(100). The coverage independence of the sticking coefficient has been explained by an island formation of superstructure on the surface in this case [220]. The reaction rate of isotopic equilibration of N_2 was measured over Mo_2N (5 m^2/g) at a temperature from 725 to 774 K. The rate at 748 K and a N_2 pressure of 21 Torr is 0.85×10^{11} molec. $N_2/cm^2/s$. The ammonia synthesis rate was calculated to be 0.68×10^{12} molec.$N_2/cm^2/s$ by assuming the same intermediate state of adsorbed nitrogen. The good agreement with the experimental value (0.75×10^{12} molec.$N_2/cm^2/s$) at 748 K and 760 Torr of $N_2 + 3H_2$ suggests the rate-determining step is N_2 dissociative adsorption and desorption [189].

At lower temperatures (80 K), Re(0001) and Re(1120) adsorb molecular nitrogen with a high sticking coefficient ca. 0.9 [221]. The adsorbed dinitrogen is suggested as being bound with its molecular axis perpendicular to the surface [222]. At higher temperatures (218 K), the dissociation proceeds with a probability of 10^{-2} to 10^{-6} depending on the surface defect concentration [222], or with a probability of 4×10^{-4} (1120), 9×10^{-6} (0001), or 0.25 (polycrystalline foil) [229]. The stabler form of the two molecular species is considered to be the precursor to the dissociation [223].

The nitride-forming metals, such as vanadium and molybdenum in particular, seemingly give rise to a faster rate of N_2 chemisorption and a slower hydrogenation rate for the adsorbed nitrogen. In fact, the initial rate of N_2 chemisorption on pure molybdenum is very fast even at 583 K. However, the rate decreases rapidly with an increase in N_2 uptake, or the extent of nitride formation. Thus, the rate of N_2 chemisorption on Mo_2N is much slower than on Mo. Even with iron, the rate of ammonia decomposition on iron nitride (Fe_4N) was reported to be two orders of magnitude slower than on pure iron [224]. Since both vanadium and molybdenum are transformed into bulk nitrides under the reaction conditions of ammonia synthesis, the mechanism should be referred to a nitride surface.

Kinetic studies suggest that the chemisorption of N_2 is the rate-determining on nitrides. This was also confirmed by observing the absence of isotopic mixing in N_2 on Mo_2N [191]. In this experiment, the ^{15}N content of product ammonia collected in a trap was always lower than that in N_2, indicating that the bulk nitrogen in Mo_2N undergoes an exchange reaction with adsorbed nitrogen or ammonia. Moreover, the synthesis rate is as slow as expected [191]. The ammonia synthesis rate can be calculated from the isotopic equilibration rate assuming the same rate-determining step. Calculated data corresponds well with the observed data on Mo_2N [189].

3.3.3.3 Nickel, Platinum, Rhodium, Palladium, Iridium and Copper

The rate-determining step of ammonia synthesis and decomposition on Ni powder was studied using tracer isotopes (determination of the stoichiometric number) at 743 K. Dissociative adsorption and desorption were the slowest steps [225]. Dissociative adsorption was studied in few cases. The slight extent of chemisorption of N_2 at 523 K to 573 K was speculated to be atomic species on Ni powder [226]. A spectroscopic study of surface "nitride" on polycrystalline Ni was carried out under a N_2 pressure of 200 Torr [227].

Thermal desorption peaks at 470 (low) and 800 K (high) were attributed to dissociatively adsorbed nitrogen on Ni(110). This nitrogen was adsorbed at 591 K and gives Auger peaks at 380 eV. The sticking probability was very low [228].

Very few works have been reported on the dissociative adsorption of N_2 on Pt, Rh, Pd and Ir. N_2 is inferred to be adsorbed dissociatively on polycrystalline platinum ribbon with an apparent sticking coefficient of ca. 10^{-2}. Desorption of dissociatively adsorbed nitrogen occurs around 650 K with second-order kinetics and an apparent activation energy of 80 kJ/mol [229]. Dissociative adsorption of nitrogen has not been observed on Pt by other investigators. The use of a high-frequency discharge or the use of a hot W filament supplies activated N_2 on Pt [230, 231], on Pd [61], and on Cu [231]. The dissociative adsorption of N_2 on a Rh field emitter was suggested to occur with a significant rate at a temperature of 500 K in addition to the molecularly adsorbed state [232]. Atomically adsorbed nitrogen on Pd was formed by electron bombardment [61–63] or by reduction of adsorbed NO [51, 64] (see also Sect. 3.2.1). Nitrogen is adsorbed in the molecular form (γ-state) on Ir only at lower temperatures (below 300 K). Dissociative β-form was produced only by the use of electron beams on Ir [233]. Dissociatively adsorbed nitrogen was observed on Raney Rh, Ir and Pt in addition to Raney Fe, Co, Ni and Ru [76].

3.3.3.4 Alkali Metals, Alkaline-Earth Metal, Scandium and Lanthanides

Isotopic equilibration of N_2 has been studied over nitrides of alkali metals, alkaline-earth metals, and rare earth metals [32, 87, 234, 235]. The kinetic results including U, Fe and Mo are summarized in Table 3.13. Many nitrides are characterized by the mobility of the nitrogen atom in the nitride. More than several monolayers of nitrogen atoms can take part in the isotopic exchange with N_2 on group IIA or IIIA elements, and the entire bulk nitrogen in the solid phase takes part in the exchange even at 573 K [235]. However, it is interesting to point out that the mobility is not directly related to the equilibration activity. The activity seems, rather, to be related to the surface M–N bond strength, that is, the heat of adsorption. Since the isotopic equilibration reaction generally takes place through the dissociative adsorption and desorption, the activation energy is equal to that of the adsorption step when the coverage is low and is equal to the desorption step when the coverage is high. In the case of nitrides, the coverage is high. Thus, the activation energies in Table 3.13 represent the activation energies of desorption, which may be comparable to the heat of adsorption. Nitrides with low heats of activation (or low heat of adsorption) are active for isotopic equilibration as is seen in Table 3.13.

It should be pointed out that nitrides of Sr and Ba are quite active. The activity per unit surface may be the highest among those reported. However, the estimated weight-base activity of $SrN_{0.5}$ is lower than Raney Ru and is higher than 5% Ru–K/AC (see Fig. 3.3 and Table 3.13). The high catalytic activity of Ba nitride is suggested as being due to the presence of various nitrogen-containing complexes (Ba_2N, BaN_2, Ba_3N_4) on the Ba_3N_2 surface, which may give the best route of bond expansion into the dissociation [32].

The rates of ammonia synthesis have been measured on several nitrides, and the results are shown on Table 3.14 [87]. If there is no hydrogen effect, ammonia synthesis rate can be estimated from the rate data of isotopic equilibration (see Table 3.14 foot note). The experimental values are mostly lower than the calculated values. In case of Ba, Sr, Ca, and Li, the nitride is changed to the hydride and the activity decreases. Sr, the most active equilibration element,

Table 3.13. Isotopic equilibration of N_2 on nitrides or metals[a]

Nitrides or Metals	Exchangeable N(a) (Monolayers)	React. order	Act. Energy (kJ/mol)	log V[b] (molec/cm^2/s)	Surf. Area (m^2/g)	V[c] (μmol/h/g)
$BaN_{0.62}$	50	0.7	50	12.4	0.1[f]	550
$SrN_{0.5}$	60	0.8	55	12.5	1.1[g]	12500
$UN_{1.70}$	bulk	0.5	105	10.7	–	–
Ru[d]	–	0.7	125	8.6	–	–
Fe[e]	–	0.5	130	10.1	5[e]	82
$CaN_{0.5}$	0.7	0.7	130	9.0	2.2[i]	7.5
CeN	3	0.5	130	8.8	0.9[h]	0.74
Mo_2N[e]	bulk	0.5	155	7.0	5[e]	0.075
Li_3N	bulk	0.4	175	7.5	–	–
α-U_2N_3	bulk	0.5	226	5.1	0.6[g]	0.00021

[a] Ref. [87], LaN, PrN, and ErN have very low activities, and $MgN_{0.6}$ is inactive
[b] Rate under 1.3 kPa of N_2 at 573 K
[c] Estimated rate under 20 kPa (150 Torr) of N_2 at 588 K
[d] Ref. [236]
[e] Ref. [189]
[f] Ref. [32]
[g] Ref.
[h] Ref. [237]
[i] Ref.

Table 3.14. Ammonia synthesis on nitrides[a]

Nitrides	Temp. (K)	NH_3 yield (mole%)	NH_3 synthesis rate log (NH_3 molec./cm^2/g) exp.	calc.	Phase composition after synthesis
Ba	573	0.32	13.8	14.5	Hydride
Sr	873	Inactive	–	–	Nit.-Hydirde
$UN_{1.70}$	623	0.24	11.6	12.9	Nitride
CA-1[c]	673	0.27	11.8	–	–
Ca	773	0.08	12.3	13.3	Nit.-Hydride
Ce	673	0.28	11.4	11.5	Nitride
Li	673	0.03	11.3	12.0	Hydride
Mg	873	Inactive	–	–	Nitride
La, Pr, Nd Tb, Ho	773	0.01–0.07	9–10	–	Nitride

[a] $N_2 + 3H_2 = 98$ kPa [87]
[b] calculated as $V([N_2]/[N_2]_e - 1)$ where $[N_2] = 25$ kPa, $[N_2]_e$ is a ficticious N_2 pressure under the equilibrium with NH_3 produced, and V is isotopic equilibration rate at $[N_2]_e$
[c] Promoted Fe catalyst

turns out to be inactive, and Ba does not activate N_2 in the presence of H_2. The activity of Ba comes only from the hydrogenation of the nitride which is only formed by N_2 without H_2. For $UN_{1.70}$, the rate of isotopic equilibration has a reaction order of -0.2 with respect to H_2 [235]. Recalculation of the ammonia activity using this data gives values closer to the experimental value of the synthesis rate [87, 235]. For CeN, hydrogen has no effect on the rate of isotopic equilibration of N_2 [237].

N_2 is inferred to be adsorbed dissociatively on Dysprosium at 115 K, yielding two XPS peaks in the N_{1s} region at 396.2 and 398.2 eV, corresponding to a nitride and chemisorbed N(a), respectively. No peaks corresponding to molecularly adsorbed N_2 (400.2 eV) were observed. Upon heating the sample, the N(a) is converted into nitride species. At a warm-up temperature of 300 K, the N(a) species accounts for only ca. 10% of the total surface [238].

The rate of nitrogen sorption was studied at 10^{-4} to 10^{-6} Torr and 298 to 473 K on continuously renewed scandium films at a condensation rate of 4.8×10^{13} atom $cm^{-2} s^{-1}$. The rate of adsorption is limited by the rate of physical adsorption [239].

3.3.4 Adsorbed State of Nitrogen

3.3.4.1 Infrared Adsorption Spectroscopy

The adsorbed state of nitrogen was studied by various instrumental and spectroscopic techniques. These adsorbed species observed at room temperature may be desorbed at higher temperatures, however, they may be a short-lived intermediate of ammonia synthesis. As was shown before, N_2 must generally be dissociated before the reaction with hydrogen. Thus N(a), the main adsorbed species, is more important than N_2(a) in the mechanism. However, it gives important information, for example, how N_2 is activated by the surface and turned to the dissociated form N(a). Although Ni had been known to be inactive in the chemisorption of N_2, an IR-active adsorbed dinitrogen was found on Ni/SiO_2 which disappeared on evacuation at 303 K. The adsorbed state was confirmed to be molecular as evidenced by the isotope shift (2195, 2160 and 2123 cm^{-1} for $^{28}N_2$ $^{29}N_2$ and $^{30}N_2$, respectively) and was assigned to $Ni{-}N \equiv N^+$ [240]. It was later found that the IR-active species is formed only when the size of the nickel particles on silica is in a range of 1.5 to 7.0 nm. The effect of particle size was interpreted as that the concentration of "B5" sites (inverted square pyramid) was at a maximum when the particle diameter ranged from 1.5 to 7.0 nm [241]. Later, analogous IR-active species were found on other metals as summarized in Table 3.15. These bands were also shown to be diatomic by the isotope shift [240, 242, 243] as well as the first order kinetics of desorption [244]. The high extinction coefficient associated with the adsorption bands, comparable to the value of chemisorbed CO, has given evidence for an end-on type of coordination rather than a side-on type [245].

The weak adsorption was formerly interpreted as an interaction with the electric field on the metal surface [241]. However, later findings disclosed a chemical nature of adsorption as suggested by: 1) a fairy high initial heat of adsorption on Ni, 50 kJ/mol, 2) the fact that the shift in IR band frequency was large and 3) the effect of hydrogen which causes a shift toward higher wave numbers [240, 242, 243]. The deviation from gaseous N_2 (2331 cm^{-1}) is a kind of measure of the M–N bond strength in the case of a metal surface. On oxides, on the other hand, the N–N strength is not changed with a very weak M–N interaction, which gives an adsorption band at a wave number higher than for the normal dinitrogen molecule. Such cases are seen on ZnO [246], zeolite [247], TiO_2 [248] and ZrO_2 [249].

Although iron is well known as a dinitrogen-activating metal, analogous IR-active species observed on iron are formed by treatment with ammonia [250] or a N_2–H_2 mixture [251] at elevated temperatures. The adsorption strength, shown as the deviation of the wave number from that of gaseous N_2 (Table 3.15), is influenced by the coadsorbed hydrogen on Co/Al_2O_3 [252], by the oxidized site of Rh on Al_2O_3 [253], or by the diluent element (Ag) in Pt on SiO_2 [254].

A very stable and IR-active dinitrogen species which can be converted to ammonia by hydrogen treatment has been found on Ru-K/Al_2O_3 at 2020 cm^{-1} by N_2 treatment above 473 K [255]. $^{15}N_2$ gives rise to an isotope shift to 1997 cm^{-1}. Since the IR absorption at 2020 cm^{-1} was also observed when ammonia was decomposed on Al_2O_3-K (without Ru) at 623 K, it was tentatively assigned to a novel K–N_2 surface complex [256]. Thus, it does not seem to be the intermediate but rather a side product on the surface. The spectra which might be due to the "intermediate" complex was observed recently on active Ru catalysts. The isotope shift confirms these species are end-on adsorbed species

Table 3.15. IR data of adsorbed dinitrogen on metals

Sample	Support	Wave No (cm^{-1})	Deviation from gaseous $N_2(cm^{-1})$	Heat of adsorption (kJ/mol)	Ref.
Fe	Al_2O_3	2200	131		[250]
	MgO	2200	131		[250]
	MgO	2050	281		[250]
Co	Al_2O_3	2214	117		[252]
Ni	SiO_2	2202	129	50	[240]
	SiO_2	2202	129		[241]
	SiO_2	2195	136		[242]
	SiO_2	2202–2220	129–111		[257]
	(evaporated)	2202	129		[258]
Ru	K–Al_2O_3	2020	311		[255]
		1935	396		[256]
	SiO_2	2240	111		[259]
	Al_2O_3	2214	117		[260]
	MgO	2168	163		[260]
	CsOH/MgO	1910	421		[260]
Rh	SiO_2	2236	95		[243]
		2255	76		[261]
	Al_2O_3	2257	74	9	[253]
		2252	79		[261]
	TiO_2	2254	77	49	[261]
Pd	Al_2O_3	2260	71		[241]
Ir		2185	146	17	[262]
Pt	SiO_2	2230	101		[241, 254]
	SiO_2	2230	101	15	[244]
	(as Pt black)	–	–	18	[264]
ZnO		2337	−6	19	[246]
Na-A-Zeolite		2339	−8		[247]
η–Al_2O_3		2360	−29	46	[263]
TiO_2		2337	−6	11	[248]
		2348, 2340	−17, −9	25	[265]
ZrO_2		2341	−10		[249]
SiO_2		2329	2		[265]
BeO		2350, 2337	−19, −6		[265]
SiO_2–Al_2O_3		2357	−26	31	[265]
H-mordenite		2356, 2335	−25, −4		[265]

(Ru–N ≡ N). The wave numbers (cm^{-1}) are 1910 for Ru–CsOH/MgO, 2268 for Ru/MgO, 2214 for Ru/Al_2O_3, and 2240 for Ru/SiO_2 (Table 3.15). The deviation from the gas phase N_2 molecule is 421, 163, 117 and 111 cm^{-1}, respectively. This number is a measure of N_2 activation, N–N bond looseness, and the extent of electron donation by the Ru atom which is interacting with the support or the promoter. Interestingly, the ammonia synthesis activity is ranged in this order (Fig. 3.4). Although the life time of these species might be short at the synthesis reaction conditions, the species might be a precursor of the dissociated state. A mechanism of ammonia synthesis and the role of the support and promoter is shown in Fig. 3.10.

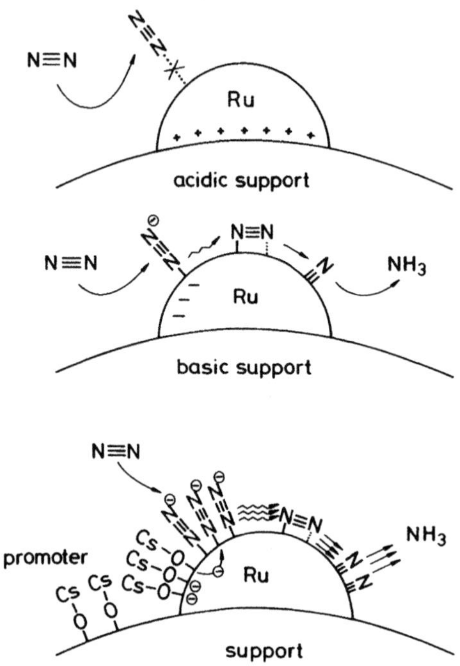

Fig. 3.10. A model of the support and promoter effect on Ru catalyst in ammonia synthesis mechanism

3.3.4.2 Electron Spectroscopy

Electron spectroscopy of adsorbed species has been developed in recent years. Both AES (Auger electron spectroscopy) and XPS (X-ray photoemission spectroscopy) can give information about the electronic state of atoms in the outermost layer of a solid surface, thus permitting the study of adsorbed species. When dinitrogen is adsorbed at 83 K on Mo or at 373 K on Fe, at which temperatures molecular adsorption is expected, a single peak is observed at around 385 eV. While, when it is adsorbed at 273 K on Mo or 673 K on Fe, three peaks are observed. Since Mo_2N gives three similar peaks, the adsorbed species obtained at higher temperatures are assigned to atomic nitrogen [218].

Analogous Auger spectra were observed on W(110) for N_2 adsorbed at 100 K [206]. The amount of adsorbed nitrogen can be monitored by the peak height of the Auger spectra as was done on W by the 390 eV peak [202] and on Fe by the 380 eV peak [266].

Dinitrogen adsorbed on W, Ni or Fe at 100 K or lower gives two peaks in XPS at around 400 and 406 eV, which were interpreted in different ways: 1) N_2(a) for 406.6 eV and N(a) for 397.6 eV on W because the latter remained while the former disappeared at 300 K [267]; 2) Two types of N_2(a) [268], end-on and bridge [269]; and 3) two nitrogen atoms of an end-on type dinitrogen, $W \equiv N \equiv N$ [206].

There has been agreement that the atomically adsorbed nitrogen gives rise to a peak of around 400 eV, e.g. 397.2 and 398.0 eV for N(a) on W at 300 K [267];

397.2 eV for N(a) on Fe at 290 K [269] and 397.5 eV for N(a) on promoted iron at 673 K [270]. Surface reconstruction by nitrogen adsorption has been observed on various surfaces [219]. Surface science reviews have been published [3, 4, 271].

The valuable comments of Professor Emeritus A. Ozaki are acknowledged by one of the authors.

3.4 References

1. Timm B (1984) The Ammonia Synthesis and Heterogeneous Catalysts, A Historical Review in "Proc. 8th Intern. Congr. Catal.," DECHEMA ed., Verlag Chemie, Weinheim, Vol 1 p 7
2. Mittasch A (1951) Geschichte der Ammoniak Synthese, Verlag Chemie, Weinheim
3. Grunze M (1982) In: King DA, Woodruff DP (eds) Chem Phys Solid Surf Heterog Catal, Elsevier, Amsterdam, Vol 4 p 143
4. Ertl G (1983) In: Anderson JR, Boudart M (eds) Catalysis, Science and Technology, Springer-Verlag, Berlin, Vol 4 p 209
5. Shilov AE (1989) In: Bottomley F (ed) A Treatise on Dinitrogen Fixation, John Wiley & Sons, New York, Sec. 1, p 31
6. Bottomley F (1989) In: Bottomley F (ed) A Treatise on Dinitrogen Fixation, John Wiley & Sons, New York, Sec. 1, p 109
7. Khan F, Yue P-L, Rizzuti L, Augugliaro V, Schiavello M (1981) J Chem Soc Chem Commun 1049
8. Endoh E, Leland JK, Bard AJ (1986) J Phys Chem 90: 6223
9. Khader MM, Lichtin NN, Vurens GH, Salmeron M, Somorjai GA (1987) Langmuir 3: 303
10. Uyama H, Uchikura T, Niijima H, Matsumoto O (1987) Chem Lett 555
11. Kalman J, Varga TA, Hajos R (1983) Proc 6th Intern Symp Plasma Chem Vol 3: 686
12. Sugiyama K, Akazawa K, Oshima M, Miura H, Matsuda T, Nomura O (1986) Plasma Chem Plasma Process 6: 179
13. Miyahara K (1983) Chem Lett 1971
14. Nielsen A (1970) Review of Ammonia Catalysis, In: Heinemann H (eds) Catalysis Rev, Marcel Dekker, New York., Vol 4 p 1
15. Nielsen A (1977) Fert Sci Tehcnol Ser 2: 87
16. Ammoniak NH_3-Bildung und Zerfall (1935) In: Gmelin Handbook of Inorganic Chemistry, Vol 4 Nitrogen p 320
17. Mittasch A (1949) Early Studies of Mulicomponent Catalysts, In: Frannkenburg WG et al. Adv Catalysis, Academic Press, New York, Vol 2 p 81
18. Ammonia In: (1968) Encyclopedia of Chemical Technology (R.E. Kirk, D.F. Othmer), Hark HF et al. (eds) John Wiley, New York, p 258
19. Ammoniak, Synthetisches, In: Ullmann, Vol 3 p 544
20. Vancini CA, (1971) Synthesis of Ammonia, Macmillan, London
21. Nielsen A (1968) An Investigation on Promoted Iron Catalysts for the Synthesis of Ammonia, Jul Giellerups Forlag, Copenhagen
22. Nielsen A (1981) Catal Rev -Sci Eng 23: 17
23. Boudart M (1981) Catal Rev -Sci Eng 23: 1
24. Jennings JR (1991) Catalytic Ammonia Synthesis, Fundamentals and Practice Plenum Press, New York
25. Ozaki A, Aika K (1979) The Synthesis of Ammonia by Heterogeneous Catalysis, In: Hardy RWF et. al. (eds) A treatise on Dinitrogen Fixation Sec. I and II, John Wiley, New York, p 169
26. Ozaki A, Aika K (1981) Catalytic Activation of Dinitrogen, In: Anderson JR, Boudart M (eds) Catalysis-Science and Technology, Springer-Verlag, Berlin, Vol 1 p 87
27. Emmett PH, Love KS (1933) J Am Chem Soc 55: 4043
28. Aika K, Ozaki A (1968) Bull Chem Soc Jpn 41: 2818

29. Jolly WL (1964) The Inorganic Chemistry of Nitrogen, Benjamin, New York, p 36
30. Trapnell BMW (1953) Proc Roy Soc A218: 566
31. Bond GC (1962) Catalysis by Metals, Academic Press, London New York
32. Panov GI, Boreskov GK, Kharitonov AS, Moroz EM (1981) React Kinet Catal Lett 16: 247
33. Panov GI, Boreskov GK, Kharitonov AS (1982) Kinet Ketal 23: 438
34. Schlier E, Farnsworth HE (1950) Phys Rev 78: 316
35. Toyoshima I, Takezawa N, Suzuki H (1973) J Chem Soc Chem Commun 270
36. Toyoshima I, Takezawa N, Suzuki H (1977) Proc 6th Intern Congr Catal, Bond GC et al. (eds) Chem Soc London, p 708
37. Scholten JJF, Zwietering P (1957) Trans Farad Soc 53: 1363
38. Beeck O (1950) Advan Catal 2: 151
39. Ko SM, Schmidt LD (1974) Surf Sci 42: 508
40. Beeck O, Colle WA, Wheeler A (1950) Discuss Faraday Soc 8: 314
41. Oguri T (1964) J Phys Soc Jpn 19: 77
42. Oguri T (1963) J Phys Soc Jpn 18: 1280
43. Parry AA, Pryde JA (1967) Br J Appl Phys 18: 329
44. Kisliuk P (1959) J Chem Phys 31: 1605
45. Hickmott TW, Ehrlich G (1958) J Phys Chem Solids 5: 47
46. Hill MP, Lecchini SMA, Pethica BA (1966) Trans Farad Soc 62: 229
47. Madey TE, Yates JT Jr (1966) J Chem Phys 44: 1675
48. Yates JT Jr, Madey TE (1969) J Chem Phys 51: 334
49. Bagg J, Tomkins FC (1955) Trans Faraday Soc 51, 1071
50. Urabe K, Aika K, Ozaki A (1975) J Catal 38: 430
51. Obuchi A, Naito S, Onishi T, Tamaru K (1982) Surf Sci 122: 235
52. Mimeault VJ, Hansen RS (1966) J Phys Chem 70: 3001
53. Vajo JJ, Tsai W, Weinberg WH (1985) J Phys Chem 89: 3243
54. Miyazaki E, Yasumori I (1976) Surf Sci 55: 747
55. Sachtler WMH, Van Reijen LL (1962) J Res Inst Catal Hokkaido Univ 10: 87
56. Brenann D, Hayward DO, Trapnell BMW (1960) Proc Roy Soc A256: 81
57. Roberts MW (1960) Nature 188: 1020
58. Tanaka K, Tamaru K (1963) J Catal 2: 366
59. Toyoshima I, Somorjai GA (1979) Catal Rev -Sci Eng 19: 105
60. Frankenburg WG (1955) The Catalytic Synthesis of Ammonia from Nitrogen and Hydrogen, In: Emmett PH (ed) Catalysis, Reinhold Pub, New York, 1955, Vol 3 p 171
61. Kunimori K, Kawai T, Kondow T, Onishi T, Tamaru K (1976) Surf Sic 59: 302
62. Davies PW, Lambert RM (1981) Surf Sci 110: 227
63. Kuwahara Y, Fujisawa M, Jo M, Onchi M, Nishijima M (1987) Surf Sci 180: 421, ibid 188: 490
64. Matsuo I, Nakamura J, Hirano H, Yamada T, Tanaka K, Tamaru K (1989) J Phys Chem 93: 7747
65. Mittasch A (1950) Adv Catal 2: 81
66. Stathism E (1937) Osterr Chemiker Ztg 40: 80
67. Mittasch A, Keunecke E (1931) Z Phys Chem 574
68. King DA, Sebba F (1965) J Catal 4: 253
69. Aika K, Yamaguchi J, Ozaki A (1973) Chem Lett 161
70. Spitsyn VI, Mikhailenko IE, Pokrovskaya OV (1982) Dokl Akad Nauk SSSR, 263: 656
71. Jost F (1908) Z Anorg Chem 57: 414
72. Logan SR, Kemball C (1960) Trans Farad Soc 56: 144
73. Aika K, Hori H, Ozaki A (1972) J Catal 27: 424
74. Ozaki A, Aika K, Hori H (1971) Bull Chem Soc Jpn 44: 3216
75. Urabe K, Oh-ya A, Ozaki A (1978) J Catal 54: 436
76. Ogata Y, Aika K, Onishi T (1989) Bull Chem Soc Jpn 62: 642
77. Keunecke E (1930) Z Elektochem 36: 690
78. Brill R, Osumi Y (1966) Bull Chem Soc Jpn 39: 1678
79. Zabuga V Ya, Markova GP (1965) Kataliz i Katalizatory, Akad Nauk Ukr SSR, Resp Mezhved sb 110 (CA64:4312d)
80. Chalenko VG, Tovbin MV (1964) Ukr Khim Zh 30: 1128
81. Artyukh Yu. N, Rusov MT, Boldyreva NA (1967) Kinet Katal 8: 1319
82. Smith PJ, Taylor DW, Dowden DA, Kemball C, Taylor D (1982) Appl Catal 3: 303
83. Komarov VS, Efros MD, Rabina PD, Dmitrenko LM, Rozin AT, Kuznetsov LD, Lemeshonok GS, Khachaturyan IG, Mantseva GM, Eremenko SI (1983) SU 988327 (CA98:114516b)

84. Urabe K, Yoshioka T, Ozaki A (1978) J Catal 54: 52
85. Hikita T, Aika K, Onishi T (1990) Catal Lett 4: 157
86. Hikita T, Kadowaki Y and Aika K (1991) J Phys Chem 95: 9396
87. Panov GI, Kharitonov AS (1984) Proc 9th Intern Congr Catal DECHEMA (eds) Verlag Chemie, Veinheim, Vol 3 p 371
88. Takeshita T, Wallace WE, Craig RS (1976) J Catal 44: 236
89. Wallace WE, France J, Shamsi A (1982) Rare Earth Mod Sci Technol 3: 561
90. Kirch G, Schwab E, Wicke E, Zuchner H (1984) Proc 8th Intern Congr Catal DECHEMA (ed) Verlag Chem, Weinheim, Vol 4 p 209
91. Baiker A, Schlogl R, Armbruster E, Guntherodt HJ (1987) J Catal 107: 221
92. Shiflett WK, Dumesic JA (1982) J Catal 77: 57
93. Spencer ND, Schoonmaker RC, Somorjai GA (1982) J Catal 74: 129
94. Aika K, Ohya A, Ozaki A, Inoue Y, Yasumori I (1985) J Catal 92: 305
95. Aika K (1986) Angew Chem 25: 558
96. Ohya A, Aika K, Ozaki A (1984) J Chem Soc Chem Commun 321
97. Urabe K, Aika K, Ozaki A (1974) J Catal 32: 108
98. Urabe K, Aika K, Ozaki A (1976) J Catal 42: 197
99. Ichikawa M, Kondo T, Kawase K, Sudo M, Onishi T, Tamaru K (1972) J Chem Soc Chem Commun 176
100. Croft RC (1956) Austr J Chem 9: 184
101. Volpin ME, Novikov Yu N, Postnikov VA, Shur VB, Bayerl B, Kaden L, Wahren M, Dmitrenko LM, Stukan RA, Nefedev AV (1977) Z Anorg Allg Chem 428: 231
102. Sudo M, Ichikawa M, Soma M, Onishi T, Tamaru K (1969) J Phys Chem 73: 1174
103. Nishiyama S, Matuura S, Morita H, Tsuruya S, Masai M (1985) Appl Catal 15: 185
104. Nishiyama S, Yoshioka K, Yoshida T, Tsuryua S, Masai M (1988) Surf Sci 33: 1081
105. Asscher M, Somorjai GA (1984) Surf Sci 143: L389
106. Asscher M, Carraza J, Khan MM, Lewis KB, Somorjai GA (1986) J Catal 98: 277
107. Stoltze P, Structure and Surface Chemistry of Industrial Ammonia Synthesis Catalysts, chapter 2 in this book
108. Amariglio H, Rambeau G (1977) Proc 6th Intern Congr Catal, Bond GC et al (eds) Chem Soc London, p 1113
109. Strongin DR, Bare SR, Somorjai GA (1987) J Catal 103: 289
110. Ozaki A, Urabe K, Shimazaki K, Sumiya S (1979) Preparation of Catalysts II, Elsevier, Amsterdam, p 381
111. Lyubchenko YuA, Sergeeva AN, Dmitrenko LM, Tkachenko ZhI, Pavlenko LI (1981) SU 484 718 (CA 96:037844M)
112. Johnson MM, Tabler DC, Nowack GP (1982) US 4309311
113. Sydney A, Smith P, Yarm C (1986) EP 201205
114. Aika K, Shimazaki K, Hattori Y, Ohya A, Ohshima S, Shirota K, Ozaki A (1985) J Catal 92: 296
115. Lear AM, McCarroll JJ, Pippard DA, Tennison SR (1982) GB 2087746
116. McCarroll JJ, Tennison SR (1983) GB 2109361
117. Aika K, Takano T, Murata S (1992) J Catal 136: 126
118. McCarroll JJ, Tennison SR, Wilkinson NP (1986) US 4600571
119. Aika K, Kawahara T, Murata S, Onishi T (1990) Bull Chem Soc Japan 63: 1221
120. Murata S, Aika K, Onishi T (1990) Chem Lett 1067
121. Murata S, Aika K (1992) J Catal 136: 118
122. Kadowaki Y, Murata S, Aika K (1993) New Frontiers in Catalysis, Proc, 10th Intern Congr Catal Budapest, Guczi L et al eds, Elsevier p 2055
123. Aika K, Ozaki A (1969) J Catal 13: 232
124. Ozaki A, Taylor HS, Boudart M (1960) Proc Roy Soc (London) 258: 47
125. Boudart M, Oyama ST, Leclercq L (1981) Proc 7th Intern Cogr. Catal 1980, Seiyama T, Tanabe K (eds) Elsevier, Tokyo, p 578
126. Boudart M, Oyama ST (1985) US 4515763
127. McCandlish LE, Turaew LW, Wrisht FJ, Kugler EL (1982) EP 53018
128. Oyama ST, Boudart M (1980) J Res Inst Catal Hokkaido Univ 28: 305
129. Schulz-Ekloff G, Barsesel D, Sarholz W (1976) J Catal 43: 353
130. Kuznetsov BN, Kuznetsov VL, Ermakov Yu I (1975) Kinet Katal 16: 915
131. Temkin MI, Pyzhev VM (1940) Acta Physicochim (USSR) 12: 327
132. Kiperman S, Temkin MI (1946) Acta Physicochim (USSR) 21: 267
133. Kiperman S, Temkin MI (1946) Zh Fiz Khim 20: 623

134. Sidrov IP, Livshits VD (1947) Zh Fiz Khim 21: 1177
135. Peters C, Krabetz R (1956) Z Elektrochem 60: 859
136. Dumesic JA, Topsoe H, Khammouma S, Boudart M (1975) J Catal 37: 503
137. Kiperman S (1947) Zh Fiz Khim 21: 1435
138. Kiperman S, Granovskaya V (1951) Zh Fiz Khim 25: 557
139. McGill WJ, Sebba F (1963) J Catal 2: 104
140. Kunsman CH (1928) J Am Chem Soc 50: 2100
141. Tamaru K (1961) Trans Faraday Soc 57: 1410
142. Jungers JC, Taylor HS (1935) J Am Chem Soc 57: 679
143. Barrer RM (1936) Trans Faraday Soc 32: 490
144. Hinshelwood CN, Burk RE (1925) J Chem Soc (London) 127: 1105
145. Hailes HR (1931) Trans Faraday Soc 27: 601
146. Shindo H, Egawa C, Onishi T, Tamaru K (1979) Z Naturforsch 34a: 96
147. McGeer JP, Taylor HS (1951) J Am Chem Soc 73: 2743
148. Logan SR, Moss RL, Kamball C (1958) Trans Faraday Soc 54: 922
149. Schwab GM, Krabetz R (1956) Z Electrochem 60: 855
150. Sidrov IP, Livshits VD (1952) Zh Fiz Khim 26: 538
151. Love KS, Emmett PH (1941) J Am Chem Soc 63: 3297
152. Takezawa N, Toyoshima I (1966) J Phys Chem 70: 594
153. Tamaru K, Tanaka K, Fukasaku S, Ishida S (1965) Trans Faraday Soc 61: 765
154. Friedlander AG, Courty PhR, Montarnal RE (1977) J Catal 48: 322
155. Amano A, Taylor H (1954) J Am Chem Soc 76: 4201
156. Apel'baum LO, Temkin MI (1959) Zh Fiz Khim 33: 2697
157. Robertson AJB, Willhoft EMA (1967) Trans Faraday Soc 63: 476
158. Dixon JK (1931) J Am Chem Soc 53: 2071
159. Schwab GM, Schmidt FH (1929) Z Elektrochem 35: 605
160. Elod E, Banholzer W (1926) Z Elektrochem 32: 555
161. Dixon JK (1931) J Am Chem Soc 53: 1763
162. Rusov MI, Pevsner Ts V (1954) Zh Fiz Khim 28: 1765
163. Brill R, Tauster S (1962) J Chem Phys 36: 2100
164. Emmett PH, Kummer JT (1943) Ind Eng Chem 35: 677
165. Bokhoven C, van Raayen W (1954) J Phys Chem 58: 471
166. Aika K, Kumasaka M, Oma T, Kato O, Matsuda H, Watanabe N, Yamazaki K, Ozaki A, Onishi T (1986) Appl Catal 28: 57
167. Holzman PR, Shiflett WK, Dumesic JA (1980) J Catal 62: 167
168. Morikawa Y, Ozaki A (1971) J Catal 23: 97
169. Baris H, Glinski M, Kijenski J, Wokaun A, Baiker A (1986) Appl Catal 28: 295
170. Shiflett WK, Dumesic JA (1981) Ind Eng Chem Fundam 20: 246
171. Temkin MI, Morozov NM, Shapatina EN (1963) Kinet Ketal 4: 260
172. Tamaru K (1965) Proc 3rd Intern Congr Catal, Sachtler WHM et al (ed) Amsterdam, North Holland Pub p 664
173. Tamaru K (1988) Acc Chem Res 21: 88
174. Shindo H, Egawa C, Onishi T, Tamaru K (1980) J Chem Soc Faraday Trans 1 76: 280
175. Egawa C, Nishida T, Naito S, Tamaru K (1984) J Chem Soc Faraday Trans 1 80: 1595
176. Loeffler DG, Schmidt LD (1976) J Catal 44: 244
177. Loeffler DG, Schmidt LD (1976) J Catal 41: 440
178. Lotz CR, Sebba F (1957) Trans Faraday Soc 53: 1246
179. Segal N, Sebba F (1967) J Catal 8: 105
180. Kummer JT, Emmett PH (1951) J Chem Phys 19: 289
181. Morikawa Y, Ozaki A (1968) J Catal 12: 145
182. Schulz G, Schaefer H (1969) Z Phys Chem (N.F.), 64: 333
183. Schulz-Ekloff G (1971) Ber Bunsenges Phys Chem 75: 110
184. Guyer WRF, Joris GG, Taylor HS (1941) J Chem Phys 9: 287
185. Gorbunov AI, Boreskov GK (1960) In: Problemy Kinet Katal Akad Nauk (USSR), No. 10: pp 192
186. Joris GG, Taylor HS (1939) J Chem Phys 7: 893
187. Gasser RPH, Lowrence CP, Newman DG (1965) Trans Faraday Soc 61: 1771
188. Gasser RPH, Hale A, Marsay CJ (1967) Trans Faraday Soc 62: 1789

189. Boreskov GK, Kolchanova VM, Rachkovskii EE, Filimonova SN, Khasin AV (1975) Kinet Katal 16: 1218
190. Moore GE, Unterwald FC (1968) J Chem Phys 48: 5393
191. Aika K, Ozaki A (1969) J Catal 14: 311
192. Rambeau G, Amariglio H (1981) J Catal 72: 1
193. Rambeau G, Amariglio H (1981) Appl Catal 1: 291
194. Rambeau G, Jorti A, Amariglio H (1982) Appl Catal 3: 273
195. Rambeau G, Jorti A, Amariglio H (1982) J Catal 74: 110
196. Danielson LR, Dresser MJ, Donaldson EE, Dickinson JT (1978) Surf Sci 71: 599
197. Danielson LR, Dresser MJ, Donaldson EE, Sandstrom DR (1978) Surf Sci 71: 615
198. Feulner P, Menzel D (1982) Phys Rev B25: 4295
199. Menzel D, Pfnur H, Feulner P (1983) Surf Sci 126: 374
200. Rigby LJ (1965) Can J Phys 43: 532
201. Yates JT Jr, Madey TE (1965) J Chem Phys 43: 1055
202. Joyner RW, Rickman J, Roberts MW (1974) J Chem Soc Faraday Trans 1 70: 1825
203. Delchar TA, Ehrlich G (1965) J Chem Phys 42: 2686
204. Goymour CG, King DA (1973) J Chem Soc Faraday Trans 1 69:749
205. Yates JT Jr, Klein R, Madey TE (1976) Surf Sci 58: 469
206. Fuggle JC, Menzel D (1978) Vak Tech 27: 130
207. Wilf M, Folman M (1975) Surf Sci 52: 10
208. Adams DL, Germer LH (1971) Surf Sci 26: 109
209. King DA, Wells MG (1974) Proc Roy Soc (London) Ser A 339: 245
210. Singh-Boparai SP, Bowker M, King DA (1975) Surf Sci 53: 55
211. Tamm PW, Schmidt LD (1971) Surf Sci 26: 286
212. Adams DL, Germer LH (1971) Surf Sci 27: 21
213. Clavenna LR, Schmidt LD (1970) Surf Sci 22: 365
214. Han HR, Schmidt LD (1971) J Phys Chem 75: 227
215. Tompkins FC, (1978) Chemisorption of Gases on Metals, Academic Press, London, p 26.
216. Egawa C, Naito S, Tamaru K (1983) Surf Sci 131: 49
217. Pasternak RA, Endow N, Bergsnov-Hansen B (1966) J Phys Chem 70: 1304
218. Mahnig M, Schmidt LD (1972) Z Phys Chem (N. F.) 80: 71
219. Kunimori K, Kawai T, Kondow T, Onishi T, Tamaru K (1976) Surf Sci 54: 525
220. Egawa C, Naito S, Tamaru K (1983) Surf Sci 125: 605
221. Haase G, Asscher M (1987) Surf Sci 191: 75
222. Grunze M, Golze M, Fuhler J, Neumann M, Schwarz E, Proc 8th Intern Congr Catal 1984, DECHEMA (ed) Verlag Chemie, Weinheim Vol 4 p 133
223. Haase G, Asscher M (1987) Chem Phys Lett 142: 241
224. Khrizman IA, Korneyiuchuk G (1943) Acta Physicochim (URSS) 18: 420
225. Kazusaka A (1971) J Res Inst Catal Hokkaido Univ 19: 42
226. Kokes RJ, Emmett PH (1958) J Am Chem Soc 80: 2082
227. Honda F, Hirokawa K (1977) J Electron Spectrosc Relat Phenom 10: 125
228. Grunze M, Driscoll RK, Burland GN, Cornish JCL, Pritchard J (1979) Surf Sci 89: 381
229. Wilf M, Dawson PT (1976) Surf Sci 60: 561
230. Schwaha K, Bechtold E (1977) Surf Sci 66: 383
231. Kiss J, Berko A, Solymosi F (1981) Magy Kem Foly 87: 566
232. Hendrickx HACM, Hoek A, Nieuwenhuys BE (1983) Surf Sci 135: 81
233. Gorodetskii VV, Sobyanin VA (1980) Proc 7th Intern Congr on Catal, Seiyama T, Tanabe K, (eds) Elsevier, Amsterdam, 1981, p. 566.
234. Kharitonov AS, Boreskov GK, Panov GI, Pankrotiev Yu D (1983) React Kinet Catal Lett 22: 309
235. Panov GI, Boreskov GK, Kharitonov AS, Moroz EM, Sobolev VI (1984) Kinet Katal 25: 123
236. Magomedkov EP, Kasatkina LA (1978) Trudy MKhTl im Meneleeva DI No 99: 60
237. Panov GI, Boreskov GK, Kharitonov AS (1980) Dokl Akad Nawk, SSSR 252: 646
238. Schreifels JA, Deffeyes JE, Neff LD, White JM (1982) J Electron Spectrosc Relat Phenom 25: 191
239. Varnakova RG, Morozova LV, Khamidova Kh Kh (1982) Zh Fiz Khim 56: 1533
240. Eischens RP, Jacknow J (1965) Proc 3rd Intern Congr Catal, Sachtler WHM et al (ed) Amsterdam, North Holland Pub pp 627

241. Hardeveld RV, Montfoort AV (1966) Surf Sci 4: 396
242. Hardeveld RV, Montfoort AV (1969) Surf Sci 17: 90
243. Borodko YuG, Lyutov VS (1971) Kinet Katal 12: 238
244. Egerton TA, Sheppard N (1974) J Chem Soc Faraday Trans 1 70: 1357
245. Eischens RP (1972) Acc Chem Res 5: 74
246. Chang CC, Kokes RJ (1973) J Phys Chem 77: 2640
247. Cohen de Lara E, Delaval Y (1978) J Chem Soc Faraday Trans 2 74: 790
248. Sakata Y, Kinoshita N, Domen K, Onishi T (1987) J Chem Soc Farad Trans 1 83: 2765
249. Sakata Y, Abe H, Kondo J, Maruya K, Domen K, Onishi T (1989) Chem Lett 711
250. Okawa T, Onishi T, Tamaru K (1977) Z Phys Chem (N. F.) 107: 239
251. Brill R, Jiru P, Schulz G (1969) Z Phys Chem (N. F.) 64: 215
252. Kinoshita K, Kido K, Domen K, Aika K, Onishi T (1986) J Chem Soc Farad Trans 1 82: 2269
253. Wang HP, Yates JT Jr (1984) J Phys Chem 88: 852
254. DeJong KP, Bongenaar-Schlenter BE, Meima GR, Verkerk RC, Lammers MJJ, Geus JW (1983) J Catal 81: 67
255. Oh-kita M, Aika K, Urabe K, Ozaki A (1976) J Catal 44, 460; (1975) J Chem Soc Chem Commun 147
256. Aika K, Midorikawa H, Ozaki A (1982) J Phys Chem 86: 3263
257. Nakata T, Matsushita S (1965) J Catal 4: 631
258. Bradshaw AM, Pritchard J (1970) Surf Sci 19: 198
259. Lyutov VS, Borodko YuG (1971) Kinet Katal 12: 1566
260. Kubota J, Aika K (1991) J Chem Soc Chem Commun 1544; (1994) J Phys Chem 99:
261. Wey JP, Worley CG, Neely WC, Worley SD (1992) J Phys Chem 96: 7088
262. Ravi A, King DA, Sheppard N (1968) Trans Faraday Soc 64: 3359
263. Kubsov SA, Borovkov VYu, Kazansky VB, Gagarlin SG (1984) Chem Phys Lett 107: 337
264. Rissman EF, Parry JM (1975) J Phys Chem 79: 1975
265. Zverev SM, Smirnov KS, Tsyganenko AA (1988) Kinet Katal 29: 1439
266. Bozso F, Ertl G, Weiss M (1977) J Catal 49: 18
267. Madey T, Yates JT Jr , Erickson NE (1974) Surf Sci 43: 526
268. Brundle CR (1976) J Vac Sci Tech 13: 301
269. Kishi K, Roberts MW (1977) Surf Sci 62: 252
270. Ertl G, Thiele N (1979) Appl Surf Sci 3: 99
271. Rao CNR, Ranga Rao G (1991) Surf Sci Report 13: 221
272. Walker AP, Rayment T, Lambert RM (1989) J Catal 117: 102

Chapter 4

Kinetics of Ammonia Synthesis and Decomposition on Heterogeneous Catalysts

John Bøgild Hansen
Haldor Topsoe A/S Copenhagen, Denmark

Contents

Nielsen (Ed.), Ammonia

4.1 Early Developments

Despite the vast practical importance of ammonia synthesis on heterogeneous catalysts, more than 25 years passed from when these catalysts were introduced until the first workable rate equation was proposed in the literature.

The ground work had been laid by the publication of the rates of the ammonia synthesis as a function of operating conditions. Larson and Tour [1] published in 1922 data of a study on doubly promoted (i.e. Al_2O_3 and K_2O) catalysts between 400 and 475 °C and 1–100 atm. This was followed in 1926 by the Almquist and Crittenden paper [2].

Benton [3] attempted to develop a rate equation from these data. He noted the retarding effect of ammonia on the synthesis rate and explained this by the accumulation of ammonia on the surface of the catalyst.

Winter [4] determined the rate of NH_3 decomposition on an iron foil at atmospheric pressure in the temperature range 500 to 700 °C. The experiments were carried out in the presence of excess H_2 in order to avoid the formation of iron nitride. Many earlier studies had used pure NH_3, thereby transforming the catalyst into a nitride phase and making it drastically different from the catalyst under synthesis conditions.

Winter fitted the data to the following rate equation:

$$-\frac{dP_{NH_3}}{dt} = k\frac{P_{NH_3}^{0.9}}{P_{H_2}^{1.5}} \quad (1)$$

Here P_{NH_3} and P_{H_2} are the partial pressures of NH_3 and H_2, t is the contact time.

To rationalize his findings, Winter assumed that the steps leading to adsorbed nitrogen in the atomic state on the catalyst surface were so fast that the equilibrium

$$NH_3(gas) \rightleftarrows N(ads) + 1.5\ H_2(gas) \quad (2)$$

is established. This equilibrium can be obtained by combining the two equilibria

$$NH_3(gas) \rightleftarrows N(gas) + 1.5\ H_2(gas) \quad (3)$$

and

$$N(gas) \rightleftarrows N(ads) \quad (4)$$

From equilibrium (3) one gets

$$K = \frac{P_{H_2}^{1.5}\, P_N}{P_{NH_3}} \quad (5)$$

Here P_N, the pressure of nitrogen atoms in the gas, is a virtual pressure and serves as an auxiliary quantity only.

Assuming low coverage and Henry's law to be valid one obtains for the coverage θ_N of nitrogen far from equilibrium

$$\theta_N = a\,P_N \tag{6}$$

Inserting P_N from Eq. (5) one arrives at

$$\theta_N = a\,K\,\frac{P_{NH_3}}{P_{H_2}^{1.5}} \tag{7}$$

Winter further assumed that nitrogen is desorbed as atoms, which recombine in the gas phase according to the scheme

$$N(ads) \rightleftarrows N(gas) \tag{8}$$

$$2\,N(gas) \rightleftarrows N_2(gas) \tag{9}$$

These assumptions lead to the rate of decomposition being proportional to θ_N, i.e.

$$-\frac{dP_{NH_3}}{dt} = a\,K\,\frac{P_{NH_3}}{P_{H_2}^{1.5}} \tag{10}$$

which is very close to rate equation (1).

The idea that adsorption of N_2 is the rate determining step in the synthesis reaction was beginning to take hold, as discussed by Taylor [5], Frankenburger [6], and Emmett et al. [7]. Indirect evidence supporting this theory began to accumulate.

Emmett and Love [8] found that hydrogenation of iron nitride takes place at temperatures below that of synthesis, which suggests that the hydrogenation of adsorbed nitrogen is fast.

Emmett and Brunauer [9] determined the rate of N_2 adsorption. It was found to be comparable to the rate of NH_3 formation or sufficiently fast to be the first step in the synthesis mechanism.

Taylor and Jungers [10] demonstrated that isotopic exchange between NH_3 and D_2 takes place even at room temperature. NH_2D, NHD_2 and ND_3 were detected on the doubly promoted catalyst used in the study. They concluded that

$$N_2(gas) \rightarrow 2\,N(ads)$$

must be the rate determining step in the ammonia synthesis.

Kozhenova and Kagan [11] demonstrated that the hydrogenation of adsorbed nitrogen proceeds much faster than N_2 adsorption.

Data on the rate of synthesis and decomposition continued to be published. Finkelstein studied NH_3 decomposition on promoted and unpromoted iron and the synthesis on iron [12]. Chrisman [13] did experiments with doubly promoted catalyst at atmospheric pressure. Some empirical rate equations were proposed. Finkelstein and Rubanik [14] proposed that the reaction is second order and the rate determining step is

$$2\,K_x + N_2 \rightarrow 2\,K_xN$$

with the following hydrogenation steps being fast. On this basis they determined the activation energy from earlier data to be between 18.6 and 22 kcal/mol.

Roiter [15] proposed empirical rate equations for the synthesis rate, being different for subatmospheric pressures and pressures between 50–200 atm. He did not discuss the reaction mechanism.

4.2 The Temkin-Pyzhev Rate Equation

Temkin and Pyzhev used the data of Winter as the starting point for their development of the first successful rate equation for ammonia synthesis named after them [16–18]. They demonstrated, that using Winter's assumptions leads to an apparent activation energy for decomposition E_d larger than 97 000 cal/mol, clearly disagreeing with the experimentally found energy of 51 000 cal/mol. Furthermore, they pointed out that the desorption of nitrogen in the atomic form is extremely unlikely due to the large amount of energy needed.

If, however, the predominant view that the rate of ammonia decomposition is determined by the rate of

$$2\,N(ads) \rightleftarrows N_2(gas) \tag{11}$$

is accepted, then the rate should be proportional to θ_N^2 again in conflict with the experimental findings.

In order to overcome this difficulty Temkin and Pyzhev proposed to abandon the Langmuir adsorption concept, which basically assumes the surface of the catalyst to be energetically uniform. Instead they chose to use for the adsorption equilibrium the isotherm proposed by Frumkin and Slygin [19],

$$\theta = \frac{1}{f} \ln a_0 P \tag{12}$$

where θ is the coverage, P the equilibrium pressure, and a_0 and f are constants, and for the rate of adsorption,

$$r = k_a P e^{-g\theta} \tag{13}$$

an equation developed by Zeldowitsch and Roginsky [20]. Here P is the gas pressure and K_a and g are constants. For the rate of desorption they used the equation proposed by Langmuir [21]

$$w = k_d e^{h\theta} \tag{14}$$

where k_d and h are constants.

Temkin and Pyzhev pointed out that choosing the three equations above is equivalent to assuming a linear dependence of the heat of adsorption with the coverage. In accordance with Polanyi [22] the change in energy of activation should be a certain fraction α of the change in the heat of adsorption. These assumptions involve the concept of a heterogenous surface. Temkin and Pyzhev ascribed this heterogeneity to considerable repulsive forces between the adsorbed molecules. They noted that these assumptions describe the *intermediate degrees* of coverage and are valid for adsorption of both molecules and atoms. The constants f, g and h are related as follows

$$\alpha = \frac{g}{f} \tag{15}$$

$$\beta = \frac{h}{f} \tag{16}$$

$$\alpha + \beta = 1 \tag{17}$$

Another assumption for the Temkin-Pyzhev equation is that the nitrogen adsorption is not influenced by hydrogen and ammonia, an assumption which seems to be corroborated by the experimental data of Emmett and Brunauer [7].

Following Winter, Temkin and Pyzhev assumed that the amount of adsorbed nitrogen is determined by the equilibrium with hydrogen and ammonia in the gas phase, i.e., the nitrogen equilibrium pressure is given by

$$(P_{N_2})_{eq} = k' \frac{P^2_{NH_3}}{P^3_{H_2}} \tag{18}$$

Inserting into Eq. (12) one finds

$$\theta = \frac{1}{f} \ln a_0 k' \frac{P^2_{NH_3}}{P^3_{H_2}} \tag{19}$$

Using Eq. (14) the rate of desorption becomes

$$w = k_d (a_0 k')^{\beta} \left(\frac{P^2_{NH_3}}{P^3_{H_2}} \right)^{\beta} = k_2 \left(\frac{P^2_{NH_3}}{P^3_{H_2}} \right)^{\beta} \tag{20}$$

In order to get Eq. (20) to agree with Winter's experimental findings, Temkin and Pyzhev had to choose

$$\alpha = \beta = 0.5$$

Using data from Emmett and Brunauer [7], Temkin and Pyzhev were able to demonstrate that the activation energy for decomposition on the average should be 46 500 cal/mol, in good agreement with Winter's results.

To arrive at an expression for the synthesis rate equation, Eq. (19) is inserted in Eq. (13) to get the net rate as the difference between the forward and reverse processes,

$$\frac{dP_{NH_3}}{dt} = k_1 P_{N_2} \left(\frac{P^3_{H_2}}{P^2_{NH_3}} \right)^{\alpha} - k_2 \left(\frac{P^2_{NH_3}}{P^3_{H_2}} \right)^{\beta} \tag{21}$$

Assuming $\alpha = \beta = 0.5$, Eq. (21) reduces to

$$\frac{dP_{NH_3}}{dt} = k_1 \frac{P_{N_2} P_{H_2}^{1.5}}{P_{NH_3}} - k_2 \frac{P_{NH_3}}{P_{H_2}^{1.5}} \tag{22}$$

The retarding effect of ammonia on the synthesis rate observed by Benton [3] is also borne out by Eq. (21), but the underlying reason is not the accumulation of NH_3 on the surface, but rather an increase in the amount of nitrogen on the surface due to the equilibrium

$$NH_3(\text{gas}) \rightleftarrows N(\text{ads}) + 1.5\ H_2(\text{gas}) \tag{23}$$

Equation (22) can be rewritten by considering that at equilibrium

$$\frac{dP_{NH_3}}{dt} = 0$$

and therefore

$$\frac{k_1}{k_2} = \frac{P_{NH_3}^2}{P_{N_2} P_{H_2}^3} \tag{24}$$

or, using the usual equilibrium constant K_p, one has

$$\frac{k_1}{k_2} = K_p^2 \tag{25}$$

If $(P_{NH_3})_{eq}$ denotes the partial pressure of ammonia in equilibrium with hydrogen and nitrogen, given by

$$(P_{NH_3})_{eq}^2 = \frac{k_1}{k_2} P_{N_2} P_{H_2}^3 \tag{26}$$

then Eq. (22) reduces to

$$\frac{d\dfrac{P_{NH_3}}{(P_{NH_3})_{eq}}}{dt} = \frac{k_2}{P_{H_2}^{1.5}} \left[\frac{(P_{NH_3})_{eq}}{P_{NH_3}} - \frac{P_{NH_3}}{(P_{NH_3})_{eq}} \right] \tag{27}$$

At low pressures, or low P_{NH_3}, $\dfrac{P_{NH_3}}{(P_{NH_3})_{eq}}$ is a good approximation for the efficiency η, defined as the ratio of the actual ammonia yield to the yield under equilibrium conditions at a given temperature and pressure. With this approximation, Eq. (27) can now be integrated to give

$$k = 0.5\,\gamma^{1.5} P^{0.5} V \ln(1 - \eta^2) \tag{28}$$

Here V is the space velocity, the ratio of the volume of gas at standard conditions to the volume of catalyst per hour, and γ characterizes the deviation from stoichiometric gas composition

$$\gamma = \frac{P_{H_2}}{0.75\,P} \tag{29}$$

and k is given by

$$k = \frac{273\,\phi k_2}{0.75^{1.5}T} \tag{30}$$

where ϕ is the catalyst bed void volume ratio.

Temkin and Pyzhev tested the rate equation by applying Eq. (28) to a series of experiments at atmospheric pressure with a doubly promoted catalyst, first by varying the H_2/N_2 ratio at 400 °C using a space velocity of 30 000 h^{-1}. As predicted by Eq. (22), the H_2/N_2 ratio giving the highest yield at low conversion was 1.5.

A second set of experiments was carried out at 400 and 450 °C. They showed a satisfactory constancy of k at each temperature as expected if Eq. (28) were correct. The apparent activation energy for the decomposition rate constant k_2 was calculated to be 40 000 cal/mol.

The decomposition data of Chrisman [13] were evaluated with Eq. (20), and an activation energy of 39 700 cal/mol was found.

Temkin and Pyzhev derived the following integrated form of their rate equation without making any simplifying assumptions, except assuming H_2/N_2 to be 3.

$$k = P^{0.5}V_0 \int_0^z \frac{z(1-z)^{1.5}dz}{(1+z)\,[L^2(1-z)^4 - z^2]} \tag{31}$$

where z is the mole fraction of NH_3 in the gas mixture, given by

$$z = \frac{P_{NH_3}}{P} \tag{32}$$

and V_0 the space velocity at the entry to the reactor. L is defined by

$$L = \frac{z_{eq}}{(1-z_{eq})^2} = \left(\frac{1}{4}\right)^{0.5}\left(\frac{3}{4}\right)^{0.5} PK_p \tag{33}$$

Temkin and Pyzhev made an error in deriving Eq. (31), as pointed out first by Emmett and Kummer [23]. Emmett and Kummer introduced a factor of $(1+z)^3$ instead.

The correct first denominator term, $(1+z)^2$ was first worked out by Kodama et al. [24]. Temkin also arrived at $(1+z)^2$ in [25]. The error introduced by a wrong denominator is normally small as Z is small.

Temkin and Pyzhev used Eq. (31) to test their rate equation with the data of Larson and Tour [2] at 10, 31.6 and 200 atm and at 420 and 450 °C. They concluded that the data agreed well with Eq. (21), however, the rate constants calculated for the 200 atm data were lower than the ones calculated for the lower pressures. Temkin and Pyzhev attributed this difference to the fact that the measurements at high pressures were carried out with different equipment than those at low pressures. Temkin and Pyzhev later pointed out that at pressures above 300 atm fugacities should be used instead of pressures and the constants k_2 and k_1 should be assumed to depend on the pressure.

By differentiating Eq. (21) with respect to temperature, the optimum synthesis conditions (maximum rate) is found at a temperature at which the equilibrium constant is equal to

$$\frac{P_{NH_3}}{P_{N_2}^{0.5} P_{H_2}^{1.5}} = \sqrt{\frac{E_d}{E_s}} \tag{34}$$

where E_d and E_s are the apparent activation energies for decomposition and synthesis, respectively.

Love and Emmett [26] studied the decomposition of ammonian over both singly (10.2% Al_2O_3) and doubly promoted (10.2% Al_2O_3 1.59% K_2O) catalysts at atmospheric pressure.

The doubly promoted catalyst showed the rate equation

$$-\frac{dP_{NH_3}}{dt} = k\frac{P_{NH_3}^{0.6}}{P_{H_2}^{0.85}} \tag{35}$$

which is very close to Temkin and Pyzhev's Eq. (20) if β is set at 0.29. The apparent activation energy for decomposition was found to be 45600 ± 2000 cal/mol.

The data obtained on singly promoted catalysts, however, could not to be fitted to the Temkin-Pyzhev equation. Their performance depended in a highly complex manner on the temperature and the partial pressures and the activation energy depended on the temperature and even displayed hysteresis effects. These deviations from Temkin-Pyzhev kinetics could be removed by impregnating the catalyst with a KOH solution although the activity dropped considerably.

The decomposition rate found was approximately independent of the ammonia and hydrogen partial pressures at 390 and 450 °C. At 429 °C, it was proportional to $P_{H_2}^{1.2}/P_{NH_3}^{0.84}$ i.e., showing a negative β value in the Temkin-Pyzhev equation.

Love and Emmett suggested that this abnormal behaviour could be due to the presence of substantial amounts of NH and NH_2 groups on the surface on singly promoted catalysts, whereas on doubly promoted catalysts, such groups would be "driven off" by the alkali.

This interpretation was supported in some chemisorption experiments with H_2 at 100 °C on singly and doubly promoted catalysts previously exposed to N_2 at 450 °C (Brunauer and Emmett [27]).

Frankenburger later explained the phenomena by NH, NH_2 and NH_3 being adsorbed on the acidic Al_2O_3 groups in singly promoted catalysts. These acidic functions could be neutralized by K_2O [28].

4.3 The Rate Equations of Brunauer, Love, and Keenan

Brunauer et al. [29] extended the range of the Eqs. (12), (13), (14), which are the basis for the Temkin-Pyzhev equation. In order to develop an adsorption isotherm, Brunauer et al. subdivided the surface into elements each following (the original) Langmuir isotherm, so that

$$a_1(1-\theta)p = b_1\theta e^{-q/RT} \tag{36}$$

where q is the heat of adsorption, and a_1 and b_1 are constants. Assuming q to be linear function of the surface area covered,

$$q = q_0 - As \tag{37}$$

where q_0 is the heat of adsorption at $\theta = 0$, one finds

$$\theta = \int_0^1 \frac{a_0 e^{-As/RT} P \, ds}{1 + a_0 e^{-Bs/RT} P} \tag{38}$$

Integration gives

$$\theta = \frac{RT}{A} \ln \frac{1 + a_0 P}{1 + a_0 e^{-a/RT}} \tag{39}$$

where

$$a_0 = \frac{a_1}{b_1} e^{q_0/RT} \tag{40}$$

if $a_0 \gg 1 \gg a_0 e^{-A/RT}$, then Eq. (39) gives the Frumkin and Slygin isotherm (Eq. (12)) used by Temkin and Pyzhev, if

$$\frac{1}{f} = \frac{RT}{A} \tag{41}$$

The Temkin-Pyzhev equation is thus only valid at medium surface coverage as already pointed out by the authors.

Brunauer et al. also developed equations for the rates of desorption and adsorption using as starting points the equations proposed by Taylor [5],

$$w = b\theta e^{-E_d/RT} \tag{42}$$

$$z = aP(1-\theta)e^{-E_a/RT} \tag{43}$$

where w and z are rates of desorption and adsorption, respectively, and b and a are constants.

They assumed the activation energies to vary linearly with coverage:

$$E_d = E_d^0 - Bs \tag{44}$$

$$E_a = E_a^0 - Js \tag{45}$$

and, furthermore, divided the surface into two parts, an almost completely covered part and an almost completely bare part, which is equivalent to assuming B and J to be sufficiently large. Integrating only over the relevant part of the surface gives

$$w = be^{-E_d^0/RT}(RT/B)\,(e^{B\theta/RT} - 1) \tag{46}$$

$$z = aPe^{-E_a^0/RT}(RT/J)\,(e^{-J\theta/RT} - e^{-J/RT}) \tag{47}$$

Equation (46) gives $w = 0$ for $\theta = 0$ and the maximum value of w for $\theta = 1$. Equation (47) gives $z = 0$ for $\theta = 1$ and a maximum value of z for $\theta = 0$. For large values of θ, Eq. (46) reduces to the Langmuir desorption rate equation (14) and for small values of θ, Eq. (47) gives the Zeldotwitsch equation with

$$h = B/RT \tag{48}$$

$$g = J/RT \tag{49}$$

Brunauer et al. also deduced rates for adsorption and desorption if the surface is homogeneous, but strong attractive and repulsive forces exist between the adsorbed molecules or atoms. They assumed the activation energy to be linear with coverage

$$w = be^{-E_d^0/RT}\,e^{B\theta/RT} \tag{50}$$

$$z = aPe^{-E_a^0/RT}(1 - \theta)e^{-J\theta/RT} \tag{51}$$

and found the adsorption isotherm formula by equating Eqs. (50) and (51)

$$P = \left[\frac{\theta}{a_0}(1 - \theta)\right]e^{A\theta/RT} \tag{52}$$

Here A, B, J are again related by the equations

$$B + J = A \tag{53}$$

$$E_d^0 - E_a^0 = q_0 \tag{54}$$

At intermediate surface coverage Brunauer et al. used Eqs. (46) and (47) (which then reduces to the Temkin Eqs. (13) and (14)) to calculate the adsorption rate of nitrogen

$$\frac{dV}{dt} = k_a PV_m e^{-JV/V_mRT} - k_d V_m e^{BV/V_mRT} \tag{55}$$

where V_m is adsorbate volume in a monolayer. Equation (55) contains four constants $k_a V_m$, $k_d V_m$, J/V_m, and B/V_m which can be derived from the experimental nitrogen adsorption rate curve.

Using Emmnett and Brunauer's data [7] on doubly promoted catalyst, Brunauer et al. calculated the four constants and obtained a close fit to the experimental rate curve. At equilibrium

$$\ln P = \ln\frac{k_d}{k_a} + \frac{B + J}{V_m RT} \tag{56}$$

By using the constants, determined in the adsorption rate experiment, Brunauer et al. achieved an excellent fit to the adsorption isotherm determined experimentally with the same catalyst.

Substituting the β term in the Temkin-Pyzhev equations

$$\beta = \frac{B}{B + J} = \frac{800}{2900} = 0.276 \tag{57}$$

into Eq. (20) gives, using the β determined from N_2 adsorption,

$$-\frac{dP_{NH_3}}{dt} = k_2 \frac{P_{NH_3}^{0.55}}{P_{H_2}^{0.83}} \tag{58}$$

in good agreement with what Love and Emmett had with the same catalyst referred to before. An activation energy of 49000 cal/mol could also be estimated in good agreement with the experimental value.

In 1943, Emmett and Kummer [23] presented the results of high pressure experiments on a doubly promoted catalyst (3.02% Al_2O_3, 0.94% K_2O) at 33.3, 66.6 and 100 atm, H_2/N_2 ratios: 3/1, 1/1 and 1/3 and space velocities from 25000 to 125000 h^{-1} at 370, 400 and 450 °C were used. The data were analysed with the Temkin-Pyzhev equation using $\alpha = \beta = 0.5$. The rate constant k was constant with variations in space velocity, except at 370 °C where it decreased with an increase in space velocity of 5. Apparent activation energy for decomposition was found to be from 45000 to 53000 cal/mol.

Although the Temkin-Pyzhev equation seemed to fit the data reasonably well with respect to variations in space velocity and temperature, all data sets exhibited a clear decrease in the rate constant with increasing pressure.

Using a β value of approx. 0.3 as found by Love and Emmett [26] an attempt was made to calculate the data by the Temkin-Pyzhev equation with $\alpha = 0.67$, but this gave a poorer agreement. With respect to changes in gas composition the agreement with $\alpha = 0.67$ was fair at 370 and 400 °C, but the rate constant varied by a factor 2 at 450 °C between the 3:1 and 1:3 H_2/N_2 mixtures.

Emmett [30] presented a review on the ammonia decomposition reaction in 1946.

In 1947, Temkin and Kiperman [31] gave a general discussion of the Temkin-Pyzhev equation and its usefulness for both the synthesis reaction and the decomposition. They pointed out that it would not be valid at low ammonia partial pressures because the equilibrium

$$*N_2 + 3\,H_2 \rightleftarrows 2\,NH_3 \tag{59}$$

would be displaced to the right leading to low surface coverages, where the Temkin-Pyzhev equation is no longer valid. The rate r should be independent of coverage and given by

$$r = k' P_{N_2} \tag{60}$$

Sidorov and Livshits [32] carried out experiments at 300 atm in the temperature range 450–500 °C. They obtained a good fit of the data by using the Temkin-Pyzhev equation with $\alpha = 0.5$ and an activation energy of 176 kJ/mol. They still used $(1 + z)^3$ in the denominator in the integration of the Temkin-Pyzhev equation introduced by Emmett and Kummer [23].

Some data on the synthesis rate were given by Kobayashi and Kubota in [33].

The less-than-satisfactory agreement with experimental data at higher pressures of the Temkin-Pyzhev rate equation has been mentioned. In 1950, Temkin modified the original equation by introducing fugacities instead of partial pressures [34]. Furthermore, the original equation, Eq. (21), was multiplied by a term

$$\frac{dP_{NH_3}}{dt} = \left[k_1 P_{N_2}\left(\frac{P_{H_2}^3}{P_{NH_3}^2}\right)^{\alpha} - k_2\left(\frac{P_{NH_3}^2}{P_{H_2}^3}\right)^{1-\alpha}\right] e^{\frac{(\alpha V_s - V_a)P}{RT}} \tag{61}$$

to take high pressures into account. Here V_a is the partial molar volume of the activated complex of the nitrogen adsorbed. V_s the partial molar volume of adsorbed N_2, and P the total pressure. Temkin suggested using the molar volume of solid nitrogen of 27 cm^3/mol both V_s and V_a.

In a series of experiments with triply promoted catalyst at 1 atm, 400 °C and SV = 30000 at different H_2/N_2 ratios Nielsen [35] found that maximum conversion was obtained at a H_2/N_2 ratio of 1.5 as predicted by the Temkin-Pyzhev equation (Eq. (22)) when the backward reaction can be ignored. At 330 atm, 450 °C, and SV = 15000 the decomposition rate constant (k in Eq. (31)) was however found to increase with approximately a factor of two when the H_2/N_2 was increased from 1/1 to 6/1.

Further discussions by Temkin et al. of the use of the Temkin-Pyzhev equation can be found in [36].

Brill [37] carried out synthesis experiments on pure iron, singly promoted (3% Al_2O_3) and doubly promoted catalyst (3% Al_2O_3, 2% K_2O) at atmospheric pressure. The catalyst was reduced within 20–80 h at 450 °C with a space velocity of 4000 h^{-1}. The experiments were carried out at 314–320 °C, because abnormal reaction rates were observed at 450 °C, attributed to diffusion effects. When the reaction rate data were fitted to the Temkin-Pyzhev equation taking $\alpha = 0.5$ a dependency on space velocity became apparent, k being smaller at low space velocities than at high ones. This trend could be removed by using different α values. The best fit was obtained with α values 0.7 for pure Fe, 0.7 or somewhat larger for Fe–Al and 0.65 for doubly promoted catalyst.

Annable was the first to use the Temkin-Pyzhev equation to analyze operating data from industrial ammonia units [38]: an adiabatic reactor operating at 245 atm and a multibed quench reactor operating at 300 atm.

For a given inlet gas composition, pressure, and temperature, numerical integration was carried out throughout the reactor. The rate constant k_2 in the Temkin-Pyzhev equation was adjusted so that the calculated temperature profiles matched the measured ones. Radial gradients and axial heat conduction were ignored. The catalyst particles were assumed to have the same temperature as the gas. The variance between the actual and calculated ammonia outlet concentrations was between 0 and −7%. The accuracy of k_2 was estimated to be 10% for the adiabatic reactor and 16% for the quench reactor. The temperature covered the range 374–555 °C and the efficiencies the range 0.09 to 0.96. The results in the form of k_2 as a function of temperature were plotted as Arrhenius plot. Although the experimental scatter was considerable (around 30%), the Temkin-Pyzhev equation simulated the plant performance well.

Activation energies for decomposition reaction of 45 800 cal/mol and 47 900 cal/mol were found at 245 atm and 300 atm, respectively, in good agreement with Emmett and Kummer's results [23]. A trend in k_2 with pressure could be observed. Whereas Emmett and Kummer found that k_2 was proportional to $p^{-0.65}$, Annabel found k_2 to be proportional to $P^{-0.5}$.

The Temkin-Pyzhev equation was also used to analyze the performance of a catalyst charge with a feed gas containing larger amount of poisons than the experiments above. The activity was only one third of the activity of a catalyst operating with a relatively pure gas. A bend in the Arrhenius plot could be observed around 450 °C. The Temkin-Pyzhev equation with the as-found constants was eventually used to find the optimum temperature curve in an ideal as well as a practical reactor.

Kiperman and Granovskaya [39] studied the kinetics at low ammonia partial pressures and efficiencies below 1%. They found the rate to be proportional to the nitrogen partial pressure in agreement with the Temkin equation (Eq. (60)).

Sidorov and Livshits used the special high-pressure version of the Temkin equation (Eq. (61)) to study the kinetics on doubly promoted catalyst at 500 °C [40]. They found the rate constant to be independent of pressure between 10 and 400 atm with $\alpha = 0.5$. The total pressure range studied was 10–500 atm.

Adams and Comings made an extensive study on the ammonia synthesis rate on an industrial ammonia catalyst (2.84% Al_2O_3, 1.04% K_2O, 0.26% SiO_2) [41]. They used an isothermal reactor (within 1 to 3 °C) holding 1 ml of catalyst, in a tubular catalyst chamber measuring 0.95 cm (inside diameter) by 1.25 cm (length).

A total of 300 runs were made at 350, 400, 450, 475 and 500 °C and 100, 200 and 300 atm. The H_2/N_2 ratios studied were 3/1, 1/1 and 1/3 and the space velocities ranged from 10 000 to 125 000 h^{-1}.

They used the original Temkin-Pyzhev equation with $\alpha = 0.5$ to correlate the data, but agreement between theory and experiment was even less satisfactory than that obtained by Emmett and Kummer [23].

Partial pressures were replaced with fugacities calculated according to Lewis and Randall rules, and the data sets were recalculated. The rate constant k' still showed considerable variations with operating conditions. Below 400 °C it decreased with space velocity and above 400 °C it increased. It decreased with both pressure and H_2/N_2 ratio. In view of this, Adam and Comings resorted to a graphical approach to reactor design.

Bokhoven et al. later pointed out that the experimental data might have been influenced by back diffusion in the reactor due to the shallow bed used [42].

Frankenburger in a comprehensive review article [28] on the ammonia synthesis extensively discussed the physical meaning of α and β in the Temkin-Pyzhev equation.

Furthermore, he discussed the unresolved problems in interpreting the kinetics of ammonia synthesis. In principle, it should be possible by theoretical means to calculate, for instance, the rate constant k_1. Emmett and Brunauer [7] had already pointed out that only approx. 1 out of 10^6 N_2 molecules striking the catalyst surface and possessing the necessary activation energy are actually chemisorbed on the surface. Frankenburger pointed out that such calculations ignore the entropy changes associated with the chemisorption step. The entropy effects are linked to the probability that a molecule will actually be chemisorbed. The entropy of the chemisorbed state is mostly lower than that of the gaseous state.

Frankenburger suggested a two-step mechanism for N_2 adsorption involving a short lived "physical" adsorption on the surface followed by nitrogen dissociation. Another reason for the heterogeneous behaviour of the ammonia catalyst surface, in addition to the two already mentioned, was suggested by Frankenburger: "previously adsorbed particles influence the entire catalyst and particularly its surface in such a way that it exerts forces of attraction or repulsion toward N_2 molecules from the gas phase different from those exerted by a completely bare surface."

Bokhoven et al. in a review article [42] critically discussed the application of the Temkin-Pyzhev equation to fit the data published before 1955. They quote their own unpublished results of synthesis experiments at 1 atm on a doubly promoted catalyst at 350 °C with $H_2/N_2 = 3.0$. The best fit is obtained with $\alpha = 0.6$.

Concerning high pressure experiments, Bokhoven does not discuss the data of Larson and Tour [1], because these were influenced by diffusion effects. The data of Adams and Comings [41] were not used either because of the possible influence of back diffusion in these experiments.

Bokhoven et al. used data from Emmett and Kummer [23], Sidorov and Livshits [32], [42] and Nielsen [35]. They pointed out that the data of Emmett and Kummer are restricted to a narrow efficiency range. Regarding the data of Nielsen, they only compare those at an intermediate space velocity, because non-isothermal conditions could be more or less pronounced with higher or lower space velocity.

The calculated rate constants from the data of Nielsen and Sidorov and Livshits are influenced by the H_2/N_2 ratio and are decreasing with pressure; in the latter case there is only a small decrease, however, k_2 is independent of efficiency.

Bokhoven et al. cited some of their own, unpublished results obtained with a differential reactor at 350 °C whereby the exponents of the partial pressures were directly determined. The rate equation was found to be

$$\frac{dP_{NH_3}}{dt} = k\, P_{N_2}^{1.1} \frac{P_{H_3}^{2.0}}{P_{NH_3}^{1.2}} \tag{62}$$

which is close to the Temkin-Pyzhev equation if $\alpha = 2/3$. The pressures were $P_{N_2} = 2.7 - 10$ atm, $P_{H_2} = 9.4 - 22$ atm and $P_{NH_3} = 0.2 - 1.1$ atm. A higher value of α removed some of the inconsistencies in the literature data surveyed, but on the other hand it made the pressure dependencies of k_2 worse.

Bookhoven attributed the different pressure dependencies of k_2 in the various studies to differences in the catalysts used. The results on doubly promoted catalyst are best fitted by the Temkin-Pyzhev equation. It was noted that K_2O addition to singly promoted catalysts lowers its activity at 1 atm, but increases it at 200 atm.

Bokhoven also discusses the decomposition kinetics, especially the results of Love and Emmett [26]) on singly promoted catalysts. He repeated their experiments with a gas containing 6%NH_3 and 94% H_2 and found that the singly promoted catalysts obeyed the Temkin-Pyzhev equation with an activation energy of 60 kcal/mol up to 411 °C, whereafter the activation energy became much lower. Hysteresis effects were also observed, although less pronounced than in Love and Emmett's work. When the catalyst was impregnated with KOH, the abnormal behaviour disappeared and the activation energy in the temperature range 400–450 °C was constant (59 kcal/mol).

Bokhoven interpreted the results obtained on singly promoted catalysts of Love and Emmett to be due to nitrogen dissolution in iron. The necessary high nitrogen fugacities are due to the relatively high ammonia concentrations used. This could also explain the observed hysteresis effect as being due to slow adjustment of the adsorption equilibrium. The activity of the catalyst is expected to depend on the amount of N_2 dissolved in the iron, which could explain the increase in decomposition rate with increasing hydrogen pressure.

Peters and Krabetz [43] reported data on both singly and doubly promoted catalysts for the synthesis reaction at 1 atm. At 324 °C they found $\alpha = 0.5$ and for 372–378 °C they found $\alpha = 0.5$–0.6 for singly promoted catalysts.

Scwab and Krabetz [44] studied the decomposition of NH_3 on pure iron in the temperature range 380–500 °C and the pressure range 0.013–0.106 atm. The decomposition rate was described by

$$-\frac{dP_{NH_3}}{dt} = k\frac{P_{NH_3}^{0.9-1.0}}{P_{H_2}^{1.4-1.5}} \tag{63}$$

corresponding to a β value in the Temkin-Pyzhev equation of 0.45–0.50.

Kawamura et al. [45] measured the rate of synthesis for doubly promoted and triply promoted (Al_2O_3, K_2O, CaO) catalysts. The operating parameters were: temperature = 400–550 °C, pressure 100 or 200 atm, and space velocities between 5000 and 400 000 h^{-1}. They tried without success to fit the data to the published rate equations.

Shishkova et al. [46] confirmed the validity of the Temkin-Pyzhev equation at pressures up to 300 atm using a commercial catalyst in a recycle reactor.

Logan et al. [47] studied the decomposition of NH_3 at low pressures ($P = 0.01$–0.044 atm) from 335 to 470 °C and found the rate law

$$-\frac{dP_{NH_3}}{dt} = k\,\frac{P_{NH_3}^{0.5}}{P_{H_2}^{0.7}} \tag{64}$$

i.e., $\beta \cong 0.25$ in the Temkin-Pyzhev equation. The activation energy was found to be 38.8 kcal/mol.

Mills and Bennett made an extensive kinetic study at 100–1000 atm and 400 and 450 °C on doubly promoted catalysts [48]. H_2/N_2 ratios of 3/1, 1/1 and 2/1 were used and the space velocities ranged from 2000 to 230 000 h^{-1}. The reactor used held 1.625 g or 0.602 cm^3 of catalyst and had the dimensions 4.76 cm in diameter and 5 cm long. The maximum temperature difference recorded was 3 °C and in most cases it was less than 1 °C. Approximate calculations demonstrated that the experimental data should be free of axial or radial dispersion of heat or mass. The size of the catalyst used was 2 mm diameter, and intraparticle diffusion effects were thought to be absent. The lowest effectiveness factor estimated was 0.88. Reduction of the catalyst was done according to the procedures of Nielsen [35].

The experimental data were presented in the form of curves showing the space time yield (cm^3 NH_3 produced per cm^3 per hour) as a function of space velocity. All data were evaluated with the Temkin-Pyzhev equation modified for high pressure (Eq. (61)) according to Temkin [34]. The ideal solution method was selected for calculating the fugacities.

The Temkin-Pyzhev equation in form of Eq. (61) was found to describe the variations with space velocity not too close to equilibrium fairly well, but k_2 was clearly decreasing with pressure. To fit the data, the partial molar volumes of nitrogen would have to be in the order of 200 or 500 cm^3/mol instead of the more probable 20–50 cm^3/mol. The activation energies for decomposition were calculated at the different pressure levels.

Pressure atm	Energy of Activation cal/mol
100	36000
400	44000
1000	53000

4.4 Rate Equations Derived from the Langmuir Isotherm

If one assumes the catalyst surface to be homogeneous energetically, then the classical Langmuir isotherm can be applied. This is equivalent to assuming q to be constant in Eq. (36) in the treatment by Brunauer et al. [28].

Assuming the rate-determining step to be dissociative adsorption of nitrogen, the rate of synthesis becomes

$$r = k\,P_{N_2}(1-\theta)^2 \tag{65}$$

If the most abundant surface species is assumed to be nitrogen, the coverage θ can be determined from the equilibrium

$$N(ads) + 1.5\,H_2 \rightleftarrows NH_3 \tag{66}$$

with the adsorption equilibrium constant K. According to Langmuir, θ is determined by

$$\theta = \frac{K\left(\dfrac{P_{NH_3}}{P_{H_2}^{1.5}}\right)}{1 + K\left(\dfrac{P_{NH_3}}{P_{H_2}^{1.5}}\right)} \tag{67}$$

and the rate then becomes

$$r = \frac{k' P_{N_2}}{\left\{1 + K\left(\dfrac{P_{NH_3}}{P_{H_2}^{1.5}}\right)\right\}^2} \tag{68}$$

If instead of adsorbed nitrogen, NH is assumed to be the most abundant surface species, while N_2 adsorption is still rate-determining, one arrives at

$$r = \frac{k' P_{N_2}}{\left\{1 + K_a'\left(\dfrac{P_{NH_3}}{P_{H_2}}\right)\right\}^2} \tag{69}$$

because the coverage is determined by the equilibrium

$$(NH)_{ads} + H_2 \rightleftarrows NH_3$$

and K_a' is the equilibrium adsorption constant for this equilibrium.

If nitrogen is assumed adsorbed in the molecular form and this is the most abundant surface species, then one obtains

$$r = \frac{k'' P_{N_2}}{1 + K\left(\dfrac{P_{NH_3}^2}{P_{H_2}^{1.5}}\right)} \tag{70}$$

Kubota and Shindo [49] carried out the analysis above and reevaluated the data given in Refs. [23, 32, 33, 41, 50, 51]. They used fugacities instead of partial pressures and included terms for the reverse reaction. The data were well fitted. The results obtained on doubly promoted catalysts were best fitted by Eq. (68), whereas Eq. (69) worked best for singly promoted catalysts. The following activation energies were found:

95 kJ/mol	Emmett and Kummer [27]
77–95 kJ/mol	Kobayashi and Kubota [31]
76 kJ/mol	Uchida and Kuraishi [50]

Kubota and Shindo pointed out that the data could also be correlated by Eq. (70); thus, kinetics alone cannot be proof of a mechanism. Boudart [52] as early as 1956, had noted that it should be possible to describe the ammonia synthesis kinetics by the Langmuir isotherm. This was also discussed by Stelling and von Krusenstierna [53].

Both Temkin [54] and Kwan [55] showed that assuming a simple Freundlich isotherm can also lead to the Temkin-Pyzhev equation.

Ozaki et al. [56] carried out experiments at and below atmospheric pressure with two types of doubly and triply promoted catalysts (A = 0.85% Al_2O_3, 0.27% K_2O and B = 2.6% Al_2O_3, 1.6 K_2O and 1.6% SiO_2). Amounts of approx. 0.8 g were used. Reduction was carried out in pure H_2 for 100 hours at 400 °C. The experiments were carried out at 218, 251, 278, and 302 °C and the pressure was kept at 1/3, 1/2 and 1 atm. Stoichiometric gas was obtained by cracking NH_3 and ND_3.

The experimental data were first fitted to the Temkin-Pyzhev equation. It was noted that the value of α decreased from 0.8 to 0.4 with increasing efficiency. This can be attributed to low coverage by nitrogen, thus making the assumptions for the Temkin-Pyzhev equations invalid as noted by Brunauer [27] and also Temkin [17], [31]. At a fixed temperature and at relative high efficiencies, k still depended on pressure. Ozaki et al. correlated the data with Eq. (68) and found k independent of pressure but K dependent on pressure. A successful fit, however, was obtained using Eq. (69) with both k′ and K_a' independent of pressure.

The isotope effect K_H/K_D also was different from the one calculated from partition functions if Eq. (68) was used. Good agreement between the theoretical and experimental isotope effect was obtained with Eq. (69). The rate constants k are identical for H_2 and D_2 strongly indicating that nitrogen adsorption is the rate determining step.

Ozaki et al. also extended the Temkin-Pyzhev equation by using the isotherm developed by Brunauer et al. [27].

$$\theta = \frac{2}{f} \ln \frac{1 + K_0\left(\frac{P_{NH_3}}{P_{H_2}^{1.5}}\right)}{1 + e^{-0.5f} K_0\left(\frac{P_{NH_3}}{P_{H_2}^{1.5}}\right)} \tag{71}$$

and noting that if $1 \gg e^{-0.5f} K_0 \frac{P_{NH_3}}{P_{H_2}^{1.5}}$

then

$$\theta = \frac{2}{f} \ln \left\{ 1 + K_0 \left(\frac{P_{NH_3}}{P_{H_2}^{1.5}} \right) \right\} \tag{72}$$

Using the Zeldowitsch equation (Eq. (13)), one arrives at the rate equation

$$r = \frac{k_0 P_{N_2}}{\left\{ 1 + K_0 \left(\frac{P_{NH_3}}{P_{H_2}^{1.5}} \right) \right\}^{2\alpha}} \tag{73}$$

with $\alpha = \frac{g}{f}$.

The different possible rate equations with the underlying assumptions are summarized for the main surface species as follows:

N_a		$(NH)_a$	
$\frac{k' P_{N_2}}{\left\{ 1 + K \left(\frac{P_{NH_3}}{P_{H_2}^{1.5}} \right) \right\}^2}$	(74)	$\frac{k' P_{N_2}}{\left\{ 1 + K' \left(\frac{P_{NH_3}}{P_{H_2}} \right) \right\}^2}$	(75)
$\frac{k_0 P_{N_2}}{\left\{ 1 + K_0 \left(\frac{P_{NH_3}}{P_{H_2}^{1.5}} \right) \right\}^{2\alpha}}$	(76)	$\frac{k_0 P_{N_2}}{\left\{ 1 + K'_0 \left(\frac{P_{NH_3}}{P_{H_2}} \right) \right\}^{2\alpha}}$	(77)
$k_1 P_{N_2} \left(\frac{P_{H_2}^3}{P_{NH_3}^2} \right)^{\alpha}$	(78)	$k_1 P_{N_2} \left(\frac{P_{H_2}^3}{P_{NH_3}^2} \right)^{\alpha}$	(79)

For Eqs. (74) and (75) a uniform surface is assumed, whereas for Eqs. (76) and (77) a heterogeneous surface according to Brunauer is considered. Equations (78) and (79) are associated with a heterogeneous surface with the Temkin-Pyzhev approximation. Ozaki et al. pointed out that, at sufficiently high efficiencies, Eqs. (76) and (77) reduce to Eqs. (78) and (79). If $\alpha = 1$, Eqs. (76) and (77) are identical with Eqs. (74) and (75). It is noted that Eqs. (78) and (79) can be regarded as approximations to Eqs. (74) and (75), in which case α will be without any physical meaning.

As to the reason for the presence of NH instead of N on surface, Ozaki et al. pointed out that the catalysts may not have been completely reduced. Since N_{ads} is more thermodynamically stable than NH_{ads} the reason for the dominance of NH_{ads} must be a kinetic one.

Brill and Tauster [57] carried out experiments on doubly promoted catalysts. No experimental details about catalysts or reduction procedures were given. The constant α in the Temkin-Pyzhev equation was increased from 0.11 to 0.7 for doubly promoted catalyst between 182 and 369 °C and from 0.47 to 0.7 for singly promoted catalyst between 242 and 286 °C. There was no change in

α with efficiency. Brill pointed out that the change in α observed by Ozaki et al. [56] could be due to temperature changes rather than efficiency changes. He also noted that it is impossible to calculate the activation energy from the Temkin-Pyzhev equation, when α is changing with temperature, because k is a function of temperature according to

$$k = k_1^{1-\alpha} (k_2 K_p)^\alpha e^{-E/RT} \tag{80}$$

Using a Langmuir equation Eq. (74), i.e. N_{ads} being the most abundant reaction intermediate, Brill obtained a good fit to the data and calculated the activation energy on the doubly promoted catalyst to be 17.4 kcal/mol.

Krabetz and Peters [58] used both the Temkin-Pyzhev equation and the Langmuir formulation to test data from synthesis experiments on singly and doubly promoted catalysts at atmospheric pressure and temperatures between 280 and 370 °C. For the doubly promoted catalyst, the Temkin-Pyzhev equation fitted the data well if $\alpha = 0.8$ at 333 °C and $\alpha = 0.7$ at 372 °C. The Langmuir equation (Eq. (80)) with N as most abundant reaction intermediate, also gave a satisfactory fit yielding an activation energy of 84–105 kJ/mol. For the singly promoted catalyst the best fit was obtained with the Ozaki-Taylor-Boudart equation Eq. (75), i.e. NH as most abundant reaction intermediate. The activation energy was 59 kJ/mol.

Nielsen et al. [59] carried out an extensive kinetic study on a commercial triply promoted KMIR (K_2O, CaO, Al_2O_3) catalyst. It had been prereduced in the size range 3–6 mm but was tested in the size range 0.3–0.7 mm. The sample was reduced again at 150 atm up to 400 °C after which the pressure was increased to 300 atm and the temperature to 480 °C. The reactor had an internal diameter of 5 mm and was equipped with three thermocouples of 1 mm in outer diameter in the catalyst bed. Total catalyst volume was 2.5 cm^3.

The operating conditions studied were as follows:

Pressure : 149–309 atm abs
Temperature : 330–495 °C
Space velocity : 13200–105600 vol/h
H_2/N_2 ratio : 6.23–1.15

The inlet gas did not contain any inerts or ammonia. A total of 35 runs at different conditions were carried out. Throughout the experiment the stability of the catalyst activity was checked by measurements at standard conditions. The influence of different departures from ideal conditions on the reliability of the measurements is discussed in the article. In most of the experiments, the temperature was held within 5 °C, but in runs at high pressures and space velocities, temperature differences up to 17 °C were observed. A weighed mean temperature was used in the analysis of the experiments. Computer calculations using a stepwise kinetic integration, simulating the actual temperature variations, showed that using such a weighed average temperature as the assumed isothermal operating temperature removed any error from the runs where the temperatures differences were below 5 °C and the difference between calculated and measured ammonia outlet percentages were small even with the 17 °C temperature difference.

The effect of diffusion was also investigated by means of computer calculations. A run at rather extreme conditions, 450 °C, 317 atm, and space velocity = 32 100 vol/h, had an effectiveness factor of 0.5 and 0.8 at 10 and 20% into the catalyst bed respectively. In the main part of the bed, the effectiveness factor was very close to 1.0 as in the majority of the runs. Flow conditions in the experiment were in the viscous region or in the transition region between viscous and turbulent flow. The non-ideal conditions due to flow and wall effects were believed to have minor effects. The fact that the catalyst was reduced in the 3–6 mm size range affect the intrinsic activity due to the self-poisoning by water diffusing out of the pores during the reduction process.

Nielsen et al. analyzed their data using the rate Eqs. (74), (75), (76) and (77) developed by Ozaki et al. [56]. They combined the four equations into

$$r = \frac{k_1 P_{N_2}}{\left\{1 + K_3\left(\frac{P_{NH_3}}{P_{H_2}^{\omega}}\right)\right\}^{2\alpha}} \tag{81}$$

Introducing the equilibrium constant K_{a2};

$$K_a^2 = \frac{a_{NH_3,eq}^2}{a_{H_2,eq}^3 a_{N_2,eq}} \tag{82}$$

and taking the reciprocal reaction (ammonia decomposition) into account and substituting activities for partial pressures, they arrived at:

$$r = \frac{k_2^0\left\{a_{N_2}K_a^2 - \left(\frac{a_{NH_3}^2}{a_{H_2}^3}\right)\right\}}{\left\{1 + K_3\left(\frac{a_{NH_3}}{a_{H_2}^{\omega}}\right)\right\}^{2\alpha}} \tag{83}$$

Expressing

$$k_2^0 = k_{20}\,e^{-E_2/RT} \tag{84}$$

$$K_3 = K_{30}\,e^{-E_3/RT} \tag{85}$$

it can be seen that Eq. (83) contains six unknown coefficients k_{20}, E_2, K_{30}, E_3, ω and α. The best values for the constants were found by computer calculations to be

$K_{30} = 3.07 \times 10^{-2}$ $\quad$ $E_3 = -19\,361$ cal/mol

$^1k_{20} = 1.06 \times 10^{13}$ $\quad$ $E_2 = 17\,429$ cal/mol

$\omega = 1.564$ $\quad$ $\alpha = 0.640$

---

[1] A printing error in [59] giving k_{20} as 2.12×10^{13} or twice the correct value was later corrected by Nielsen in [60].

Equation (83) fitted the data at 370, 410, 450 and 490 °C quite well when these values are used, whereas it predicted too high reaction rates at 330 °C compared to the measured ones. In this connection it should be remembered that even minute amounts of oxygen-containing compounds will severely retard the reaction rate, especially at low temperatures, as will be discussed later.

The authors pointed out that the second term in the denominator bracket far exceeds 1 under most circumstances. In fact, only when the ammonia content is down to around 0.1%, is it comparable to 1. This means that Eq. (83) is reduced to the original Temkin-Pyzhev equation. For the Temkin-Pyzhev equation, Nielsen et al. derived an apparent activation energy of 42 300 kcal/kmol with $\alpha = 0.64$. The rate constant was found to be independent of pressure, which was ascribed to the use of fugacities instead of partial pressures.

Temkin et al. [61] studied the rate of ammonia synthesis far from and near the equilibrium on a doubly promoted catalyst at and below atmospheric pressure. The catalyst, of 0.1 to 0.2 mm in size, was reduced in the synthesis gas at a space velocity of 100 000 h^{-1} and a final temperature of 550 °C. Batches of 0.02 to 0.05 g catalyst were used in the experiments. In one series of experiments the partial pressure of nitrogen was fixed at 100 mm Hg and the hydrogen pressure varied from 100 to 600 mm Hg. In another series the opposite was done. Temperatures were 450 °C and 350 °C in both series and the space velocities used ranged from 138 000 h^{-1} to 2 640 000 h^{-1}, which ensured an ammonia percentage less than 0.01 mol %, i.e. far from equilibrium.

The data could be reproduced with a kinetic equation of the form

$$r = k' P_{H_2}^{0.5} P_{N_2}^{0.5} \tag{86}$$

Another series of experiments were carried out at 350 and 450 °C at atmospheric pressure, but varying the H_2/N_2 ratio from 8.5 to 0.15. The data from this experiment could also be fitted to Eq. (86). The apparent activation energy for k was found to be 11.5 kcal/mol.

The authors interpreted the applicability of Eq. (92) as an indication that, at low ammonia concentrations, the equilibrium between adsorbed nitrogen and ammonia plus hydrogen in the gas phase is no longer established.

Another set of experiments was carried out with stoichiometric gas near or at an intermediate distance from equilibrium. The pressure was fixed at 1 atm and the temperature varied in steps of 25 °C from 325 to 550 °C. The space velocity ranged from 61 000 h^{-1} to 542 000 h^{-1}.

The original Temkin-Pyzhev equation (Eq. (21)) with $\alpha = 0.5$ reproduced the data above 400 °C reasonably well. The activation energy for k_2 was found to be constant at 37.0 kcal/mol above 400 °C. This corresponds to an activation energy for k_1 in Eq. (21) of 12.0 kcal/mol which is close to the 11.5 kcal/mol found for k′ in Eq. (86). At the temperatures of 325 and 350 °C, k_2 in Eq. (22) declined at lower ammonia concentrations indicating that Eq. (22) is no longer valid because the conditions are far from equilibrium.

In a later communication [62], Temkin et al. developed a rate equation covering the complete range of operating conditions including the transition region between Eq. (21) and Eq. (86). They assumed an associative mechanism for the synthesis as follows:

$$\text{Step 1} \quad N_2 + * \underset{\omega_{-1}}{\overset{\omega_1}{\rightleftarrows}} N_2* \tag{87}$$

$$\text{Step 2a} \quad N_2* + H_2 \underset{\omega_{-1}}{\overset{\omega_{2a}}{\rightleftarrows}} N_2H_2* \tag{88}$$

$$\text{Step 2b} \quad N_2H_2* + 2H_2 \rightleftarrows 2NH_3 + * \tag{89}$$

Only step 2b is assumed to be sufficiently rapid so that equilibrium is established. Another assumption is that adsorbed N_2 covers the surface to a moderate degree, whereas N_2H_2, N_2H_4 etc. only cover a small fraction of the surface. Furthermore, it is assumed that the adsorption energy for N_2 varies linearly with surface coverage as discussed earlier. For N_2H_2 the adsorption energy is assumed to be independent of coverage. Using these assumptions the following rate equation for the synthesis is obtained:

$$\omega = \frac{1}{f} \frac{\Pi}{\sin \alpha\Pi} \frac{\chi_1^0 P_{N_2} \chi_{2a}^0 P_{H_2} - \chi_{-1}^0 \dfrac{\chi_{-2a}}{K_{2b}} \dfrac{P_{NH_3}^2}{P_{H_2}^2}}{\left(\chi_1^0 P_{N_2} + \dfrac{\chi_{-2a}}{K_{2b}} \dfrac{P_{NH_3}^2}{P_{H_2}^2}\right)^{\alpha} (\chi_{-1}^0 + \chi_{2a}^0 P_{H_2})^{1-\alpha}} \tag{90}$$

where

$$f = \frac{c}{RT}$$

and χ_1^0, χ_{-1}^0 and χ_{2a}^0, χ_{-2a}^0 are the rate constants at zero surface coverage for the reactions described by step 1 and 2a. These are related to the equilibria constants

$$K_1^0 = \frac{\chi_1^0}{\chi_{-1}^0} \tag{91}$$

$$K_{2a}^0 = \frac{\chi_{2a}^0}{\chi_{-2a}^0} \tag{92}$$

and

$$K_1^0 K_{2a}^0 K_{2b} = K \tag{93}$$

Introducing the substitutions

$$k_{\pm} = \frac{1}{f} \frac{\Pi}{\sin \alpha\Pi} (\chi_1^0)^{1-a} (\chi_{-1}^0)^{\alpha} \tag{94}$$

and

$$1 = \frac{\chi_{-1}^0}{\chi_{2a}^0} \tag{95}$$

together with the equilibrium relations, one obtains

$$\omega = \frac{k_{\pm} P_{N_2}^{1-\alpha}\left(1 - \frac{1}{K}\frac{P_{NH_3}^2}{P_{N_2}P_{H_2}^3}\right)}{\left(\frac{1}{P_{H_2}} + \frac{1}{K}\frac{P_{NH_3}^2}{P_{N_2}P_{H_2}^3}\right)^{\alpha}\left(\frac{1}{P_{H_2}} + 1\right)^{1-\alpha}} \tag{96}$$

It can be seen that

$$\frac{1}{P_{H_2}} = \frac{\omega^{-1}}{\omega_{2a}} \tag{97}$$

If the probability of hydrogenation of adsorbed nitrogen is much higher than the probability of nitrogen desorption, it follows that

$$\frac{1}{P_{H_2}} \ll 1 \tag{98}$$

which obviously requires that the hydrogen partial pressure is not negligible. Close to equilibrium, the term

$$\frac{1}{K}\frac{P_{NH_3}^2}{P_{N_2}P_{H_2}^3}$$

will be close to 1, and Eq. (96) approaches the normal Temkin-Pyzhev equation (Eq. (21)). If the system is very far from equilibrium, then

$$\frac{1}{K}\frac{P_{NH_3}^2}{P_{N_2}P_{H_2}^3} \ll \frac{1}{P_{H_2}} \tag{99}$$

Equation (96) then converts into Eq. (86) deduced in Ref. [61]. Calculations showed, however, that under industrial ammonia synthesis conditions the original Temkin-Pyzhev equation is invariably applicable. However, it was necessary to apply Eq. (96) in order to get constant values of K_{ad} at 350 °C and 325 °C for the data reported in [61]. Additional data are given at 450 °C in [62]. When the total pressure was varied from 0.5–1.0 atm, the hydrogen/nitrogen ratio from 0.5–5, and the space velocity from 200000 to 526000 h^{-1}, the relative yields Z/Z_{eq} were in the range 0.13 to 0.35. The calculated k_+ values were satisfactorily constant, although a tendency towards lower k_+ values with higher yields could be observed.

Bridger and Snowdon [63] cited tests from both differential and integral reactors without giving any experimental details. They found the extended Temkin-Pyzhev equation, Eq. (96), to describe the experimental data best with $\alpha = 0.465$ and $E_2 = 40.4$ kcal/mol.

Aika and Ozaki [97] did experiments on 10.4 g of unpromoted catalyst at 305 °C at and below atmospheric pressure. The catalyst had been reduced at 400 °C for 200 h and stabilized for approx. two weeks. They tested three kinetic equations for the best fit to the experimental data.

1) The Temkin-Pyzhev equation Eq. (21)
2) A Langmuir-type equation with NH as the main adsorbed species.
3) A Langmuir-type equation with N as the main adsorbed species.

The least sum of squared standard deviations between measured and calculated rate constants

$$S^2 = \sum_1^n \frac{\frac{(k - k_{av})^2}{k_{av}^2}}{N - 1} \tag{100}$$

was found with equation type 3.

Aika and Ozaki thus found N the most abundant surface species at 305 °C, in contrast to Ozaki et al. who found it to be NH. Aika and Ozaki attributed this discrepancy to the absence of promoters in the catalyst they used. The fact that Nielsen et al. [59] also found N to be the most abundant species for a doubly promoted catalyst, and the Temkin-Pyzhev equation best, they ascribed to the higher temperatures used in the study of Nielsen et al. compared to the low temperature of 305 °C used by Aika and Ozaki.

Logan and Philip [65] analyzed the data of Ozaki et al. again. They used the extended isotherm and rate of N_2 adsorption developed by Brunauer et al. [27] as did Ozaki et al. The quality of the fit was assessed by the variance as defined by

$$S = \left\{ \sum_1^n \left(\frac{y_a - y'}{y_a} \right)^2 \right\}^{0.5} \tag{101}$$

where y_a and y denote the measured and calculated NH_3 mole fractions respectively. Ozaki selected the best linear log-log plot of y_a vs 1/v, where v is the flow rate.

However, Logan and Philip did not as Ozaki et al. limit the value of $g/f = \alpha$ to 1.0 and indeed found a best fit with α around 0.8. Furthermore, they found a more normal isotope effect of $K_h/K_d = 3.8$ at room temperature consistent with N being the most abundant species on the catalyst surface and adsorbed H participating in the slow step of nitrogen dissociation.

Based on certain circumstantial evidence, Carra and Ugo [66] proposed a mechanism where the rate-determining step is the hydrogenation of a half-hydrogenated surface nitrogen, dehydroimide (N_2H). This leads to the following rate equation using the Langmuir adsorption isotherm

$$r = \frac{K P_{N_2} (b_h P_{H_2})^{0.5}}{1 + (b_H P_{H_2})^{0.5} + b_{NH_3} + P_{NH_3}} \tag{102}$$

which can be approximated by

$$r = \frac{kKb_H^{0.5}}{b_{NH_3}} \frac{P_{H_2}^{1.5} P_{N_2}}{P_{NH_3}} \cong k_{eff} \frac{P_{H_2}^{1.5} P_{N_2}}{P_{NH_3}} \tag{103}$$

which is identical with the original Temkin-Pyzhev equation with $\alpha = 0.5$.

Takezawa [67, 68] determined the rate of ammonia decomposition on a triply promoted catalyst (4.72 Al_2O_3, 0.31% K_2O and 0.05% silica). The pressure was kept constant at 1 atm with helium and the ammonia partial pressure was varied between 0.25 and 0.88 atm, the nitrogen partial pressure between 0 and 0.49 atm. They found the Temkin-Pyzhev equation

$$v = k_2\left(\frac{P_{NH_3}}{P_{H_2}^{1.5}}\right)^{\beta} \tag{104}$$

to be valid at 420 °C with $\beta = 0.48$.

Above 479 °C they found however

$$v = k_2\left(\frac{P_{NH_3}}{P_{H_2}^{0.5}}\right)^{0.75} \tag{105}$$

to be valid. At slight inhibitive effect of nitrogen could be observed with a reaction order around 0.1. They interpreted the findings by a change in the rate-determining step from nitrogen desorption to dehydrogenation of adsorbed amino radical NH_2 (a).

In order to determine the effect of potassium they eluted part of the potassium oxide by immersing the catalyst in water, which gave a catalyst containing 0.25 wt% K_2O [69]. The rate of ammonia decomposition was measured again at 378 °C, and the following rate dependency was found:

$$v = k_2\left(\frac{P_{NH_3}}{P_{H_2}^{0.5}}\right)^{0.44} \tag{106}$$

which indicates that dehydrogenation of an adsorbed amino group may be rate-determining. Takezawa concluded that the change in rate determining step at low temperatures was due to the lower potassium content of the catalyst.

Brill [70] applied the following rate equation proposed by Ozaki et al. [56]

$$\frac{dP_{NH_3}}{dt} = kP_{N_2}(1-\theta)^2 - k_2\theta \tag{107}$$

with

$$\theta = \frac{K\left(\frac{P_{NH_3}}{P_{H_2}^{1.5}}\right)}{1 + K\left(\frac{P_{NH_3}}{P_{H_2}^{1.5}}\right)} \tag{108}$$

(i.e. assuming a uniform surface with N adsorbed as the most abundant reaction intermediate) to experimental data at low temperatures (approx. 265–340 °C), presumably atmospheric pressure, and a stoichiometric gas mixture using pure iron as catalyst. Brill also applied the rate equation

$$\frac{dP_{NH_3}}{dt} = kP_{N_2}(1 - \theta) - k_2\theta \tag{109}$$

with

$$\theta = \frac{cK^{-1}\left(\frac{P_{NH_3}}{P_{H_2}^{1.5}}\right)}{1 + cK^{-1}\left(\frac{P_{NH_3}}{P_{H_3}^{1.5}}\right)} \tag{110}$$

where K is the equilibrium constant for ammonia synthesis. This is equivalent to assuming a uniform surface and hydrogenation of molecularly adsorbed nitrogen to be rate determining.

Both Eq. (107) and (109) in their integrated form represented data at 340 °C equally well. Equation (109) could, however, be adapted better to an Arrhenius plot than Eq. (107).

Brill noted that Eq. (109) gives a better approximation to the integrated rate equation,

$$k_1 = \left(\frac{x}{t}\right)\frac{P_{NH_3,eq}}{P_{N_2}} \tag{111}$$

than Eq. (107) at yields below 1% (whether x is the equilibrium yield $Z_{NH_3}/Z_{NH_3,eq}$).

Kazarnovskaya et al. [71] made experiments in a recycle reactor at 400–450 °C, and 200, 250, 300 atm with different particle sizes at four space velocities. For the smallest particle size (0.25–0.5 mm), the rate constant in the Temkin-Pyzhev equation with $\alpha = 0.5$ was calculated. The calculated rate constants increased systematically with increasing space velocity, i.e. decreasing ammonia concentration.

Dyson and Simon [72] used the data of Nielsen et al. to estimate the parameters in the original Temkin-Pyzhev equation,

$$r = 2k\left(K^2 a_{N_2}\left[\frac{a_{H_2}^3}{a_{NH_3}^2}\right]^{\alpha} - \left[\frac{a_{NH_3}^2}{a_{H_2}^3}\right]^{1-\alpha}\right) \tag{112}$$

They calculated the fugacity coefficients in a manner slightly different from Nielsen et al. and estimated k for $\alpha = 0.5$ and $\alpha = 0.75$.

Both values of α gave reasonably good fits, and Dyson and Simon chose $\alpha = 0.5$ as this gave the simplest rate equation. With $\alpha = 0.5$ they found:

$$2k = 1.7698 \times 10^{15} e^{-\frac{40765}{RT}}$$

Dyson and Simon found, as Nielsen et al. did, the rate constant k to be independent of pressure.

Ferraris and Donati [73] reevaluated the data of Nielsen et al. using both the Temkin equation Eq. (112) modified by Dyson and Simon and Eq. (83) used by Nielsen et al. They used the same thermodynamic data as Dyson and Simon. A non-linear parameter estimation program was used to calculate the constants in the two equations minimizing the sum of squares defined by:

$$SSQ = \sum_{i=1}^{N} (Z_{calc} - Z_{meas})^2 \tag{113}$$

where Z_{calc} and Z_{meas} are the calculated and measured ammonia mole fractions.

For the Dyson and Simon equation, Ferraris and Donati found $SSQ = 9.30 \times 10^{-3}$ with $\alpha = 0.5$. The best fit was obtained with $\alpha = 0.73$ and $2k = 1.30 \times 10^{15} e^{\frac{-43230}{RT}}$. This gave $SSQ = 6.33 \times 10^{-3}$.

Using the values of Nielsen et al. Ferraris and Donati found $SSQ = 6.87 \times 10^{-3}$. Reestimation of the parameters gave:

$K_{30} = 2.1496$ $\quad E_3 = -18\,458$ cal/mol
$K_{20} = 1.349 \times 10^{15}$ $\quad E_2 = 16\,480$ cal/mol
$W = 1.574$ $\quad \alpha = 0.6923$

The SSQ became 6.26×10^{-3}.

In a later article [74], Ferraris et al. again used the data by Nielsen et al. to test 23 different rate equations for ammonia synthesis. Some of the rate equations considered both associative and dissociative mechanisms. Others were purely empirical. They concluded that a large number of the models represented the data equally well, in fact better than the Temkin-Pyzhev equation. This again proves that it is extremely difficult, and in fact usually impossible, to draw definite conclusions about the mechanism from kinetic data alone.

Cappeli and Collina [75] investigated the kinetics of five different commercial ammonia synthesis catalysts. The following rate equations were tested:

1) The Temkin-Pyzhev equation modified by Dyson and Simon (112)
2) the equation proposed by Nielsen et al. (83)
and 3) a new equation:

$$r = \frac{k_1 a_{N_2}^{0.5}\left(1 - \dfrac{a_{NH_3}}{a_{H_2}^{1.5} a_{N_2}^{0.5} K_e}\right)}{1 + k_2 \dfrac{a_{NH_3}}{a_{H_2}^{1.5}} + k_3 \dfrac{a_{NH_3}}{a_{H_2}^{1.5}} + k_4 a_{NH_3}} \tag{114}$$

The new equation was claimed to give the best description of the experiments. The $k_4 a_{NH_3}$ term was found to be insignificant compared to the other terms in the denominator. Industrial data from a third bed (adiabatic) with intermediate cooling were also well described by Eq. (114).

Guacci et al. [76] tested eight different commercial ammonia catalysts under industrial conditions. They used the Temkin-Pyzhev equation modified by Dyson and Simon to analyze their results. The α values found for the different catalysts ranged from 0.426 to 0.687

Altenburg et al. [77] carried out kinetic studies on a commercial, promoted catalyst (Al_2O_3 2.90, K_2O 1.30, ZrO_2 0.45 and SiO_2 0.45 wt%). The potassium content was varied from 0.1 to 3.8 wt% by elution and reimpregnation of prereduced catalyst with KOH. Three temperature levels of 350, 400 and 450 °C were used. The pressure was varied from 5 to 150 atm. Altenburg et al. used the modified Temkin-Pyzhev equation

$$r = k_1 a_{N_2} \left(\frac{a_{H_2}^{\omega}}{a_{NH_3}} \right)^{2\alpha} \left(1 - \frac{a_{NH_3}^2}{a_{H_2}^3 a_{N_2} K_e} \right) \tag{115}$$

to analyze the results. k_1, ω, and α were estimated by minimizing

$$SSQ = \sum \left(1 - \frac{k_{1\,meas}}{k_{1\,calc}} \right)^2 \tag{116}$$

at each temperature level. The α values varied from 0.48 to 0.79 in a nonsystematic manner, not correlated with the potassium content. The reaction order of hydrogen through the value of ω did, however, increase systematically from 0.71 to 1.55 with increasing potassium content.

Altenburg et al. speculated that this trend could be explained by assuming a fast reaction to NH, and other partly hydrogenated species between adsorbed N and H from polarized hydroxylic groups on the oxidic part of the catalyst, followed by a reaction between hydrogen and the O groups. This pathway is blocked by potassium either by reaction with the acidic OH groups or by screening of adsorbed N from these groups. They noted that the promoting effect of potassium was most pronounced at higher pressures.

Lu Zhenming [78] tested two different commercial ammonia catalysts using the Temkin-Pyzhev equation for analysing the results. The pressure was fixed at around 148 atm and the temperature was varied from 338–550 °C. The catalyst size was 0.2–0.3 mm, and a stoichiometric feed gas was used. The optimum α value was 0.5, and activation energies of 42.1 and 44.8 kcal/mol were found.

Using slightly different mechanistic assumptions than Temkin et al., Huan Kahui [79, 80] derived a kinetic equation similar to the one proposed by Temkin et al. Liu De-ming et al. [81] tested Huan Kahui's equation on data obtained with an industrial catalyst and compared the fits with other rate equations proposed by Capelli and Collina [75], Dyson and Simon [72], Buzzi Ferraris et al. [75], and Nielsen et al. [60]. The modified Temkin equation was found to give the best fit.

4.5 Kinetics Based on Surface Science Techniques

Bowker et al. [82] used computer calculations to extrapolate the results obtained by Ertl et al. [83] at low pressures and coverages to industrially relevant conditions around 450 °C and above 100 atm. They used the reaction mechanism proposed by Ertl et al.

$$\begin{aligned} N_2 + * &\rightarrow N_2\text{-}* \\ N_2\text{-}* + * &\rightarrow 2N\text{-}* \\ H_2 + * &\rightarrow H_2\text{-}* \\ H_2\text{-}* + * &\rightarrow 2H\text{-}* \\ N\text{-}* + H\text{-}* &\rightarrow NH\text{-}* + * \\ NH\text{-}* + H\text{-}* &\rightarrow NH_2\text{-}* + * \\ NH_2\text{-}* + H\text{-}* &\rightarrow NH_3* + * \\ NH_3\text{-}* &\rightarrow NH_3 + * \end{aligned} \tag{117}$$

and assumed iron surface area for a commercial ammonia synthesis catalyst of $1.5\,m^2/g$ and an energetic homogeneous surface where single-site Langmuir isotherms describe the adsorption processes. To each of the elementary reactions a preexponential factor and an activation energy were assigned. The preexponential factors were estimated from transition state theory. All adsorption and desorption reactions of molecular H_2, N_2 and NH_3 were assigned a preexponential factor of 10^{13}, all other reactions a factor of about 10^{21}. The experimentally available activation energies at low surface coverages were taken from the work of Ertl et al. Knowing the overall heat of reaction ($50\,kJ\,mol^{-1}$) and assigning an activation energy of $21\,kJ\,mol^{-1}$ to all the dehydrogenation steps, the complete potential energy diagram could be constructed.

Using these parameters the performance of a back mix reactor was calculated at conditions similar to the experimental data of Nielsen et al. [59]. The model predicted ammonia percentages which were 1.5×10^5 lower than those experimentally found.

Lowering the preexponential factors for both the forward and reverse of step 2 by a factor of 10^7 only improved the fit slightly. However, the model predicted that the surface was totally hydrated instead of being totally nitrided as in the case of "normal" preexponential factors of 10^{21}. Using the energies determined at high surface coverages had virtually no effect on the model predictions. The authors attributed the very poor agreement with the experimental findings to an energy well in the model too deep for adsorbed nitrogen atoms in the model.

Reasonable fits could be obtained (within 10% of experimental values) if activated dissociation of nitrogen was invoked as, found by Scholten et al. [85] on singly promoted catalysts. The authors point out, however, that there is no surface science experimental evidence for activated nitrogen dissociation on potassium promoted iron.

Stoltze and Nørskov [85–88] used almost the same reaction scheme as Bowker et al. i.e.:

$$\begin{aligned}
&\text{Step 1} \quad N_2(g) + * \rightarrow N_2\text{-}* \\
&\text{Step 2} \quad N_2\text{-}* + * \rightarrow 2N\text{-}* \\
&\text{Step 3} \quad N\text{-}* + H\text{-}* \rightarrow NH\text{-}* + * \\
&\text{Step 4} \quad NH\text{-}* + H\text{-}* \rightarrow NH_2\text{-}* + * \\
&\text{Step 5} \quad NH_2\text{-}* + H\text{-}* \rightarrow NH_3\text{-}* + * \\
&\text{Step 6} \quad NH_3\text{-}* \rightarrow NH_3(g) + * \\
&\text{Step 7} \quad H_2(g) + 2* \rightarrow 2H\text{-}*
\end{aligned} \tag{118}$$

the only difference being that a molecularly held precursor state of H_2 is not considered for the dissociation of hydrogen. The following assumptions were made: the gas is considered to be ideal and all surface sites are assumed to be equal (homogeneous surface). Adsorbate-adsorbate interaction is thus considered to be absent until the coverage reaches one adsorbate atom per two iron atoms. Above this coverage adsorbate-adsorbate repulsion excludes further adsorption. Hydrogen and nitrogen as well as reaction intermediates are assumed to chemisorb competitively. The active area of the catalyst is taken to be that of the CO chemisorption area. Finally, all the reaction steps are assumed to be in equilibrium except step 2 i.e., this step is assumed to be the sole rate-determining step.

By generalizing Fowler and Guggenheim's [89] statistical mechanical description of adsorption that covers competitive gas adsorption, the equilibrium constants were calculated. For instance, for step 1:

$$K_1 = \frac{Z_{N_2*}}{Z^0_{N_2}} \tag{119}$$

where Z_{N_2*} is the partition function for N_2(adsorbed) and $Z^0_{N_2}$ the partition function for N_2 (gas) at the thermodynamic reference pressure P_0 taken to be 101.325 kPa. The partition functions were calculated from

$$Z_x = (\Pi Z_{trans,i})(\Pi Z_{vib,i})(\Pi Z_{rot,i})e^{-\frac{E_x}{RT}} \tag{120}$$

Z_{trans}, Z_{vib}, Z_{rot} are the partition functions for translational, vibrational, and rotational degrees of freedom, whereas E_x is the ground-state energy for the species. The equilibria can be specified as follows:

$$K_1\left(\frac{P_{N_2}}{P_0}\right)\theta* = \theta_{N_2}* \tag{121}$$

$$K_3\theta_N*\theta_H* = \theta_{NH}*\theta* \tag{122}$$

$$K_4\theta_{NH}*\theta_H* = \theta_{NH_2}*\theta* \tag{123}$$

$$K_5\theta_{NH_2}*\theta_H* = \theta_{NH_3}*\theta* \tag{124}$$

$$K_6\theta_{NH_3}* = \frac{P_{NH_3}}{P_0}\theta* \tag{125}$$

$$K_7\left(\frac{P_{H_2}}{P_0}\right)(\theta*)^2 = (\theta_N*)^2 \tag{126}$$

where θ_x is the surface coverage by species X, and $\theta*$ is the fraction of the surface not covered by any surface intermediate. For step 2 the rate can be calculated from

$$r_2 = k_2\theta_{N_2}\theta* - k_{-2}\theta_N^2* \tag{127}$$

where

$$k_2 = A_2 e^{-\Delta H_2^{\ddagger}/RT} \tag{128}$$

Noting that

$$\theta_{N_2*} + \theta_{N*} + \theta_{NH*} + \theta_{NH_2*} + \theta_{NH_3*} + \theta_{H*} + \theta_* = 1 \tag{129}$$

and solving for the different surface coverages, it is found that

$$\theta_* = \left(1 + k_1\frac{P_{N_2}}{P_0} + \frac{P_{NH_3}P_0^{0.5}}{K_3K_4K_5K_6K_7^{1.5}P_{H_2}^{1.5}} + \frac{P_{NH_3}}{K_4K_5K_6K_7P_{H_2}} + \frac{P_{NH_3}}{K_5K_6K_7^{0.5}P_{H_2}^{0.5}P_0^{0.5}} + \frac{P_{NH_3}}{K_6} + k_7^{0.5}\frac{P_{H_2}^{0.5}}{P_0^{0.5}}\right)^{-1} \tag{130}$$

The rate of ammonia synthesis is then determined by:

$$r_2 = 2k_2K_1\left(\frac{P_{N_2}}{P_0} - \frac{P_{NH_3}^2P_0}{K_gP_{H_2}^3}\right)(\theta_*)^2 \tag{131}$$

where K_g is the gas phase equilibrium constant for the synthesis.

In order to determine the input parameters for the model, parallel basic thermodynamic data were taken from [90], frustrated vibration was assumed to have a barrier for surface diffusion of 0.05 eV for N_2-* and 0.5 eV for other intermediates, and vibrational frequencies for adsorbed species were taken from electron energy loss spectra. TPD spectra for H-* and NH_3* and N-* together with the initial sticking coefficient for N_2 into 2N-* on K covered iron were used to determine the ground-state energies and A_2 and $\Delta H_2^{\ddagger}$. Values for the ground state energy of NH-* and NH_2-* were assigned to be between the values of N-* and NH_3*. The active area determined by CO chemisorption was used in the calculations.

The validity of this model was assessed by comparison with the experimental data of Nielsen [60]. The maximum deviation between observed and calculated rate constants for the ammonia synthesis is 1.4, despite the wide span of pressures from 1 to 300 atm suggests good representation of the kinetics and mechanism by the model.

The surface coverages by reaction intermediates could also be calculated by the model. It was found that N-* is by far the most abundant surface intermediate under typical synthesis conditions, but the sum of the coverages by other intermediates is also larger than the coverage by free sites. This fact is seen by the authors to be the cause for the complicated kinetics deduced by Ozaki et al. [56], rather than adsorbate-adsorbate interaction or surface heterogeneity.

The data of Temkin et al. [61] for low conversions were also analysed by this model in ref. [87] to adequately describe the observations. The failure of the normal Temkin-Pyzhev equation was found to be due to high coverage by H-*, rather than to a change in the rate-determining step. Stoltze and Nørskov also demonstrated how to deduce the apparent reaction orders and activation energies from the model.

Bowker et al. refined their model proposed in [91] by modifying the preexponential factors for the molecular nitrogen adsorption and desorption in step 2 to $10^{11.2}$ and $10^{10.2}$, respectively, and the factor for the dissociation of the molecularly held nitrogen to $10^{15.2}$, so that these agreed with the experimental findings of Ertl et al. [92, 91]. The reactor model was furthermore changed to a plug flow type as used in the experiment by Nielsen et al. [59]. The ammonia percentage was calculated to be 3.0%, while the experimental value was 13.2%.

Using the heat of dissociative nitrogen adsorption and the lower preexponential factor proposed by Stoltze and Nørskov, Bowker et al.'s model gave an ammonia exit concentration of 5.4%, which corresponds to a catalyst activity 40 times lower than that experimentally found. Bowker et al. also calculated an erroneous TPD peak position and peak width using Stoltze and Nørskov's data. The poor fit is again ascribed to too high a surface coverage of nitrogen, causing "self-poisoning" to occur.

Bowker et al. again pointed out that a net activation barrier to nitrogen dissociation would make the model predict yields closer to the experimentally observed ones. The authors believe that measurement of the nitrogen desorption energetics at high nitrogen coverage are necessary as nitrogen dissolution into the bulk phase might play a role at high nitrogen coverage.

Stoltze and Nørskov [94] in a comment to the article by Bowker et al. stated that their parameters were chosen so that the TPD spectrum for nitrogen desorption is reproduced exactly. They ascribed the indequacy of the model of Bowker et al. to unrealistic choices for preexponential factors. They pointed out that in case of hydrogen chemisorption these choices lead to an entropy of chemisorption of zero, which clearly contradicts the experimental findings.

Bowker again in Ref. [105] discussed the different kinetics proposed based on surface science studies. Bowker referred to the work of Rettner and Stein [106] in which a well defined molecular beam of nitrogen was used to measure the sticking probability on Fe(111) surfaces. Their measurement showed an increase in the sticking probability from 10^{-6} to 10^{-2} with an increase in average beam energy from 0.1 to 1.0 eV. This is consistent with an activation energy for nitrogen activation of around 35 kJ/mol in good agreement with the result of Scholten for zero coverage [84]. Bowker pointed out that Ertl et. al.

Table 4.1. Parameters of kinetic models

Kinetic model	Step	Activation energy kJ/mol	Preexponential factor s^{-1}	Ammonia yield %	Surface coverage %	
Ertl	1	0	10^{11}		H(a)	12.1
[92,93]	−1	46	10^{10}	3.0	N_2(a)	1.5
	2	31	10^{7}		N(a)	84
	−2	199	10^{13}		NH(a)	2
Modified	1	0	10^{11}		H(a)	0.6
Ertl	−1	46	10^{10}	5.4	N_2(a)	0.1
[85,86,87]	2	31	10^{10}		N(a)	98.3
	−2	144	10^{10}		NH(a)	1
Activated N_2	1	0	10^{13}		H(a)	22
dissociation	−1	46	10^{13}	14.2	N_2(a)	3
[105]	2	96	10^{13}		N(a)	54
	−2	134	10^{13}		NH(a)	8

[92, 93] in their work did not vary the gas temperature, but only the temperature of the solid. The three different kinetics proposed are summarised in table 4.1, where the different activation energies and preexponential factors for the critical steps 1 and 2 in the reaction scheme (118), i.e.

Step 1 $N_2(g) + * \rightarrow N_2\text{-}*$

Step 2 $N_2\text{-}* + * \rightarrow 2N\text{-}*$

used in the kinetics, are given together with the yield calculated from the models (experimentally found 13.2%) and the surface coverages by the predominant species.

Bowker concluded that it is necessary to invoke an activated nitrogen dissociation step in order to reproduce the experimental findings on industrial catalysts as well as the measured desorption spectra widths. It is not clear from the article, whether the surface coverages given in the table are average or typical for the exit of the reactor. Bowker discussed the differences between the different crystal planes of iron, both promoted and nonpromoted, but he still assumed in his model the surface to be homogeneous.

4.6 Concluding Remarks

This review has illustrated, that the large amount of work done on the kinetics of ammonia synthesis and decomposition has stimulated new concepts of reaction kinetics catalyzed by heterogeneous catalysts. The concept of a non-uniform surface vis-a-vis a homogeneous surface is just one example, which is still debated today. Ammonia synthesis also is one of the first reactions, where micro kinetics based on modern surface science techniques has been used with a fair degree of success. However, there is no doubt that the finer details of the ammonia synthesis reaction mechanism and kinetics still will be a controversial topic for years to come.

In summary, it is noteworthy that the first workable rate equation for ammonia synthesis was proposed before the second world war by people with very limited experimental resources. The Temkin-Pyzhev equation still is the rate equation of choice for engineering purposes. Further progress in better understanding of the fundamental rate-determining process(es) is still, however, needed in order to develop better catalysts.

4.7 Transport Phenomena

The kinetics discussed in the preceding section has been (or should have been) derived under conditions where transport processes do not restrict the measured reaction rate. In ammonia synthesis as with many other catalytic reactions it may be necessary to consider also the mass and heat transfer to and inside the catalyst particle, depending on operating conditions and catalyst particle size.

If mass and/or heat transfer is slow compared to the reaction rate, temperature and concentration gradients between bulk gas and particle center will result, thus affecting the overall reaction rate of the particle. In industrial reactors the diffusion inside the catalyst pore system is the most important transfer process. In laboratory reactors with low linear velocity and/or large particles, there may also be significant concentration and temperature gradients between particle surface and bulk gas [95, 96]. Due to the high heat conductivity of the iron catalyst, a catalyst pellet will be almost isothermal [96].

The intrinsic activity of an ammonia synthesis catalyst is also known to depend on the particle size during reduction [60, 97, 43]. This effect is due to an irreversible poisoning of the outer reduced part of the particle by water from the inner, still reducing part. The water causes the iron particles to recrystallize, resulting in a lower surface area and hence lower activity. This effect can be circumvented by a very careful reduction at high space velocities; this is possible in laboratory reactors, but normally not feasible in industrial converters.

Thus the lower activity of large particles does not necessarily mean that the reaction is diffusion limited at the operating conditions; diffusion retardation can be recognized, however, by a lower activation energy for the larger particles.

The activity of a catalyst particle is often expressed by the effectiveness factor η defined as the ratio of the actual reaction rate R to the reaction rate R_{kin} at conditions on the outer surface of the particle.

$$R = \eta R_{kin} \tag{132}$$

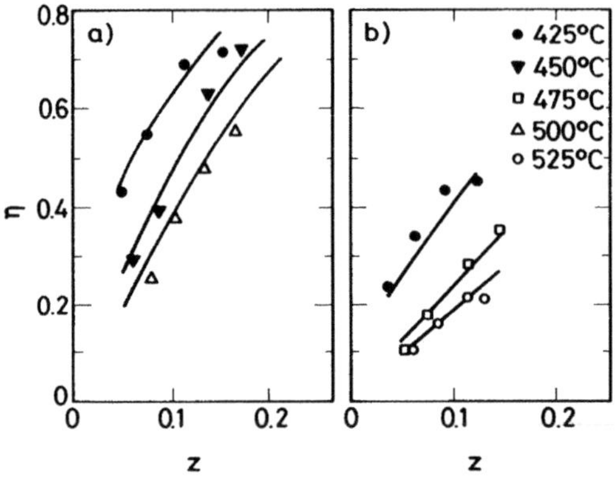

Fig 4.1. Dependence of the effectiveness factor, on the ammonia concentration z at different temperatures and two different sizes of the catalyst pellets, a) 4.5 mm b) 10 mm. SA-1 catalyst, P = 300 atm (from [98])

The effect of the operating conditions and particle size on η is illustrated in Fig. 4.1 [98]. It is mainly the concentration of ammonia that varies throughout the pellet, whereas the relative changes in the N_2 and H_2 concentrations are much smaller. Therefore the effectiveness factor depends mainly on the diffusion rate of ammonia.

If the diffusion coefficient in a straight cylindrical pore is denoted by D, the effective diffusion coefficient D_{eff} in the catalyst is given by

$$D_{eff} = \frac{\theta}{\tau} D \tag{133}$$

where θ is the porosity of the catalyst particle and τ the so-called tortuosity factor depending on the geometry of the pores.

D can be calculated by the methods outlined by Satterfield [99]. Under industrial conditions, i.e. at high pressures, diffusion occurs mainly in the bulk and D is thus the bulk diffusion coefficient of the components in the mixture. At low pressures Knudsen diffusion will also play a role, and the pore volume distribution must be known in order to calculate D.

4.7.1 Mathematical Model

For a spherical catalyst grain, the concentration C_i at radial distance r of the catalyst is given by

$$r^2 \frac{d}{dr}\left[\frac{D_i}{r^2}\frac{dC_i}{dr}\right] + v_i R = 0 \tag{134}$$

with the boundary conditions

$$\frac{dC_i}{dr} = 0 \text{ at } r = 0 \tag{135}$$

and

$$C_i = C_s \text{ at } r = r_p \tag{136}$$

where D_i and v_i are the effective diffusion coefficient and stoichiometric coefficient of component i. R is the reaction rate for a component with $v_i = 1$, and r_p is the particle radius. If the particle is not spherical, the proper particle dimension is

$$r_p = 3\,V_g/A_g \tag{137}$$

V_g/A_g is the ratio of surface to volume of the catalyst grain. Values of the tortuosity factor reported in different investigations are shown in Table 4.2.

In solving (134) D_i is usually considered constant, although D_i depends on the gas composition. This approximation seems reasonable considering the uncertainty in τ. For simple reaction kinetics, the solution to (134) is given in terms of the Thiele modulus [102]. The more complicated kinetics for ammonia synthesis does not give a simple analytical solution. Generally, a numerical integration has to be carried out.

Bokhoven and Rayen [97] measured reaction rates on 0.5–0.7 mm and 2.4–2.8 mm particles at 1, and 30 atm and 325–550 °C. The effective diffusion coefficient of NH_3 was calculated from the results of O_2 diffusion measurements on a catalyst with the same surface area and porosity as the catalyst used in the activity measurements. The authors approximated the reaction rate by a pseudo first-order reaction suggested by Wagner [103] to calculate the effectiveness factor. Good agreement between measured and calculated data was obtained. However, the approximation above is only good, if the reaction is near equilibrium.

Peters and Krabetz [43] measured reaction rates at 1 atm on Al_2O_3 promoted iron catalysts with varying particle size (0.5–5.2 mm) and mean pore radius (110–310 Å). The effective diffusion coefficient of NH_3 was also measured in a Wicke-Kallenbach experiment. They also found good agreement between the measured and calculated data using a pseudo first-order approach.

Dyson and Simon [72] calculated η for a 6–10 mm size industrial catalyst (using rate data reported by Nielsen [59]) by numerical integration of the

Table 4.2. Tortuosity factors for different catalysts and the method by which they were determined

Catalyst	θ	τ	Reference	Method
Fe + MgO, SiO	0.52	3.8	Bokhoven and van Rayen [97]	Diffusion Measurement
KMI	0.45	2.2	Nielsen [100]	Electron microscope study of pore system
GK-1	0.22	16	Rusov [101]	Rate measurements
GK-2	0.30	6.5		
SA-1	0.12	3.0		

differential equation. Their model also includes bulk flow due to volume change during the reaction. An expression for the effectiveness factor as a function of conversion and temperature was given at 150, 225, 300 atm.

Nielsen [100] also solved the simultaneous differential equations numerically and reported the results for a hypothetical reactor operating at 450 °C, 214 atm, and with a 3:1 H_2/N_2 feed gas containing 3% ammonia and 12% inerts. The concentration profiles calculated at the bed inlet and outlet are shown in Figs. 4.2 and 4.3. Nielsen used $\tau = 2.2$, based on a detailed, electron-microscopic study of the pore system. Figs 4.4 and 4.5 show the calculated ammonia concentration in the bulk at half pellet radius and the effectiveness factor for particle diameters of 5.7 mm and 1.5 mm as a function of axial distance. For the 1.5 mm particles, the effectiveness factor very quickly approaches 1. The higher reaction rates of the 1.5 mm particles are utilized in modern radial flow converters.

Rusov et al. [101] investigated a complicated model that also included molar flux due to the volume change during the reaction and different diffusibilities of the various components. They concluded, that using a simple model with constant $D_i = D_{NH_3}$ for all components was almost as accurate as the complicated model. The reaction rates for five different catalysts were measured in an ideal back-mix reactor operating at 400–525 °C and P = 300, 450 atm. The kinetic parameters in the Temkin-Pyzhev equation were estimated on crushed particles. D_{NH_3} and τ could be calculated from rate measurements on large particles. The tortuosity of the investigated catalysts varied between 3 and 15. It should be noted that the porosity of these investigated catalysts was rather low.

In the above-mentioned investigation α changed from 0.35 to 0.9 upon increasing the temperature. Beskov et al. [98] showed that the reaction rates on

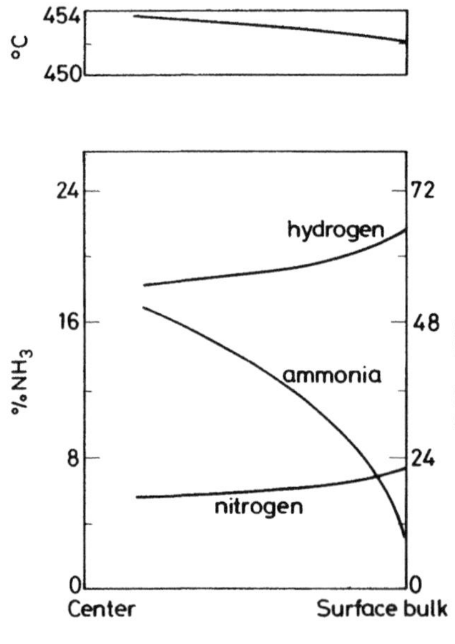

Fig 4.2. Radial concentration profiles in 5.7 mm ammonia catalyst particles located at the bed inlet. Pressure 214 atm, temperature 450 °C, inerts 12% (from [100])

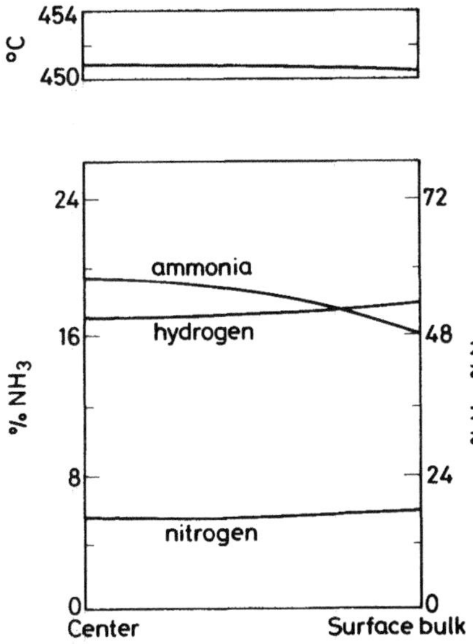

Fig 4.3. Radial concentration profiles in 5.7 mm ammonia catalyst particles located at the bed outlet, Pressure 214 atm, temperature 450 °C, inerts 13.7% (from [100])

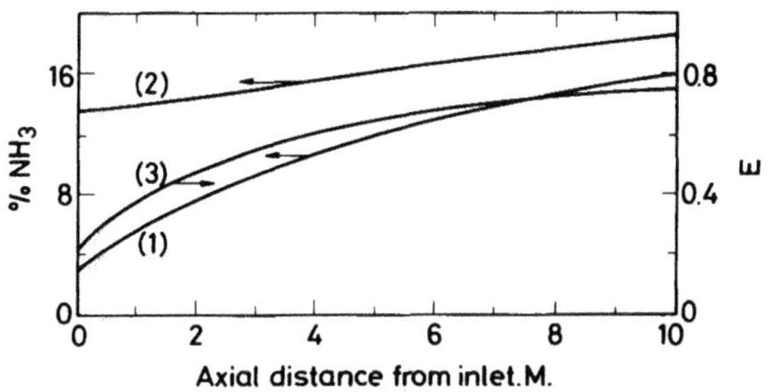

Fig 4.4. Axial profiles of NH_3 in the bulk (*1*) and at half pellet radius (*2*) as well as effectiveness factor E (*3*) for particle diameters of 5.7 mm. Pressure 214 atm, temperature 450 °C. inerts 12%, NH_3 3%, N_2 21.25%, H_2 63.75%, feed rate 6700 kmol/h (from [100])

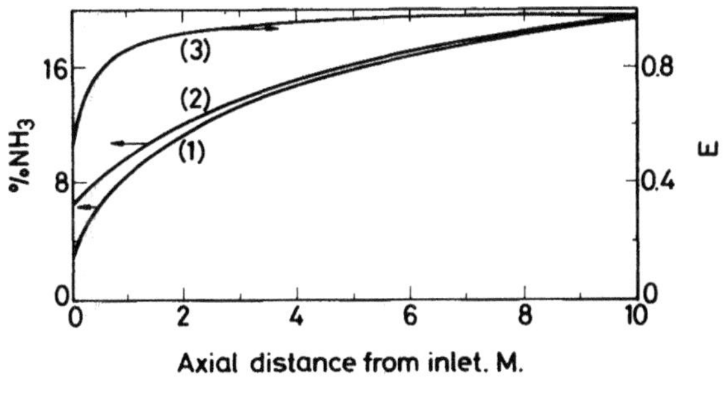

Fig 4.5. Axial profiles of NH_3 in the bulk (*1*) and at half pellet radius (*2*) as well as effectiveness factor E(*3*) for particle diameters of 1.5 mm. Other conditions as in Fig. 4.4 (from [100])

large particles could also be fitted to the Temkin-Pyzhev equation, but with lower values of α. In the temperature range 450–525 °C, α changed only slightly ($\alpha = 0.4$–0.5) for the particle sizes 4–5 mm and 10 mm.

Kazarnovskaya et al. [71] reported a lot of data from an experiment in a recycle reactor with four different particle sizes. The reactor operated at

400–550 °C, P = 200, 250, 300 atm and four different space velocities. For the largest particle size (7–10 mm) they found an apparent activation energy of 20 kcal/mol. This was half the activation energy for the small particle size (0.25–0.5 mm).

Sokolinskii et al. [104] found the following approximate expression for η in reasonable agreement with the experimental data reported in [71] for high values of η:

$$\eta = 1 - \frac{1}{15}\frac{r_p^2 R_{obs}}{D_{NH_3} c} \frac{1 + \left(\frac{c}{c_{eq}}\right)^2}{1 - \left(\frac{c}{c_{eq}}\right)^2} \tag{138}$$

where R_{obs} is the observed reaction rate, and c and c_{eq} are the concentration and equilibrium concentration of ammonia, respectively. This expression can be used to determine whether the measured rate data are diffusion-limited.

4.8 References

1. Larson AT, Tour RS (1922) Chem Met Eng 26
2. Almquist JA, Crittenden ED (1926) Ind Eng Chem 18: 1307
3. Benton AF (1927) Ind Eng Chem 19: 494
4. Winter E (1931) Z Physik Chem B 13: 401
5. Taylor HS (1931) J Am Chem Soc 53: 578
6. Frankenburger W (1933) Z Elektrochem 39: 45,97,269
7. Emmett PH, Brunauer S (1934) J Am Chem Soc 56: 35
8. Emmett PH, Love KS (1933) J Am Chem Soc 55: 4043
9. Emmett PH, Brunauer S (1933) J Am Chem Soc 55: 1738
10. Taylor HS, Jungers JC (1935) J Am Chem Soc 57: 660
11. Kozhenova KT, Kagan M Ya (1940) J Phys Chem USSR 14: 1250
12. Finkelstein WS (1934) Acta Physicochim USSR 1: 521
13. Chrisman IA (1936) Acta Physicochim USSR 4: 899
14. Finkelstein WS, Rubanik MJ (1935) J Phys Chem USSR 6: 1051
15. Roiter VA (1940) J Phys Chem USSR 14: 1229
16. Temkin MI, Pyzhev V (1939) J Phys Chem USSR 13: 851
17. Temkin MI, Pyzhev V (1935) Acta Physicochim USSR 12: 327
18. Temkin MI (1940) J Phys Chem USSR 14: 1241
19. Slygin A, Frumkin A (1935) Acta Physicochim USSR 3: 791
20. Zeldowitsch Ya, Roginsky SZ (1934) Acta Physicochim USSR 1: 449
21. Langmuir I (1932) J Am Chem Soc 54: 2798
22. Evans MG, Polanyi M (1936) Trans Farad Soc 32: 1333
23. Emmett PH, Kummer JT (1943) Ind Eng Chem 35: 677
24. Kodama S, Fukui K, Mazume A (1951) J Che Soc Japan, Ind Chem Sect 54: 157
25. Temkin MI (1957) Khim Nauka Promst 2: 219
26. Love KS, Emmett PH (1941) J Am Chem Soc 63: 3297
27. Brunauer S, Emmett PH (1940) J Am Chem Soc 62: 1732
28. Frankenburger W (1955) Catalysis Vol. III, Reinhold Publ, NY
29. Brunauer S, Love KS, Keenan RG (1942) J Am Chem Soc 64: 751
30. Emmett PH (1946) Coll Chem 6: 214

31. Temkin MI, Kiperman S (1947) Zhur Fiz Khim 21: 927
32. Sidorov IP, Livshits VD (1947) Zhur Fiz Khim 21: 117
33. Kobayashi H, Kubota H (1949) Rep Faculty Eng, Hokkaido Univ 3: 136
34. Temkin MI (1950) Zhur Fiz Khim 24: 1312
35. Nielsen A (1950) An Investigation on Promoted Iron Catalysis for the Synthesis of Ammonia, Ed 1, Gjellerup, Copenhagen
36. Temkin MI, Kiperman SL, Lukjanova LI (1951) Doklady Akad Nauk (USSR) 74: 763
37. Brill R (1951) J Chem Phys 19: 1047
38. Annable D (1952) Chem Eng Sci Vol 1,4: 145
39. Kipermann S, Granovskaya VS (1952) Zhur Fiz Khim 26: 1615
40. Sidorov IP, Livshits VD (1952) Zhur Fiz Khim 26: 538
41. Adams RM, Comings EW (1953) Chem Eng Prog 49: 359
42. Bokhoven C, van Heerden C, Westrik R, Zwietering P (1955) Catalysis vol III, Reinhold, NY
43. Peters C, Krabetz R (1956) Z Elektrochem 60: 859
44. Scwab GM, Krabetz R (1956) Z Elektrochem 60: 855
45. Kawamura M, Irie T, Yokota H (1957) 60: 1239
46. Shiskova VV, Sidorov IP, Temkin MI (1957) Tr GIAP 7: 62
47. Logan SR, Moss RL, Kemball RL (1958) Trans Faraday Soc 54: 922
48. Mills AK, Bennett CO (1959) Jour AIChE 5: 539
49. Kubota H, Shindo M (1959) Kagaku Kogaku 23: 242
50. Emmett PH ((⊠) Fixed Nitrogen, Reinhold, NY
51. Uchida H, Kuraishi M (1955) Bull Chem Soc Japan 28: 106
52. Boudart M (1956) Jour AIChE 2: 62
53. Stelling PO, Krusenstierna O (1958) Acta Chem Scand 12: 1095
54. Temkin MI (1949) Probl Kinet Katal 6: 54
55. Kwan T (1956) J Phys Chem 60: 1033
56. Ozaki A, Taylor HS, Boudart M (1960) Proc Roy Soc (London) A 258: 47
57. Brill R, Tauster S (1962) J Chem Phys 36: 2100
58. Krabetz R, Peters C (1963) Ber Bunsen-Ges Phys Chem 67: 381
59. Nielsen A, Kjær J, Hansen B (1964) J. Catal 3: 68
60. Nielsen (1968) An Investigation on Promoted Iron Catalysts for the Synthesis of Ammonia, Ed 3, Gjellerup, Copenhagen
61. Temkin MI, Morozov NM, Shapatina EN (1963) Kinetics Katal 4: 260
62. Temkin MI, Morozov NM, Shapatina EN (1963) Kinetics Katal. 4: 565
63. Bridger GW, Snowdon (1970) Catalyst Handbook, Chapter 7, Wolfe Scientific, London
64. Aika K, Ozaki A (1969) J Catal 13: 232
65. Logan SR, Philip J (1968) J Catal 11: 1
66. Carra S, Ugo R (1969) J Catal 15: 435
67. Takezawa N, Toyoshima I (1966) J Phys Chem 70: 594
68. Takezawa N, Toyoshima I (1966) J Res Inst Catalysis, Hokkaido Univ 14: 41
69. Takezawa N, Toyoshima I (1966) J Catal 6: 145
70. Brill R (1970) J Catal 16: 16
71. Kazarnovskaya DB, Atamonovskaya RM, Bomshtein EI (1973) Zh Fiz Khim 47: 1445
72. Dyson DG, Simon JM (1968) Ind Eng Chem Fundam 7:605
73. Buzzi Ferraris G, Donati G (1970) Ing Chim Italiano 6: 11
74. Buzzi Ferraris G, Donati G, Rejna F, Carra S (1974) Chem Eng Sci 29: 1621
75. Cappelli A, Collina A (1972) I Chem E Symp Series 35, 5:10
76. Guacci U, Traina F, Buzzi Ferraris G, Barisone R (1977) Ind Eng Chem 16, No 2: 166
77. Altenburger K, Bosch H, van Ommen JG, Gellings PJ (1980) J Catal 66: 326
78. Zhenming L. (1981) J Chem Ind Eng (China) 1: 39
79. Huang KH (1981) Stud Surf Sci 7: 554
80. Huang KH (1980) Scientia Sinica 24, No. 6: 800
81. De-ming L, Shi-liang Z, Zi-xing H, Xi-jun W (1979) J Chem Ind Eng (China) 2: 133
82. Bowker M, Parker IB, Waugh KC (1982) Appl Catal 14: 101
83. Ertl G (1983) Catalysis, Science and Technology, Vol 4, Anderson JR, Boudart M (Ed) Springer Berlin Heidelberg New York, p 273
84. Scholten JJF, Zwietering P, Konvalinka JA, de Boer JH (1959) Trans Faraday Soc 55: 2166
85. Stolze P, Nørskov JK (1985) Phys Rev Lett 55, No 22: 2501
86. Stolze P, Nørskov JK (1987) J Vac Sci Technol A 5(4), 581
87. Stolze P (1987) Phys Scripta 36: 824

88. Stolze P, Nørskov JK (1988) J Catal 110: 1
89. Fowler R, Guggenheim EA (1960) Statistical Thermodynamics, Cambridge Univ Press, Cambridge
90. JANAF Thermochemical Tables, US National Bureau of Standards, US GPO, Washington DC (1971)
91. Bowker M, Parker I, Waugh KC (1988) Surf Sci 197; L223
92. Ertl G, Lee SB, Weiss M (1982) Surf Sci 114: 515
93. Ertl G, Huber M, Lee SB, Weiss M (1982) Surf Sci 114: 527
94. Stolze P, Nørskov JK (1988) Surf Sci 197: L230
95. Nielsen A (1970) Catal Rev 4: 1
96. Nielsen A (1977) Ammonia Part III, Fertilizer Science and Technology Series Vol 2, Slack AV, Russell James G (Ed) Marcel Dekker, NY
97. Bokhoven C, van Rayen W (1954) J Phys Chem 58: 471
98. Beskov VS, Malinovskaya OA, Slinko MG, Streltsov OA, Rusov MT (1971) Teor Osn Khim Tekhnol 5: 537
99. Satterfield CN (1970) Mass Transfer in Heterogeneous Catalysis, MIT Press, Cambridge, Massachusett
100. Nielsen A (1969) Chim Ind (Milan) 51: 1052
101. Rusov MT, Slinko MG, Streltsov OA, Malinovskaya OA, Beskov VS, Loza AN (1971) Teor Osn Khim Tekhnol 5: 382
102. Thiele EW (1939) Ind Eng Chem 31: 916
103. Wagner C (1943) Z Phys Chem 193: 1
104. Sokolinskii YA, Kazarnovskaya DB, Atamonovskaya RM (1973) Kinet Katal 14: 1351
195. Bowker M (1992) Catal Today 12: 153
106. Rettner C, Stein H (1987) Phys Rev Letts 58, 2768

Chapter 5

Poisoning of Ammonia Synthesis Catalysts

P. E. Højlund Nielsen
Haldor Topsoe A/S Copenhagen, Denmark

Contents

Nielsen (Ed.), Ammonia

5.1 Introduction

An ammonia catalyst is poisoned when its catalytic activity is reduced due to the presence of certain elements. These elements may be present in certain gaseous compounds contained in the synthesis gas or solid and as such introduced to the catalyst during the manufacturing process. With respect to the latter, reference is made to Chap 2 by P. Stoltze.

Concerning gaseous catalyst poisons, a distinction should be made between permanent poisons causing an irreversible loss of catalytic activity and temporary poisons which lower the activity while present in the synthesis gas. For a review of poisoning of synthesis catalysts, see [1, 2]. Permanent poisons accumulate on the surface and may be detected by chemical analysis of the poisoned catalysts, whereas temporary poisons cause a partial coverage of the catalyst surface. Since oxygen is the most common temporary poison, it is difficult to detect the amount on the spent catalyst by analysis since the promoter phases are difficulty reducible metal oxides like alumina, magnesia, silica, and potassium oxide.

5.2 Permanent Poisons

A permanent poisoning is caused by a compound whose presence in the synthesis gas causes an activity decline, and the activity is not regained when the compound is removed from the synthesis gas. The poisoning action may be due to the formation of surface and/or bulk compounds with iron or one of the promoters, in particular potassium, or the deposition of solid materials in the pore system. For a discussion of permanent poisons, see [1, 2].

Among these permanent poisons, sulfur is the most common. In some plants based upon gasification of coal or oil, traces of COS and H_2S may reach the synthesis converter. In natural-gas-based plants, the Cu-based low temperature shift catalyst and the Ni-based cleanup methanation catalyst act as efficient guards. In most cases the sulfur poisoning is caused by lubricating oil. Thus, it has been considerably reduced by the use of low sulfur oil and centrifugal compressors instead of piston compressors [1].

Sulfur is chemisorbed very strongly by the iron surface, as discussed in Stoltze's chapter. Studies of sulfur poisoning of the synthesis catalyst have been carried out [5–8], whereas in [3, 4], the interaction between single crystal surfaces of iron and sulfur was discussed.

It was found that H_2S adsorption led to an increase in the work function of iron as was also the case for CO, CO_2 and H_2O [5]. But contrary to the latter, heating in H_2 did not lower the work function to its unpoisoned level, indicating an irreversible poisoning. Experiments on iron and single (Al_2O_3) promoted catalysts were carried out [6–8]. They found that about 0.2 mg S/m^2 catalyst surface completely deactivated the catalyst. The maximum uptake of S was found to be 0.4–0.5 mg/m^2 catalyst surface. Chemisorbed sulfur could not be removed by heating the synthesis gas up to 900 K. Catalysts to which sulfur was added during activation were just as poisoned as the ones where sulfur was added along with the synthesis gas. A very high level of sulfur was measured, and bulk FeS was found on the spent catalyst. An amount of 6300 ppm of sulfur completely deactivated the catalyst, whereas amounts as low as 200 ppm gave a partial deactivation to the order of 15% [8]. Analysis of spent industrial catalyst from the inlet layer reveals from 1000 to 5000 ppm of sulfur on catalysts KMI and KMII [1]. Considerable poisoning by exposure to 20–30 ppm(vol) H_2S for 24 h at a space velocity of 10000 h^{-1} was observed [1].

5.3 Other Permanent Poisons

In general, any compound that can lower the surface tension of iron may be considered as a poison [9]. Phosphorus and arsenic compounds are also known as permanent poisons [1]. However, in the natural gas-based plants they are rare and only found in exceptional cases. A catalyst prepared with 1–3 wt% P_2O_5 was examined and found to be severely poisoned [10]. Chlorine [1] is a serious poison that reacts with the potassium promoter and forms potassium chloride which is slightly volatile and consequently removed from the catalyst surface. It is expected that other halogens behave similarly [1]. Calculations show that ppb amounts in the synthesis gas should result in a marked deactivation [2].

5.4 Temporary Poisons

A temporary poison lowers the activity of an ammonia synthesis catalyst while it is present upon its surface.

If there is no poison in the synthesis gas, the poison will be removed from the surface and the catalyst will regain its activity. An oxygenic compound reacts in that manner with the catalyst surface [1]. Unsaturated hydrocarbons may also react as a reversible poison [1], but usually oxygenic compounds are the most important, temporary poisons for ammonia synthesis catalysts.

The most common oxygenic compounds encountered in ammonia synthesis are H_2O, CO, CO_2 and O_2. The equivalence of these has been known since the early studies on ammonia synthesis catalysts. This is due to the methanation activity of the ammonia synthesis catalysts,

$$3H_2 + CO \rightarrow CH_4 + H_2O$$

CO_2 will undergo a similar reaction and O_2 will become hydrogenated.

Although the effect of an oxygenic compound is temporary in the sense that the catalyst regains its activity when the poison is removed from the gas stream, a severe poisoning for an extended period of time may result in a permanent loss of activity, probably due to recrystallization, see [1] and [11].

5.5 Early Studies

Studies of ammonia catalyst poisoning by oxygenic compounds date back to 1922 when Larson carried out experiments with 400–1000 ppm of H_2O at pressures up to 100 atm [12]. Experiments were also carried out with CO, and the equivalence of the various oxygenic compounds was recognized. Experiments with up to 1600 ppm O_2 in the synthesis gas at ambient pressure were carried out, see [13]. The amount of O_2 taken up by the catalyst was determined. The promoted catalysts were found to take up a larger amount of oxygen. The thermodynamic stability of the surface oxide was evaluated by [14] and found to be more stable by 12000 cal per gram atom of iron than a gram atom of iron in magnetite. Poisoning experiments at high pressure were carried out. The results were used to suggest that the amount of oxygen take up by a catalyst varied with $[pH_2O/pH_2]^{1/2}$. The validity of this result is questioned, however, as the experiments were carried using a "pure" gas with about 100 ppm of oxygenic impurities, see [2]. An interesting set of experiments were carried using predetermined volumes of CO, CO_2, O_2 and SO_2, see [15]. These volumes of CO, CO_2, O_2 and SO_2 were injected, which resulted in a release of ammonia in amounts exceeding that of thermodynamic equilibrium; the experiment was conducted at 1 atm. The ammonia release was followed by an activity loss, which in the case of SO_2 was irreversible. Injection of 80 cm^3 of CO resulted in an almost complete poisoning of 10 cm^3 of catalyst. About 10 cm^3 of NH_3 was released during the poisoning. It was suggested that the release of NH_3 was due to a replacement of chemisorbed N with oxygen. In this case it would mean that N occupies around 1/8 of the available surface sites.

The iron catalyst was examined using chemisorption. Through this method it was found that the active surface area was a much smaller fraction of the total surface area that was determined by nitrogen condensation, see [16]. However, the conclusions may be severely affected by the limited ultrahigh vacuum technology available in 1940.

5.6 Later Studies

High pressure studies carried out using O_2, CO, CO_2 as poisoning agents [1] demonstrated that 50 ppm O_2 is just as poisonous as 100 ppm CO, and that a 6-day exposure to 100 ppm of CO at 450 °C did not harm the catalyst. Experiments were conducted using radioactive tracer methods with CO and CO_2 as poisons [17]. A slight amount of CO_2 was found to be irreversibly absorbed; the potassium promoter was thought to be responsible for that. Studies were carried out with a number of promoted catalysts at temperatures between 300 and 450 °C, a space velocity of 5600, and a pressure of 100 kg/m^2 (9.8 MPa) [18]. At lower temperatures, some of the CO was not converted. At higher temperatures, the poisoning was found to be reversible. The experiments at lower temperatures were not continued long enough to determine if the poisoning was reversible.

Changes in work function originating from the chemisorption of oxygenic compounds and the H_2S upon the catalyst were studied [5]. H_2S was found to cause a permanent increase in the work function, whereas the oxygenic compounds caused a temporary increase. Experiments were conducted at ambient pressure using a singly promoted catalyst with water as poisoning agent [19]. Between 1.2–32.3 mg of oxygen was found chemisorbed onto the catalyst which has a total surface area of 250 m^2. However, the very low space velocities employed and the duration of the experiment were not sufficient to reach steady state. Experiments were conducted at ambient pressure [20–21] which suggested a kinetic expression, something which had been done earlier based on theoretical consideration [22]. In [21] the rate expression was applied to data obtained at elevated pressure. The expression being

$$\omega = \frac{k_+ P_{N_2} - k_- \dfrac{P_{NH_3}^2}{P_{H_2}^3}}{\left[\dfrac{P_{NH_3}^2}{P_{H_2}^3} + C \cdot \dfrac{P_{H_2O}}{P_{H_2}}\right]^{\alpha}}$$

where k_+ is the rate constant for the synthesis reaction; k is the rate constant for the decomposition reaction and α is a constant, see also the chapter by Bøgild Hansen.

The expression assumes a displacement reaction

$$N_2\,(ads) + H_2O + 2H_{H_2} = O\,(ads) + 2NH_3$$

The values of C depend on temperature [see 21]:

		400 °C	425 °C	450 °C	475 °C	500 °C
C_1	$Fe + Al_2O_3 + K_2O$	0.74	0.46	0.28	0.20	0.14
C_2	$Fe + Al_2O_3$	0.63	0.39	0.24	0.17	0.12

$$C_1 = 1.30 \times 10^{-6} \exp 8930/T$$

$$C_2 = 1.15 \times 10^{-6} \exp 8930/T$$

A satisfactory description of earlier results including [12] can be found in [21]. It should be recalled, however, that experiments using a pure synthesis gas could not be carried out at that time. Examination of poisoning reported in [23] using H_2O vapour amounts between 1.8–95.8 ppm at 1 kg/cm^2 (98.1 kPa), demonstrated that the activation energy for ammonia synthesis does not depend on the amount of poison. He could, however, neither describe his results by means of Smirnov's theory, nor by means of Ozaki, Taylor and Boudart's theory [24]. It was found that oxygen reduces the rate of nitrogen adsorption [25]. By doing studies of isotope exchange, the amount of N_2 chemisorbed was reduced only 1.8 times, whereas the rate of homo-exchange was reduced 17 times, contrary to the theory adopted in Ref. [21].

Recent work has been carried out by P. E. Højlund Nielsen [26], describing the poisoning by multiplying the rate expression with a correction factor, 1-θ, where θ is given by

$$\theta = a + bT + cT \ln \left[\frac{H_2O}{H_2} \right]$$

By varying the H_2/N_2 ratio, it was found that the expression

$$\theta = a + bT + c_1 T \ln X_{H_2O}$$

where X_{H_2O} is the molar fraction of H_2O, gave a better overall description [27].

This apparent independence of the hydrogen pressure was explained in Refs. [28] and [29]. By using surface science data, kinetics valid at high pressure were established. From these data it was established that during synthesis the surface was covered by N-species; see also the old result in Ref. [15]. By assuming that a chemisorbed O also influenced chemisorption upon neighbouring sites, an explanation of the results of Ref. [15] could be established. Accordingly, Stoltze and Nørskov [28, 29] simply added an equilibrium reaction to their set of equations, see the chapter by Bøgild Hansen upon kinetics, the equation being

$$H_2O(g) + 3* \leftrightharpoons 2H - * + 0 - *$$

where $*$ denotes a surface site.

By doing so, a satisfactory description of the results reported in Refs. [26, 27] could be obtained.

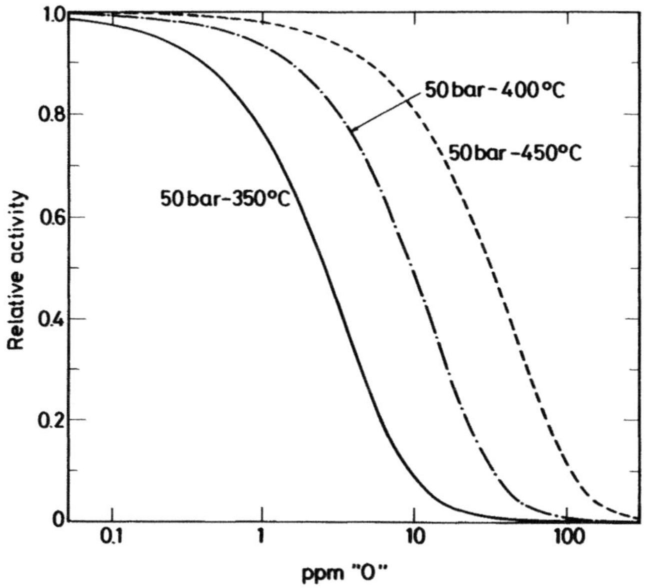

Fig. 5.1. Poisoning of ammonia synthesis catalyst

Recently reported data [30] confirm primarily the equivalence of H_2O, CO, O_2 and CO_2. The poisoning was studied at 350–400 and 450 °C, see Fig. 5.1. It was found [30] that the results could be described by adding an adsorption factor for oxygen to the rate equation, a procedure in accordance with Refs [28] and [20].

The results of the latest studies are a sort of complimentary situation, since poisoning can be described either by multiplying the rate equation by a correction term, [26, 27], or modifying the rate equation by including an adsorption term for oxygen. At low temperatures the latter treatment would be in disagreement with the results given in Ref. [23] since the adsorption of oxygen is, in practice, irreversible under these conditions. However, for all practical purposes, the two approaches should give an equally valid description of reversible oxygen poisoning.

5.7 References

1. Nielsen A (1968) An Investigation on Promoted Catalysts for the Synthesis of Ammonia, Jul. Gjellerup, Copenhagen, pp 126
2. Højlund Nielsen PE, (1991) Catalytic Ammonia Synthesis, ed. by J. R. Jennings, Plenum New York
3. Benard J, Oudar J, Barbouth N, Margot E, Berthier Y (1979) Surf Sci 88: L35

4. Grabke HJ, Paulitschke W, Tauber G, Viefhaus H (1977) Surf Sci 63: 377
5. Enikeev E, Krylova AV (1962) Kinet Katal 3: 139
6. Brill R, Tauster S (1963) Ber Bunsen-Ges phys Chem 67: 390
7. Tauster S, PhD Thesis, Polytechnic Institute of Brooklyn, June 1964, available university microfilms, 64–10, 728
8. Brill R, Schaefer H, Zimmerman G (1968) Ber Bunsen-Ges Phys Chem 72: 1218
9. Rostrup-Nielsen JR, Højlund Nielsen PE (1985) In: Oudar J, Wise H (eds) Deactivation and Poisoning of Catalyst Dekker, New York, p 259
10. Marakhovets et al. (1972) Izv Vyssh Ucheb Zaved Khim Tekhnol p 735
11. Catalyst Handbook, ed. M. V. Twigg, Wolfe London 1989, p 407
12. Larson AT, Tour RS (1922) Chem Met Eng 26: 647
13. Almquist JA, Black CA (1926) J Am Chem Soc 48: 2814
13. Almquist JA (1926) J Am Chem Soc 48: 2820
14. Emmett PH and Brunauer S (1930) J Am Chem Soc 52: 2682
15. Ussatschew PW, Tarakanowa WJ, Komarov WA (1934) Z Elektrochem 40: 647
16. Brunauer S, Emmett PH (1940) J Am Chem Soc 62: 1732
17. Bokhoven C (1954) Proceedings of the Second Radiosotope Conference, Oxford Butterworths, London, pp 53
18. Uchida H, Todo N (1951) Report of the Tokyo Industrial Research Institute Laboratory 46: 213
19. Royen P, Langhans GH (1962) Z Anorg Allg Chem 315: 1
20. Smirnov IA, Morozov NM, Temkin MI (1965) Kinet Katal 6: 351
21. Smirnov IA (1966) Kinet Katal 7: 107
22. Kiperman SL (1954) Zhur Fiz Khim 28: 389
23. Brill R. Hensel J, Schaefer H (1969) Ber Bunsen-Ges Phys Chem 73: 1003
24. Ozaki A, Taylor H, Boudart M (1960) Proc Roy Soc (London) Ser A258, 47
25. Boreskova EG, Kuchaev VL, Temkin MI (1984) Kinet Katal 25: 112
26. Højlund Nielsen PE; The Poisoning of Ammonia Synthesis Catalysts by Oxygenic Compounds, lecture given at the ACS Meeting, Washington, D.C. Sept. 1983.
27. Andersen S, Haldor Topsøe A/S, Private Communication
28. Stoltze P (1987) Phys Scr 36: 824
29. Stoltze P, Nørskov JK (1987) J Vac Sci Technol A5, 581
30. Kirkerød T, Skaugset P (1991) Abstracts of IV Nordic Symposium on Catalysis, Trondheim

Chapter 6

Ammonia Production Processes

Ib Dybkjaer
Haldor Topsøe A/S Lyngby, Denmark

Contents

Nielsen (Ed.), Ammonia

6.1 Introduction

The present chapter describes process technology for the production of ammonia. The first part gives a short review of early developments in ammonia technology. The second part contains a brief description of processes used in the production of ammonia synthesis gas, i.e. a mixture of hydrogen and nitrogen with or without minor amounts of impurities such as methane, argon, etc. A third part describes the conversion of synthesis gas to ammonia. This part includes discussion of the design of ammonia synthesis converters. A fourth part describes integrated processes for production of ammonia from primary raw materials such as hydrocarbons and coal. A final part deals with storage of ammonia.

Several publications are available which give an integrated treatment of important aspects of ammonia production. Some examples are [1–9, 876, 949]. Relevant patents are reviewed in [10, 11]. Historical aspects are treated especially in [12–17, 877]. Reviews of the technological status at various times are given in [1–9] and in [18–42, 949]. Catalytic reactions and catalysts relevant for ammonia production are treated in [43–46].

6.2 Historical

The main purpose of the present chapter is to describe modern technology for ammonia production. For the sake of completeness a brief description of the early development of ammonia synthesis technology will be given in the following. More details can be found in the relevant literature.

The history of modern ammonia production started in Germany just after 1900, when Fritz Haber and his assistants developed the process concept which is the basis for all ammonia production even today. Haber patented his work in two famous patents, the "circulation patent" [47] and the "high pressure patent" [48]. The claims of the patents read:

"Circulation patent": "Verfahren zur synthetischen Darstellung von Ammoniak aus den Elementen, wobei ein geeignetes Gemenge von Stickstoff und Wasserstoff kontinuerlich der Ammoniakbildung mittels erhitzter Katalysatoren und nachofolgender Ammoniakentziehung unterworfen wird, dadurch gekennzeichnet, daß hierbei unter dauerndem Druck gearbeitet und dafür gesorgt wird, daß die Wärme der ammoniakhaltigen Reaktionsgase auf das von neuem der Reaktion zu unterwerfende ammoniakfreie Gasgemisch übertragen wird".

"High Pressure patent": "Verfahren zur Darstellung von Ammoniak aus den Elementen durch Katalyse unter Druck bei erhöhter Temperatur, dadurch gekennzeichnet, daß die Vereinigung unter sehr hohen Drucken von etwa 100 Atmosphären, zweckmässig aber von 150 bis 250 Atm. und mehr vorgenommen wird."

The German company BASF had been following Haber's work through their two employees, Carl Bosch and Alvin Mittasch and eventually bought Haber's patents. Development work was continued on catalyst, process, and equipment design, and in 1913 the first commercial plant with a capacity of 30 tons/day was commissioned at Oppau/Ludwigshafen near Mannheim in Germany.

By 1916 the capacity had been increased to 250 tons/day. In 1917 a second plant was started at Leuna near Leipzig. At the end of the First World War this plant produced 240000 tons/year [14]. In 1937, the annual world capacity was 755000 tons/year, of which 72% was still concentrated in Oppau and Leuna [16].

Soon after the First World War, others started development work, partly on the basis of the German pioneering effort. A plant was started at Terni, Italy in 1920 based on technology developed by Luigi Casale. In France, a plant was constructed in 1921 at Grand Paroisse near Montereau, based on developments by M. G. Claude. Both the Casale and the Claude technologies operated at extremely high pressures. In further early developments in Europe, technology developed by Giacomo Fauser was adopted by the Italian company Montecatini, and Friederich Uhde GmbH constructed a plant to operate on coke oven gas from the Mont Cenis company. This plant used a special process later known as the Mont Cenis process. Its features were a very low operating temperature and pressure and a special catalyst based on iron cyanide.

Meanwhile, an independent development took place in the USA. As early as 1918 a plant called "US Nitrate Plant No. 1" was constructed in Muscle Shoals, Alabama. This plant was, however, not very successful, but the American development continued, and during the 1920s several successful plants were constructed, some based on American, some on European technology. In 1928, the Nitrogen Engineering Corporation (NEC) was commissioned to construct a plant in Europe [49], thereby entering the worldwide competition.

During the 1930s and during the Second World War production capacity was increasing rapidly using now well-established technologies. In 1945 about 125 plants existed with a total capacity of 4.5 million tons [49]. The most important processes were: Haber-Bosch, Casale, Claude, Fauser, NEC, and Mont Cenis. For descriptions of the technical features of these processes, see [50].

Just before the second world war it was a general belief [18] that ammonia synthesis technology was mature, and that no significant further developments could be expected. This was to some extent true, since most of the features which characterize modern synthesis technology – including catalyst type and most major converter designs – were already well proven in industrial operation. However, since that time, there has been a tremendous development in the scale of operation, and at the same time, the main type of feedstock has shifted from coke to natural gas, naphtha or fuel oil. New technology for the production of synthesis gas has been developed, and integrated, efficient processes for large scale production of ammonia from the different feedstocks have emerged.

6.3 Synthesis Gas Production

The ultimate raw material for ammonia production is synthesis gas, a mixture of hydrogen and nitrogen with or without minor quantities of impurities such as methane, argon, etc. Normally, the concept "Production of Ammonia" is, however, understood to cover the complete transformation of primary raw materials into ammonia [8]. The nitrogen part of the synthesis gas is always in some way derived from air, either in a separate process step, most often by cryogenic air separation, or in integrated processes, where air is used as a reactant in such a way that oxygen is consumed in the gas preparation process, while nitrogen remains as constituent in the synthesis gas (see below). When nitrogen is produced in a separate process by air separation, the oxygen part of the air is most often used in the process for production of the hydrogen part of the synthesis gas.

The most important primary raw materials for the hydrogen part of the synthesis gas are natural gas and other light hydrocarbons. Other raw materials include heavy hydrocarbons, solid materials such as lignite and coal, and hydrogen recovered from off-gases from other processes. Production of synthesis gas can be subdivided into the following process steps (see Fig. 6.1).

–Gas Preparation

From natural gas and light hydrocarbons by steam reforming, catalytic autothermal reforming, or partial oxidation.

From heavy oil or coal by partial oxidation or gasification.

From water by electrolysis.

From hydrogen-rich off-gases by separation of hydrogen.

–Gas Purification

In most cases, the gas purification train contains the following process steps:

- Carbon monoxide conversion
- Carbon dioxide removal
- Final purification

Final purification can be by:

- Methanation or Methanolation
- Cryogenic purification ("Nitrogen wash")
- Copper liquor wash

As an alternative to these process concepts, the gas purification including carbon dioxide removal can be done by Pressure Swing Absorption (PSA).

In the following paragraphs, each of the individual process steps is briefly discussed. Integrated process schemes are discussed in Sect 6.5.

6.3.1 Steam Reforming

Production of ammonia synthesis gas from light hydrocarbons is normally initiated by a two-step process consisting of so-called primary and secondary reforming. The feedstock, which may range from natural gas to heavy naphtha, is mixed with steam in an amount (normally expressed by the steam to carbon ratio) determined by feedstock properties and operating conditions. A typical range is 2.5–4 moles of steam per mole of carbon in the feed. The steam-hydrocarbon mixture is first passed through the primary (tubular) reformer, which is heated by burning fuel. The feed is converted to a mixture of carbon oxides, hydrogen and unconverted methane. This mixture is further reacted with air in the secondary reformer, which is an adiabatic reactor. Typical outlet temperatures from the two steps are 750–820 °C from the primary reformer and 950–1025 °C from the secondary reformer. The pressure in modern plants typically is about 25–35 bar. Before reforming, the hydrocarbon feed must be purified to remove compounds such as sulfur, which are poisonous to the reforming catalyst and downstream catalysts. The desulfurization takes place by adsorption of sulfur-containing compounds on active carbon or on molecular sieves or by catalytic hydrogenation of the organic sulfur compounds and subsequent adsorption of hydrogen sulfide on zinc oxide. References on desulfurization of reformer feedstock are [51–61]. Bulk removal of hydrogen sulfide by absorption in amine solutions and other solutions is discussed in [62, 63]. The effect of a possible mercury content in ammonia plant feedstocks is reviewed in [878]. Removal of mercury and arsenic is discussed in [879].

In the primary reforming reaction, hydrocarbon feed is converted with steam according to the following reactions:

$$C_nH_m + nH_2O \rightarrow nCO + (m/2 + n)H_2 \; (-\Delta H^0_{298} < 0) \qquad (1)$$

$$CH_4 + H_2O \rightleftarrows CO + 3H_2 \; (-\Delta H^0_{298} = -206 \text{ kJ/mol}) \qquad (2)$$

$$CO + H_2O \rightleftarrows CO_2 + H_2 \; (-\Delta H^0_{298} = 4 \text{ kJ/mol}) \qquad (3)$$

Higher hydrocarbons are decomposed irreversibly and completely with steam by reaction (1). The final gas composition is governed by the two reversible reactions, the methanation reaction (2) and the carbon monoxide conversion or shift reaction (3).

The conversion of methane is favoured by high temperature, low pressure, and a high steam content. The overall conversion is strongly endothermic, and heat must be added to the process. Therefore, the reaction is carried out in

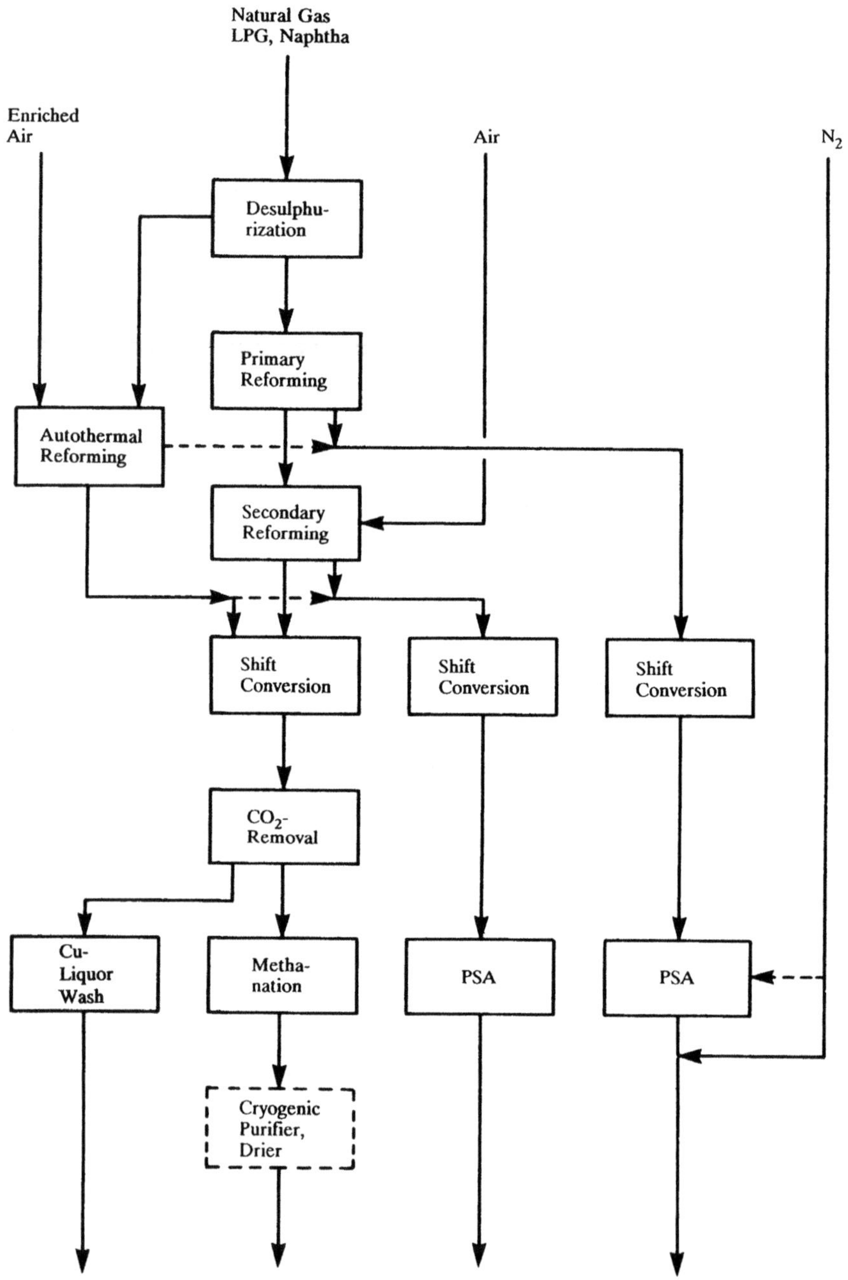

Fig. 6.1. Possible process steps in ammonia synthesis gas production

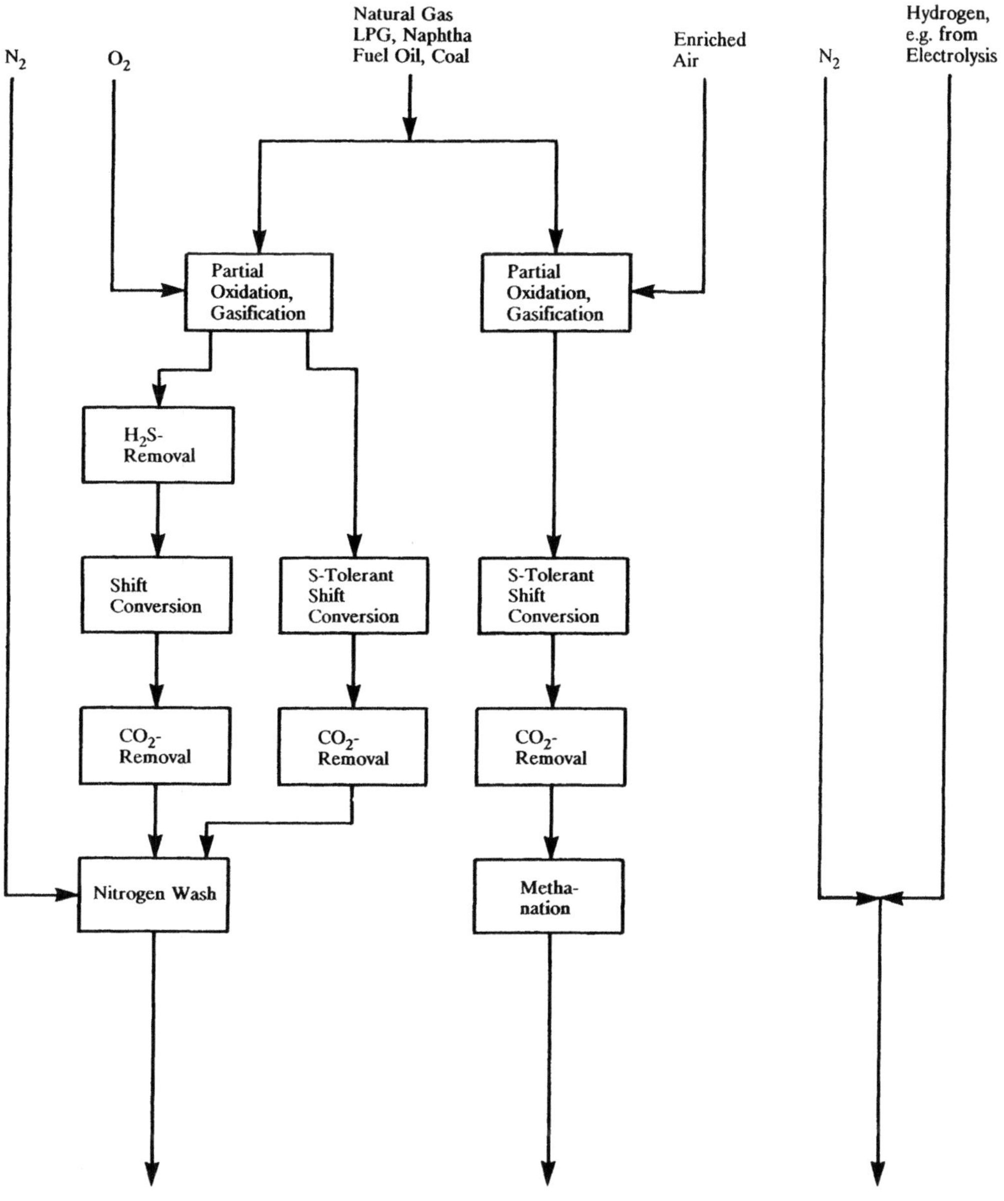

Fig. 6.1 (Continued)

reactor tubes which are heated from the outside by burning fuel. The reactor tubes are made of high alloy steel and filled with a nickel-based catalyst. A comprehensive treatise on steam reforming reactions and steam reforming catalysts is given in [64, 65]. General literature on reforming: [66–72, 880, 881];

steam reforming of naphtha (73–81). Catalyst formulations are discussed in [43, 44, 64, 65] and [82–85, 882]. Special problems in primary reformers are related to the risk of carbon formation [65, 86, 87] and catalyst poisoning, especially by sulfur [88, 883, 961]. Design of reformer furnaces and selection of tube material are discussed in [89–99, 884, 885, 962]. Creep rupture of reformer tubes is discussed in [100–104, 886]. Higher hydrocarbons (naphtha) may be converted in an adiabatic pre-reforming step [105–112, 887, 888]. Poisoning problems in pre-reformers are treated in [113]. Actual experience from installation of a pre-reformer in an ammonia plant is reported in [963].

Special reformer designs have been suggested in which the heat for the reforming reaction is supplied by heat exchange with a hot gas [114–117, 889, 890]. Application of such reformers in the production of ammonia synthesis gas is suggested in [118–124, 891]. A design where the primary reformer is split into a number of adiabatic vessels with reheat between the vessels is proposed in [125, 126].

In secondary reforming, the exit gas from the primary tubular reformer reacts adiabatically with air over a nickel-based catalyst. In this way, residual methane in the process gas is reduced to a low level and at the same time the nitrogen part of the ammonia synthesis gas is added. In most cases, the amount of air is adjusted in such a way that the stoichiometric ratio of hydrogen to nitrogen equal to 3.0 is obtained in the synthesis gas after carbon monoxide conversion and gas purification. However, there are also process schemes in which non-stoichiometric amounts of air are used; this is discussed in Sect. 6.5.

Secondary reforming is carried out in an adiabatic, refractory-lined vessel. Gas from the primary reformer typically at about 800 °C and containing 8–12 vol% (dry) residual methane is mixed in a burner with preheated air. The partially reacted mixture passes through a bed of catalyst, where the reactions are completed. The gas leaves the reactor at about 1000 °C with a residual methane content of typically 0.3–1.0 vol% (dry).

The design of secondary reformers is discussed in [90, 92, 127]. Special designs where primary and secondary reforming or primary reforming and non-catalytic partial oxidation are combined in one unit are suggested in [128–130, 964].

As an alternative to the combination of primary and secondary reforming, processes have been developed where all the conversion takes place in equipment similar to secondary reformer, but where the oxidizing agent is not air, but oxygen or oxygen-enriched air. Such so-called "catalytic partial oxidation", or "autothermal reforming" processes are described in [131–135]. A process where process air is preheated in a regenerative system to such high temperatures that stoichiometric ammonia synthesis gas can be produced from natural gas and air is described in [892]. An experimental study of catalytic partial oxidation is reported in [136]. Also non-catalytic partial oxidation of light hydrocarbons has been used for production of ammonia synthesis gas [137, 138].

Steam reforming was first used for the production of ammonia synthesis gas in the early 1930s [71, 72]. During the Second World War the process became dominant in the US ammonia industry, when natural gas became the preferred feedstock. In early plants using steam reforming, the operating pressure was low [71, 139–141]. As developments in metallurgy allowed, the pressure was gradually increased up to the present levels of about 35 bar [142–145].

6.3.2 Partial Oxidation and Gasification

Non-catalytic partial oxidation or gasification of hydrocarbons or solid feedstocks are alternatives to steam reforming and autothermal reforming for production of synthesis gas.

In contrast to the reforming processes, which accept only light hydrocarbons (up to and including naphtha) as feedstock, the partial oxidation and gasification processes are very flexible with respect to feedstock. These processes have in fact been used on all types of raw materials including natural gas, liquid hydrocarbons including naphtha and heavy oil, and coal and lignite. It is normal to refer to processes treating gaseous or liquid hydrocarbons as "*partial oxidation processes*", whereas processing of solid materials is called "*gasification processes*".

6.3.2.1 Partial Oxidation

In partial oxidation processes, a hydrocarbon feedstock is reacted with oxygen or an oxygen-containing gas at high temperature and normally high pressure to produce a gas containing hydrogen, carbon oxides, some steam, and traces of methane and of impurities such as hydrogen sulfide. The process is not sensitive to sulfur as is the reforming process, thus it is not normal practice to desulfurize the feedstock.

Reviews of the technology and of available processes are given in [146–150]. The reactions taking place may be represented by the following simplified equations:

$$C_nH_m + nH_2O \rightarrow nCO + (m/2 + n)H_2(-\Delta H^0_{298} < 0) \quad (1)$$

$$CH_4 + H_2O \rightleftarrows CO + 3H_2(-\Delta H^0_{298} = 206\ kJ/mol) \quad (2)$$

$$CO + H_2O \rightleftarrows CO_2 + H_2(-\Delta H^0_{298} = 41\ kJ/mol) \quad (3)$$

$$C_nH_m + n/2\,O_2 \rightarrow nCO + m/2\,H_2(-\Delta H^0_{298} > 0) \quad (4)$$

$$C_nH_m \rightarrow nC + m/2H_2(-\Delta H^0_{298}) \quad (5)$$

The hydrocarbon feedstock reacts with oxygen and steam according to reactions (1) and (4), and the methanation reaction (2) and the shift reaction (3) proceed to near equilibrium. It is important to avoid excessive soot formation shown by equation (5).

In general, it has been found that soot formation can be reasonably suppressed – but never completely eliminated even with light feedstocks such as natural gas – if the oxygen/carbon atomic ratio is kept above 1, and if the reaction temperature is above about 1300 °C [146]. In early plants the reaction pressure was low, but in more recent designs, pressures up to 80 bar have been used.

Two partial oxidation processes dominate the market completely, the Texaco process and the Shell process. Reaction conditions are very similar in the two cases, and the main difference is in the method of gas cooling and in the technology used for soot removal. In both cases, preheated feedstock, oxygen, and a small amount of steam are reacted in a refractory lined vessel at temperatures up to 1500 °C and at a pressure which in the Shell process typically is 60 bar and in the Texaco process up to 80 bar. The raw synthesis gas – essentially a 1 : 1 mixture of hydrogen and carbon monoxide (when the feedstock is heavy oil) with small amounts of carbon dioxide and steam and traces of methane, hydrogen sulfide, and other impurities – is cooled and treated for soot removal. The Texaco process uses quench cooling or combined cooling in a waste heat boiler and by quench, and the soot removal system consists of a Venturi scrubber followed by washing with water. The soot is recovered as a slurry in water, which is recycled to the reactor. In the Shell process the gas is cooled in a waste heat boiler, and the soot is removed by washing with light oil. The soot is recovered in the form of carbon pellets. The treated gas from the Texaco process is at about 200 °C and saturated with water, whereas the treated gas from the Shell process is cold and essentially water free. This difference is important for the layout of downstream processes. The two processes are further described in (151–158), (893). The economics of steam reforming and partial oxidation are compared in (159–161). Experience from construction and operation of plants based on partial oxidation of heavy fuel oil is reported in (162–169), (893).

6.3.2.2 Gasification

In certain situations coal and lignite are attractive as feedstocks for the production of ammonia, and plants using these raw materials have been constructed in many parts of the world. It should also be noted that coke was the most important feed- stock in the early days of the ammonia industry in plants using the water gas process (see Sect. 6.3.6). Natural gas or heavier hydrocarbons replaced coke during and after the Second World War except in special circumstances. During the oil crises of the 1970s, coal appeared to be the future feedstock for ammonia production, and many studies and much development work were carried out on processes for the production of ammonia from coal. However, later developments in oil prices made coal less competitive as a feedstock, and the expected large scale change in the industry from natural gas and heavier hydrocarbons to coal never did materialize.

Technology for the gasification of coal or lignite is reviewed in (170–183). The classical processes are two low pressure processes, the Koppers-Totzek process and the Winkler process, and two high pressure processes, the Lurgi process and the Texaco process. All these processes are used in industrial plants. A significant number of other processes have been developed to the stage where they have been proven in large demonstration units or in plants for other purposes than the production of ammonia synthesis gas.

The Koppers-Torzek process [184–188] is used in several ammonia plants today. The process operates at low pressure, and the reaction between pulverized coal and oxygen mixed with steam is carried out in an entrained bed with a residence time of less than 1 second and at a temperature of 1500–1600 °C. Ammonia plants based on this technology are described in [189–193]. A high pressure version of the process has been developed, initially in cooperation between Shell and Koppers [194–196]. Later, this process was further developed and demonstrated separately by Shell [197, 198, 894] and in a cooperation between Krupp and Koppers as the Prenflo process [199–201]. The classical Winkler process [202] is a fluid bed process operating at low pressure and at somewhat lower temperature than the Koppers-Totzek process, around 1000 °C. A high pressure Winkler process for gasification of lignite has been developed and commercialized by Rheinbraun [203, 204]. The Lurgi gasifier [205, 206] is a moving bed process. It operates at about 30 bar. Coal in the form of 5–40 mm particles is fed through a lock hopper to the top of the gasifier. Oxygen and steam is fed to the bottom, and the gasification takes place in the lower part of the gasifier. The hot gases pass countercurrently through the coal which moves slowly down through the gasifier. The hot gases degasify the coal and are at the same time cooled to an outlet temperature of about 600 °C. The gas contains impurities including about 10% methane and significant amounts of naphtha, tar etc. [207]. This high content of impurities is a major disadvantage for the Lurgi process in the production of ammonia synthesis gas. The classical Lurgi process operates as a dry ash gasifier. A further development, the British Gas-Lurgi process operates as a slagging gasifier (208).

All the above-mentioned processes use a dry coal feed. In contrast, the feed to the Texaco process [209–212] is a slurry of coal in water. The process is a further development of the Texaco process for partial oxidation of hydrocarbons. The process is an entrained bed process operating at up to 40 bar and at a temperature of about 1400 °C. The raw synthesis gas may be cooled either by direct quench or by steam production in a waste heat boiler followed by quench cooling. A large, modern ammonia plant based on gasification of coal by the Texaco process is described in [213, 214]. A widely-published commercial scale demonstration project on the retrofit of an existing natural gas-based ammonia plant to coal feed also used the Texaco process [215–219]. The Dow process is also based on slurry feed. This process has, however, until now been used only for gasification in Integrated Gasification Combined Cycle (IGCC) power plants. A review of gasification processes especially for such plants is given in [220].

6.3.3 Carbon Monoxide Conversion

The product gas from steam reforming, catalytic partial oxidation, partial oxidation, or gasification processes contains in all cases significant amounts of carbon monoxide and carbon dioxide. Since the feed gas for the ammonia synthesis loop must be completely free of these compounds, they must be removed in the gas preparation part of the plant. Carbon monoxide is removed in a conversion step by the so-called "water-gas shift" reaction or just the "shift" reaction, in some cases followed by further conversion by selective oxidation.

6.3.3.1 Shift Conversion

In the shift conversion process, carbon monoxide is converted to carbon dioxide with the formation of additional hydrogen by the following reaction:

$$CO + H_2O \rightarrow CO_2 + H_2 \, (-\Delta H^0_{298} = 41 \, \text{kJmol}) \quad (3)$$

Several catalysts and processes are available for this reaction. In the classical process, the catalyst is magnetite, Fe_3O_4, promoted with chromia and in some cases with potassium or other promoters [221–223]. This catalyst requires temperatures above about 350 °C, and the exit carbon monoxide concentration is normally about 3 vol% in the dry gas. Conversion was in some cases improved by installing a two stage unit with carbon dioxide removal between two high temperature shift reactors [224]. However, this system was rather expensive and did not gain wide acceptance.

In the early 1960s, copper-based catalysts became available for a second conversion step at temperatures about 200 °C, whereby the carbon monoxide concentration can be reduced to 0.1–0.3 vol% in the dry product gas [224–227]. This represented a major improvement in process economics, and today virtually all ammonia plants use the two step carbon monoxide conversion, or shift conversion, consisting of a high temperature and a low temperature step. Comprehensive discussions of shift conversion may be found in [44, 228]. Shift catalysts are described in [229–234, 896–898]. The formation of methanol as a by-product in low temperature shift reactors is discussed in [899]. The low temperature shift catalysts are sensitive to poisoning, especially by chlorine and sulfur. Poisoning problems are discussed in [235–240]. One way of protecting the catalyst from poisoning is to install a guard vessel upstream the main reactor [241]. The guard vessel absorbs the poisons, and it is designed for easy catalyst change. In this way the life of the catalyst in the main reactor is significantly extended.

In some cases gas with a high sulfur content coming from partial oxidation or gasification units is passed directly to shift conversion. The high sulfur content can be tolerated by magnetic type catalysts, although the activity is reduced [221]. But more efficient sulfur tolerant catalysts based on cobalt-molybdenum sulfide have become available [169, 242–245]. With such catalysts low temperature conversion can also be obtained in sulfur containing gases [244, 245].

In modern ammonia plants based on steam reforming, the tendency is to reduce the steam/carbon ratio in the reformer and thereby the steam content in the feed gas to the shift converter. At low steam contents, problems may occur in units using the classical magnetite catalyst because this catalyst is partly converted into iron carbide which promotes by-product formation [246–248, 896]. The by-products are hydrocarbons, mainly methane and lower olefins, and oxygen-containing organic compounds such as alcohols and acids. They do not only represent a loss of production, but also have a detrimental effect on the low temperature shift catalyst [249]. It has been found that adding a small amount of copper to the iron-based high temperature shift catalyst overcomes this problem, at least in some situations [250, 251]. At more extreme conditions, copper-based medium or high temperature catalysts without carbide-forming elements must be used [246, 251].

6.3.3.2 Selective Oxidation

In order to further reduce the carbon monoxide concentration in the synthesis gas, a selective oxidation unit has in some cases been installed after the low temperature shift unit. After the gas has been cooled and condensate removed, a small amount of air is introduced, and the gas is passed to a reactor containing a precious metal catalyst. Carbon monoxide reacts with oxygen, and excess oxygen is removed by reaction with hydrogen. The process is known as the "Selectoxo" process [252–254]. Use of the process in a revamp project is described in [255].

6.3.4 Acid Gas Removal

In plants based on steam reforming of light hydrocarbons, the product gas from the shift conversion contains about 18 vol% (dry basis) carbon dioxide. Gas originating from partial oxidation of heavy hydrocarbons or from gasification of solid feedstocks contains even higher concentrations of carbon dioxide, and in addition hydrogen sulfide may also be present in significant quantities depending on the sulfur content in the feed and on the type of shift conversion technology employed. In plants using standard shift systems, sulfur is most often removed from the raw gas upstream of the shift section in selective sulfur removal units, whereas the sulfur is left in the gas in plants that use special sulfur tolerant shift catalysts. In every case all sulfur compounds and all carbon oxides must be removed from the gas before it can be used for ammonia synthesis. Essentially all sulfur compounds (if present) and the bulk of the carbon dioxide are removed in so-called acid gas removal units or carbon dioxide removal units, which in all cases feature absorption of the acid gases in a suitable solvent.

The absorption, which may be physical or chemical in nature, takes place at synthesis gas pressure in an absorption vessel with trays or packing, and the rich solution is regenerated in another vessel at reduced pressure by stripping and/or heating (reboiling). Depending on the properties of the solvent and the operating conditions, part of the dissolved carbon dioxide may be released from the solvent by pressure release (flashing).

The desired carbon dioxide concentration in the treated gas depends on the downstream process used for removal of the remaining part (see Sect 6.3.6). If cryogenic gas purification is used, very low concentrations are required (most often the last traces are removed by adsorption on molecular sieves). If methanation is used for gas purification, levels of 500–1000 vol ppm carbon dioxide are typical. In old-fashioned plants using copper liquor wash for carbon monoxide removal, carbon dioxide concentration out of the absorption vessel could typically be up to 0.5 vol%.

A large number of acid gas removal processes have been developed; some are exclusively, or at least predominantly, used for natural gas sweeting, i.e. removal of hydrogen sulfide and possibly carbon dioxide from natural gas. Surveys of available acid gas removal processes may be found in [256–264, 953]. A theoretical review is given [265], and the choice between various processes is discussed in [266–272].

The solvents most used in carbon dioxide removal from ammonia synthesis gas can be characterized according to the nature of the absorption process. Chemical absorption, i.e. processes where the carbon dioxide reacts with the solvent by a chemical reaction which is reversed in the solvent regeneration stage, is most often based on the use of alkanolamines, mainly MEA (monoethanolamine) [273], or hot solutions of potassium carbonate [274] as solvents.

A series of processes based on the use of MEA with an added corrosion inhibitor called Amine Guard were developed by Union Carbide Corp. and have been extensively used under the name Amine Guard II, III, and IV. The heat requirement for solvent regeneration has been gradually decreased in this series of processes through optimization of process layout and operating conditions [275–279]. The effect of the corrosion inhibitor is to allow an increase of the MEA concentration in the solvent, thereby increasing the capacity for carbon dioxide and reducing the recirculation rate and thus the energy consumption in the process.

Several processes are based on the use of a hot solution of potassium carbonate as a chemical solvent. The processes differ mainly in the type of activator used to increase the reaction rate between carbon dioxide and the solvent. The Carbosolvant process [280, 281] uses Sulfosolvan B; the Giammarco-Vetrocoke process [282–284, 950] used initially arsenic oxide, in a later development glycine as activator, and latest a dual activator system. The Catacarb process [285] uses an amineborate activator, whereas the Benfield process [286–289] and the Carsol process [290, 291] use enthanolamines as activators. All of these processes have been improved through the years to minimize energy consumption and to reduce the risk of corrosion. Some case stories describing operational problems in hot potassium carbonate systems are reviewed in [901]. New activators have also been developed which may improve the performance of existing installations [902, 903, 950].

One of the first – if not *the* first – physical solvent to be used for carbon dioxide removal was water [292]. Other physical solvents are sulfolane (tetrahydrothiophene 1.1 dioxide) + diisopropanol in the Sulfinol process [293]; methanol in the Rectisol process [294–297]; polypropylene carbonate in the Fluor process [298–301]; methylpyrrolidone in the Purisol process [295, 302, 303]; polyethyleneglycol dimethylether in the Selexol process [304–310, 904]; and polyethyleneglycol methylisopropyl ether in the Sepasolv MPE process [311]. In the physical solvent processes no heat is required for solvent regeneration, which is done by pressure release and stripping with air or inert gas. Because of the physical nature of the solvents, a low temperature is advantageous. The Rectisol process is operated as low as $-40\,°C$, often in combination with a liquid nitrogen wash for final purification in plants based on partial oxidation or gasification [296].

Some processes combine the use of chemical and physical solvents. An example is a dual solvent process based on the use of monoethanolamine and triethanolamine (MEA/TEA) [312–314]. In other cases the process is based on the use of a solvent which exhibits at the same time some of the properties attributed to both physical and chemical solvents. This is notably true for activated methyldiethanolamine, which is used in the MDEA process developed by BASF [311, 314–320, 905, 906]. This solvent, operating at an intermediate temperature, releases a significant part of the dissolved carbon dioxide by flashing, thus reducing the energy required for reboiling and thereby the overall energy consumption.

6.3.5 Final Purification

After bulk removal of carbon monoxide in the shift conversion section and of carbon dioxide in the carbon dioxide removal section, the synthesis gas still contains typically 0.2–0.5 vol% carbon monoxide and 0.01–0.2 vol% carbon dioxide. These compounds must, together with any water present, be removed quantitatively, i.e. to low ppm levels, before the gas can be admitted to the synthesis converter, because all oxygen containing compounds are poisons to the ammonia synthesis catalysts [238]. The most important technologies for this final purification are discussed in the following.

6.3.5.1 Copper Liquor Wash

The oldest process for final purification of ammonia synthesis gas is the copper liquor wash, which was used in the world's first ammonia plant in Oppau, Germany in 1913 [321]. Descriptions of the process are given in [321–323]. The process removes carbon monoxide and carbon dioxide to the required low levels by absorption in a solution containing cuprammonium salts of acetic, formic, or carbonic acid. The chemistry involved is rather complex; both the Cu^{+}/Cu^{++}-ratio, the pH, the ammonia concentration, the total concentration in water, the temperature and pressure etc. must be controlled to obtain proper absorption and to avoid precipitation of copper salts. Reference is made to [321, 323]. The loaded solution is regenerated by pressure release and heating, and the composition is adjusted by necessary addition of ammonia, adjustment of the Cu^{+}/Cu^{++}-ratio (by addition of air if required) etc., before the solution is recycled to the absorber.

The copper liquor washing process has the advantage that the carbon oxides are removed by absorption rather than by a hydrogen-consuming reaction as in the more commonly used methanation reaction. It is, however, difficult to operate; maintenance coats are high; it has a high energy consumption; it tends to pollute the environment; and it therefore has largely been replaced by other purification processes. It is not used in any modern ammonia plants.

6.3.5.2 Cryogenic Purification

Cryogenic purification of synthesis gas for ammonia production has mainly been applied in plants based on partial oxidation of heavy oil or gasification of coal (see Sect 6.3.2). A special process unit, the so-called "Purifier", is used for final purification in the C. F. Braun ammonia process [324, 325, 954]. The purifier is installed downstream of carbon dioxide removal and methanation. In the unit, the synthesis gas is partly condensed to remove excess nitrogen which has been introduced in the secondary reformer. All water and carbon dioxide must be removed in an absorption unit upstream of the purifier to prevent freezing, and, in addition, part of the inerts (methane, argon) and removed together with the excess nitrogen in the purifier. The resulting synthesis gas is completely dry and has a low inerts content, so that energy consumption in the synthesis loop is reduced [326–331]. Operating experience from an industrial plant is reported in [332, 333].

In plants based on synthesis gas production by partial oxidation or gasification using oxygen, the final purification is often done in a "nitrogen wash" by scrubbing the gas with liquid nitrogen. In many cases the nitrogen wash is integrated with a Rectisol carbon dioxide removal unit [296] or with other carbon dioxide removal processes operating at low temperature. As in the purifier process mentioned above, it is necessary in a nitrogen was to remove all traces of carbon dioxide, water, and other impurities which could possibly cause fouling of heat exchanger surfaces by freezing. Unscheduled defrosting of a nitrogen wash requires a long down time and is therefore very undersirable. Another concern is that the gas may, at the same time, contain olefins and traces of nitrogen oxides. If these deposit together in the heat exchangers, explosions may occur [332, 333]. The necessary pretreatment of the gas is normally done by absorption of the impurities on molecular sieves and activated carbon. Investigations on the risk imposed by the simultaneous presence of nitric oxides and olefins in cryogenic units are reported in [334].

After pretreatment the gas is cooled by heat exchange with the cold product gas. Additional refrigeration is obtained by expansion of the gas and the nitrogen, or by addition of liquid nitrogen. The gas is cooled to about − 190 °C at which temperature partial condensation of carbon monoxide and methane occurs. The gas is contacted countercurrently in a washing column with liquid nitrogen. The gas leaving the top of the washing column is hydrogen containing about 10 vol% nitrogen and only traces (about 100 ppm) of argon and methane.

After addition of nitrogen to adjust the hydrogen to nitrogen ratio to 3.0 this is an excellent feed gas to a so-called "inert free" synthesis loop which may often be operated without withdrawal of a purge gas stream. The liquid nitrogen leaving the bottom of the washing column contains essentially all the carbon monoxide and methane present in the original feed gas. These compounds can be used as fuel or returned to the feed gas preparation.

The nitrogen required for the nitrogen wash is normally produced in the air separation plant which produces oxygen for the gasification or partial oxidation process. High purity nitrogen is required with less than 10 ppm oxygen. Descriptions of the nitrogen wash technology may be found in [335–340].

6.3.5.3 Pressure Swing Absorption

The Pressure Swing Absorption (PSA) technology is most often used for hydrogen purification in plants based on steam reforming or for hydrogen recovery from off gases. It has, however, also been proposed to use the PSA technology for final purification of synthesis gas in certain process concepts for the production of ammonia. In most of these cases it replaces both carbon dioxide removal unit and final purification. The schemes may be based on production of pure hydrogen which is then mixed with nitrogen produced in a separate unit [341–344], or a special version of a PSA unit may be used to add nitrogen in the unit itself, thereby improving the hydrogen recovery [345–349]. In the ICI LCA process [350–357] a special PSA unit is used for carbon dioxide removal and partial purification including removal of excess nitrogen upstream of a methanation unit. The PSA technology as such is described in [358–361].

6.3.5.4 Methanation

By far the most important process for removal of the last traces of carbon monoxide and carbon dioxide from ammonia synthesis gas in methanation. In this process, carbon oxides are reacted to extinction (less than 10 ppm) at 250–350 °C over a nickel-containing catalyst according to the following exothermic reactions:

$$CO + 3H_2 \rightarrow CH_4 + H_2O \; (-\Delta H^0_{298} = 206 \text{ kJ/mol}) \qquad (6)$$

$$CO_2 + 4H_2 \rightarrow CH_4 + H_2O \; (-\Delta H^0_{298} = 165 \text{ kJ/mol}) \qquad (7)$$

The process is simple, reliable and inexpensive both in investment and in operating cost. It causes a certain loss of hydrogen due to the reactions (6) and (7) and due to the increased content of methane in the synthesis gas, which causes an increased loss with purge gas from the synthesis loop. The effect on overall energy consumption is, however, at normal concentrations of carbon oxides (0.2–0.5 vol% carbon monoxide and 0.01–0.2 vol% carbon dioxide) rather limited, and the process is used in virtually all modern ammonia plants based on steam reforming of natural gas or other light hydrocarbons. For references on methanation, see [362–364].

The gas from a methanation unit will contain water, possibly from the carbon dioxide removal section and at least as formed by reaction (6) and (7). This water must be removed before the gas reaches the ammonia synthesis catalyst. The dehydration can be done either by adsorption on molecular sieves [365], whereafter the gas can be added to the synthesis loop just upstream of the synthesis converter, or by co-condensation and washing with ammonia. This last method is used in many ammonia plants, where the synthesis gas is added to the synthesis loop before the product ammonia is separated, see Sect 6.4.1.

6.3.5.5 Methanolation

Methanolation [907, 908] may be considered an alternative to methanation. The residual carbon oxides are converted in the methanolation process to methanol, preferably at high pressure, i.e. after the synthesis gas compressor. The methanol formed is removed by washing with water and may be returned to the feed stream to the reformer or separated as a product. Full conversion of carbon oxides is not possible, and a clean up methanation unit must be installed after the methanolation unit.

6.3.6 Synthesis Gas from Unconventional Sources

By far the largest part of the world's ammonia production is made from synthesis gas which is produced by one of the processes described above, i.e. by steam reforming of light hydrocarbons, by partial oxidation of hydrocarbons, or by gasification of solid feedstocks, in all cases followed by proper conversion, purification, and adjustment of composition.

There are, however, also other sources of synthesis gas, although their importance is limited.

In the early days of the ammonia industry the predominant feedstock was coke. Synthesis gas was produced in so-called water gas units, which were cyclic units. Coke was heated by partial combustion with air, whereby a mixture of essentially nitrogen and carbon dioxide, so-called generator gas, was produced. The hot coke was thereafter contacted with steam whereby a mixture of hydrogen and carbon oxides, so-called water gas, was produced. The two gas streams were purified and mixed, and carbon monoxide was catalytically converted to carbon dioxide in a high temperature shift unit operating at low pressure. The converted gas was compressed to, say, 25 bar, carbon dioxide was removed by washing with water, and the product gas was further compressed to, say, 200 bar, whereafter residual carbon oxides were removed in a copper liquor wash before the gas was sent to the synthesis unit. The process, which was used in the first ammonia plants in Oppau and Leuna, is described in [50, 366]. An alternative process for synthesis gas preparation, which was used mainly in Europe and Asia, was purification of coke oven gas [50, 367]. In this process, coke oven gas was first purified by the removal of tar, benzene, ammonia, sulfur-compounds etc., after which hydrogen was recovered in a cryogenic unit. Nitrogen was obtained from an air separation plant, and the pure gases were mixed and sent to synthesis.

There is still a limited production of ammonia based on coke oven gas, although its importance is dwindling. The prospects for the future are reviewed in [368, 369].

Hydrogen for ammonia production may also be produced by hydrolysis [370–374] or it may be obtained from refinery off gas [375]. Plants have been constructed from such feedstocks in the past [376, 377] and it may still be economical in special circumstances. In modern times plants have been constructed on the basis of off gases from petrochemical plants such as ethylene plants and methanol plants [378, 379], but such cases will most likely remain rare and be economical only in special circumstances. Production of ammonia from gases produced in steel plants is suggested in [380–383].

6.3.7 Synthesis Gas Compression

Whatever the production process for synthesis gas has been, the gas must be compressed before it is added to the ammonia synthesis loop. Up to around 1950, synthesis gas was generally produced at atmospheric pressure. The pressure has since then been gradually increased, and today the synthesis gas is normally available for compression at 25–30 bar in plants using steam reforming and at even higher pressure, up to about 70 bar [166] in plants based on partial oxidation of heavy oil. Synthesis pressure was typically 200–400 bar in early ammonia plants with extremes of 1000 bar in the original Claude process [384] and around 80 bar in the Mont Cenis process (385). Capacities were low compared to today's standard – increasing through the years to a maximum of 300–400 MTPD around 1960 – in part because of capacity limitations in the reciprocating compressors used in the early plants. In the 1960s the capacity of ammonia plants increased to about 1000 MTPD (metric tons per day) when centrifugal compressors became available; synthesis pressures were around 150 bar in the first ammonia plants due to limitations in pressure available from the centrifugal compressors.

The technology has since developed further, and today plants with a capacity of 1500 MTPD and a synthesis pressure of 220 bar are operating. In fact the availability of compressors is no longer imposing any limit either on capacity or on synthesis pressure in ammonia plants.

As indicated above, two types of compressors are important for the compression of ammonia synthesis gas. In old plants – and in some small modern plants – reciprocating compressors dominate, whereas centrifugal compressors are used exclusively in all large modern plants. Reciprocating compressors are described in [386, 387] and centrifugal compressors in [388–397]. Extensive discussions on the use of centrifugal compressors are given in [392–397]. Reciprocating compressors are generally driven by synchronous electric motors or in a few cases by steam turbines with special speed reduction gears. They are more expensive than centrifugal compressors, require more maintenance, and it is normal practice to install two machines in parallel so that excessive down time caused by the compressor is avoided. They are lubricated so that there is a risk for contamination of the synthesis loop with lubricating oil, they require much space and heavy foundations for installation; and their capacity limitation makes multitrain installations necessary above a certain plant size. On the other hand they are highly efficient, and the efficiency is rather independent of pressure and load.

Centrifugal compressors can handle very large gas volumes; they are reliable so that only one machine is normally installed even in very large ammonia plants; and they have the advantage of being oil free. There is a practical limit to the minimum clearance between an impeller and the stator and therefore to the minimum flow (in actual volume) which can be handled by a centrifugal compressor. Technological developments are constantly pushing this limit, but there will always be a certain capacity limit below which centrifugal compressors can not be used. Present technology (early 1990s) probably allows a flow corresponding to around 350 actual cubic meters per hour at the compressor outlet to be economically compressed in a centrifugal compressor at ammonia plant conditions. This corresponds to an upper limit in synthesis pressure of around 100 bar in a plant producing 300 MTPD ammonia.

Centrifugal compressors operate with very high rotational speed and are most often driven directly by steam turbines, although motor drivers are also used in some cases, especially for relatively small capacities. The centrifugal compressors are inherently less efficient than reciprocating compressors, although technology development may be closing the gap.

Direct steam turbine driver also means that generator loss, distribution loss, and motor and gear loss are avoided, so that the overall efficiency of a steam turbine driven centrifugal compressor may well compare favourably with the efficiency of a motor driven reciprocating compressor, at least in the upper end of the capacity ranges. Motor drives and steam turbine drivers for the large power consumers in ammonia plants are compared in [398].

Gas turbines have also been considered as drivers for compressors in ammonia plants. General discussions are given in [399, 400]. The exhaust gas from the gas turbine can be used for steam production, for preheat duties, or as combustion air in the primary reformer [401, 402]. Experience from an actual installation is reported in [909].

6.4 Synthesis of Ammonia

The formation of ammonia from hydrogen and nitrogen

$$N_2 + 3\,H_2 \rightarrow 2\,NH_3 (-\Delta H^0_{298} = 92.44\ \text{kJ/mol}) \quad (8)$$

is an exothermic reaction, which under all practical conditions is limited by equilibrium in such a way that only partial conversion can take place during the passage through a synthesis reactor. Furthermore, the product ammonia is in all practical cases separated from the unreacted gas by cooling to condense and separate the liquid ammonia from the gas; at realistic pressures relatively low temperatures are required to recover the product with reasonable efficiency. These fundamental properties of the reaction and the product, together with the properties of the synthesis catalyst and mechanical restrictions in equipment design, determine the design of the synthesis unit and the synthesis reactor.

6.4.1 The Ammonia Synthesis Loop

The synthesis of ammonia takes place in a recycle loop which always contains the following elements:

- *A reactor system* comprising one or more catalytic synthesis reactors with associated temperature control and heat recovery equipment. The design of the reactors and of reactor systems is discussed in Sect 6.4.2 below. The catalysts employed are discussed in Chapter 2. Thermodynamics of the synthesis reaction are dealt with in Chapter 1. Reaction kinetics are discussed in Chapter 4.
- Unit(s) for cooling the product gas to recover heat and to condense the product ammonia. The layout of these units, which are integrated with other parts of the synthesis loop, is dealt with in the discussion of individual processes in Sect 6.4.6.
- Unit(s) for the separation of product ammonia from unreacted gas and for the adjustment of product properties (degassing, adjustment of temperature and pressure). The gas/liquid equilibrium in the separation stage is dealt with in Chapter 1.
- Units for preheating the reactor feed. These units are most often integrated with the cooling/condensing units mentioned above.

- Equipment for the addition of make-up gas and, if the make-up gas contains inerts, for removal of purge gas from the recycling gas to prevent the build up of inerts in the loop. The addition of make-up gas and removal of purge gas may be done at various points in the recycle loop as discussed in Sect. 6.4.1.1 below.
- Equipment for the recirculation of non converted gas and make-up gas to the synthesis reactor. The recirculation equipment is in many cases integrated with equipment for compression of make-up gas to synthesis pressure as discussed in Sect 6.4.1.1 below.

6.4.1.1 Types of Ammonia Synthesis Loops

It is possible to classify ammonia synthesis loops in various ways. One viewpoint [23] is that "since the only proprietary data in commercial ammonia synthesis plants are the catalyst formulation and the design of the synthesis converter, which uses the catalyst – the rest of the process design is simply chemical engineering for the particular conditions chosen for the plant" then ammonia synthesis processes should be classified according to converter design. It is, however, also possible to classify the synthesis loops according to the arrangement of the sequence of process steps and the purge and make-up gas points [403–405]. The most important types of synthesis loops according to this classification are shown in Fig. 6.2A–E.

Fig. 6.2 A shows a loop with both recirculator and make-up gas addition point after the ammonia separator. This layout is from several points of view the most advantageous layout. Ammonia condensation and separation are done before the converter exit gas is diluted with fresh make-up gas, and consequently at the highest possible partial pressure of ammonia. Purge gas may be taken from the point in the synthesis loop where the ammonia concentration is lowest and the concentration of inert components is highest. The recycle gas from the separator is diluted with the fresh make-up gas, so that the lowest possible ammonia concentration is obtained at converter inlet. Also the volume of gas which must be recompressed in the recirculator is the lowest possible, since product ammonia has already been separated.

The disadvantage of this layout is that all impurities present in the synthesis gas or introduced by the make-up and recycle compressors are passed directly to the synthesis converter. Therefore, this layout is used mainly in cases where catalyst poisons are removed very efficiently from the synthesis gas, and where non-lubricated compressors such as centrifugal compressors are used for make-up gas compression and recirculation. This loop layout has been used in many commercial processes. Some early examples are the Fauser process [406], the Fauser Montecatini process [407] and the Casale process [408] (both with ejector recycle), and the modified Claude process [384] (in the original Claude process a series of converters with intermediate cooling and product recovery was used; this is the only process which has ever been used commercially without

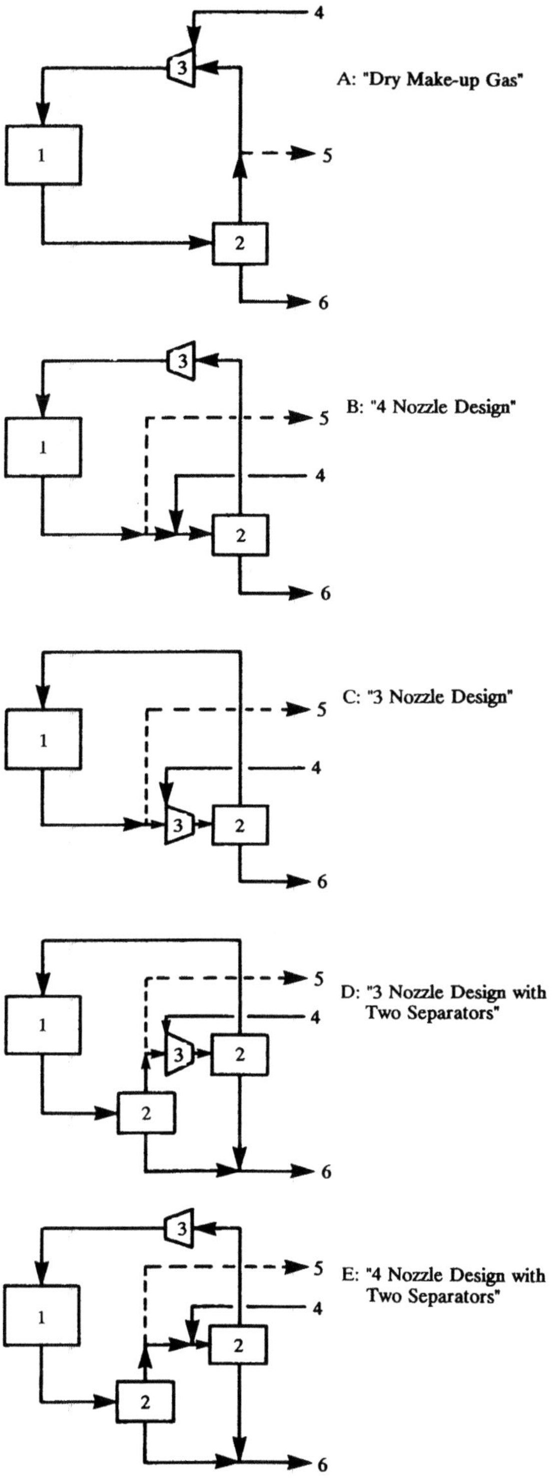

Fig. 6.2. Important types of ammonia synthesis loops (*1*: Synthesis Converter, *2*: Separator, *3*: recirculator; *4*: make-up gas, *5*: purge gas (if present), *6* ammonia)

recycle of unconverted gas). In modern processes the concept with the addition of make-up gas after the separator has been used or proposed in a number of processes where the synthesis gas is especially purified before addition to the synthesis loop. Examples are the MDF process by Humphries and Glasgow [341–343], the Linde process for small ammonia plants [344] the PARC process by KTI [345–349], and other processes using pressure swing absorption for purification of synthesis gas. Additional examples are the C. F. Braun process [326–331, 409–412], the ICI AMV process [412–418] and other processes using cryogenic units for gas purification, and the new M. W. Kellogg process [419], the LEAD process by Humphries and Glasgow (420), and other processes where the gas is dried in molecular sieve units before being added to the loop. Descriptions of all these processes are given in Sect 6.5.3.2.3.

In cases where the synthesis gas is not especially purified, and where it may therefore contain traces of water vapour and carbon dioxide, the make-up gas is added before the separator. In this way the make-up gas is in contact with liquid ammonia, which absorbes the traces of water vapour and carbon dioxide so that these poisons do not reach the synthesis converter.

This latter layout means that the ammonia is no longer condensed and separated at the highest possible concentration, and it further means that the inlet concentration of ammonia to the synthesis converter is equal to the outlet concentration from the separator, since the gas after the separator is not diluted with fresh synthesis gas. As a consequence, there is a certain increase in the energy consumption in the loop because of a slightly increased recirculation rate and/or a certain increase in the energy required for condensation of ammonia. Synthesis loops with make-up gas addition upstream of the separator are often referred to as the "3 nozzle design" and the "4 nozzle design" depending on the location of the recirculator [404, 405, 421, 422]. These designations refer to the layout of the compressor section. In the "4 nozzle design" illustrated in Fig. 6.2B, the circulator is located after the separator. This means that make-up gas and recycle gas cannot be mixed in the compressor, so that 4 nozzles are required, inlet and outlet for the make-up gas and inlet and outlet for the recycle gas. In the "3 nozzle design" illustrated in Fig. 6.2 C, the recycle compressor is located before the separator. In this way the make-up gas can be mixed with the recycle gas in the compressor, and therefore only 3 nozzles are required, inlet for make-up gas and recycle gas and outlet for the combined stream.

The "3 nozzle design" has a disadvantage compared to the "4 nozzle design" in that the volume of gas passing through the recycle compressor is increased because the product ammonia also has to be compressed here. The difference between the energy consumption in the two concepts is discussed in [404, 405].

Arrangements corresponding to the "4 nozzle design" have been used in many commercial processes including the Grande Paroisse process [423] and processes by Uhde [424, 425] and Topsoe [426, 427]. The "3-nozzle design" has e.g. been used in the Casale process (with recycle done by an ejector) [384, 408, 428] and notably by M.W. Kellogg [429–432].

In loops operating at high pressure (above about 250 bar) it has been found advantageous to install two ammonia separators (404), (405). Fig. 6.2D and 6.2E show two ways of doing this. Fig. 6.2 D is equivalent to the "3 Nozzle Design" shown in Fig. 6.2 C, whereas Fig. 6.2 E represents the corresponding "4 Nozzle Design". The characteristics of the designs with one separator are maintained in each case, and in addition energy is saved, because part of the product ammonia bypasses the last, energy consuming refrigeration stage. The "3 nozzle design with 2 separators" (Fig. 6.2 D) has been used by M. W. Kellogg [407, 433], Chemico [434], BASF [435], Fauser-Montecatini [436], Pritchard [437]. Topsøe [438], and Esso [439]. The "4 nozzle design with 2 separators" (Fig. 6.2 E) has been used in the Haber-Bosch-Mittach process [440] and the equivalent NEC-process [441], by Uhde [442, 443], and by ICI [444].

In a few cases, loop designs have been suggested which cannot be described by any of the types represented by Fig. 6.2 A–E. This is notably so for certain complicated loops with double conversion and double separation, e.g., in late versions of the Claude process [445–447]. A similar concept has been proposed by PDIL [448]. It has also been proposed [449, 450] to place two synthesis loops in series to improve overall efficiency, and to install a "preconverter" – i.e. a synthesis converter in the make-up gas stream – in order to improve conversion and to protect the synthesis loop against poisons [451]. Pre-conversion has also been used in revamp of existing synthesis loops to increase capacity [452].

6.4.1.2 Mathematical Models

The conditions in the ammonia synthesis loop can be described by a number of parameters, which may each be in some cases independent variables, and in other cases a function of other parameters. Important parameters are:

- Operating pressure
- Separator temperature
- Recycle ratio (ratio between recycle flow and make-up gas flow)
- Purge ratio (ratio between recycle flow and purge gas flow)
- Conversion per pass (fractional conversion of synthesis gas to ammonia by passage through the reactor system).

The relationship between these parameters (and other parameters such as space velocity, inert level, concentrations and temperatures at various points in the synthesis loop, etc.) may be described in mathematical models which can be used for design, simulation, and optimization. The models must contain procedures to calculate the performance of each of the elements in the synthesis loop (at least synthesis reactor and separator, and for complete models also the compressor and recirculator, the heat exchangers, etc.) as well as procedures to describe the sequence of operations and the interaction between them. Descriptions of loop models are given in [453–458]; complete mass - and energy balances are given in [453–457]; optimization studies using computer models are reported in [454–457]. Simulation and optimization of complete ammonia plants are dealt with in Sect. 6.5.3.2.2.

6.4.1.3 Effect of Individual Parameters

The *synthesis pressure* has, as mentioned above, an important influence on the performance of the ammonia synthesis loop because of its influence on the reaction equilibrium, reaction kinetics, and gas/liquid equilibrium in the product separation. A wide range of operation pressures has been used in practice, from less than 100 bar (in the early Mont Cenis process and recently in the ICI AMV-process) to 1000 bar (in the early Claude and Casale processes). The trend in modern plants has been to select operating pressure in the low to medium pressure range; typical operating parameters for modern synthesis loops at two different pressures are given in Table 6.1.

Table 6.1. Operating conditions, NH_3 loops at 220 bar and 140 bar (1000 MTPD*) [459]

Inlet pressure, bar	220	140
Inlet flow, $N\,m^3/h$	407,000	500,000
Inlet mol% NH_3	3.8	4.1
Outlet mol% NH_3	19.9	17.1
Inert level, inlet mol%	12	8
Separator temp., °C	− 5	− 5
Relative catalyst volume, %	60	100

* 1000 metric tons per day

The operating pressure is, of course, partly imposed by considerations concerning catalyst stability and activity. But the combined effects of the unfavorable equilibrium conversion at low pressure and the low temperature required to condense the ammonia product may be more important. With reference to Fig. 1.1 and 1.4 in chapter 1 this can be explained as follows:

- at a low operating pressure, a very low operating temperature is required to obtain a reasonably high concentration at the reactor outlet. Note that at the relevant temperature level, a decrease in temperature of 100 °C may cause a reduction in reaction rate of about a factor of 10 (at the same ammonia concentration).
- at a low operating pressure, a very low temperature is required to obtain reasonable recovery of ammonia. Refrigeration for product condensation is normally achieved by cooling with ammonia, and this limits the applicable separator temperature to about − 25 °C (ammonia boils at atmospheric pressure at − 33 °C).

It might be argued that one could allow a higher concentration of ammonia in the recycling gas and by increasing the concentration in the converted gas still obtain a reasonable conversion. This would, however, increase the required catalyst activity (or volume for constant activity) very significantly since the ammonia synthesis reaction on existing types of catalyst is strongly inhibited by the product ammonia (see Chapter 2).

A further point which favours an elevated operating temperature in the catalyst bed, and thereby – through the equilibrium relations – an elevated operating pressure, is the risk of poisoning. Ammonia synthesis catalysts are extremely sensitive to poisoning by oxygen-containing compounds such as H_2O and CO. The poisoning is caused by a reversible adsorption of oxygen species on the active sites, and the equilibrium is such that at temperatures below about 350 °C, almost complete coverage – and therefore almost complete deactivation of the catalyst – is obtained even at concentrations of oxygen-containing compounds about or below 1 ppm [238]. Such low concentrations are very difficult (and expensive) to achieve, and as a consequence, the risk of poisoning sets a practical lower limit to the catalyst temperature and thereby the operating pressure.

The power consumption for synthesis gas compression, for recirculation, and for refrigeration depends critically on the operating pressure. An extensive study on the influence of operating pressure (in the range 140–315 bar (2000–4500 psig)) on total power consumption is reported in [404, 405]. Loops with single and double condensation and with 3- and 4-nozzle compressor configuration (see Sect 6.3.1) are considered, and the power consumption is given for a number of cases. The main conclusions are that when realistic compressor efficiencies are used, then the total power consumption for a constant loop pressure drop depends only marginally on operating pressure in the range considered.

There seems to be a flat minimum at about 155 bar (2200 psig); a lower loop pressure drop means, of course, lower power consumption; the 4-nozzle design is slightly more energy efficient than the 3-nozzle design over the entire pressure range; and the design with double condensation (considered only at high pressure) is more efficient than single condensation. Some of the results are shown graphically in Fig. 6.3 (from [405]).

Catalyst Activity/Volume and Converter Configuration. An analysis of the effect of various parameters on overall energy consumption in an ammonia synthesis loop is reported in [460, 461]. The main purpose of the study was to determine the potential benefit from the installation of superactive catalysts (this is, of course, equivalent to the installation of a large catalyst volume). It was found that when optimum values of inert level, separator temperature, etc. are selected, then the energy consumption in the loop depends only on the equilibrium temperature which is established at converter exit. (see Fig. 6.4 (from [460]). The energy consumption at a constant equilibrium temperature was found to be almost independent of loop pressure in the range of 80–220 bar (8–22 MPa). The influence of various parameters on the conversion per pass through the synthesis converter system is discussed in Sect. 6.4.2. Various converter configurations are discussed in Sect. 6.4.3.

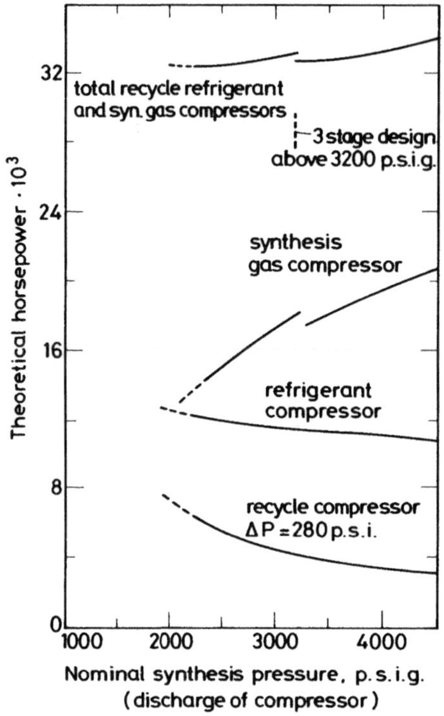

Fig. 6.3. Theoretical horsepower versus synthesis pressure for base case design (early M. W. Kellogg process) for plant capacity of 1500 short tons of ammonia per day (from [405])

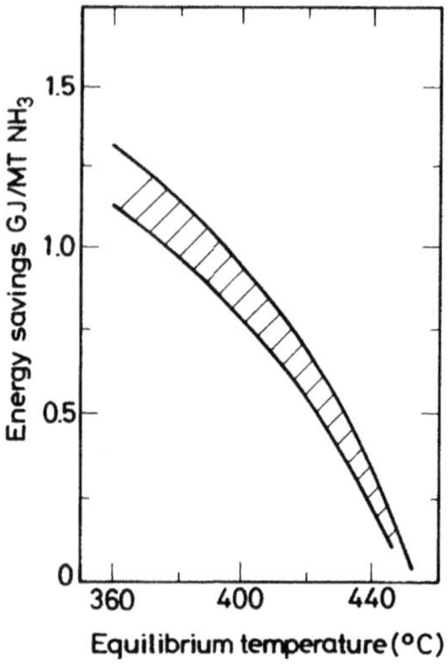

Fig. 6.4. Energy saving as function of equilibrium temperature at outlet last catalyst bed. Basis: Haldor Topøe S-200 synthesis loop (from [460])

The *separator temperature*, together with operating pressure and location of make-up gas addition point, determines the ammonia concentration at the converter inlet. A low temperature means low ammonia concentration, which again means either a low catalyst volume or high conversion. But low separator temperature can only be obtained by increased power consumption and cost of the refrigeration units, so that the choice of separator temperature is a compromise between energy consumption and capital costs in various parts of the synthesis unit. For an existing synthesis loop, a reduction in separator temperature will mean an increase in production capacity [449, 462] since the converter with its given catalyst volume will be able to make more ammonia due to the lower inlet concentration.

The *inert level* in the synthesis loop (most often measured at converter inlet) depends on the inert level in the make-up gas, the production of ammonia per unit make-up gas (the loop efficiency), and the purge rate. It may easily be seen that when two of these four parameters are given, then the two others can be calculated, if steady state (including constant pressure) is maintained in the loop. The inert level in the make-up gas is, of course, determined by the conditions in the synthesis gas preparation unit. The ammonia production is determined by conditions around the converter, the gas flow (which may be expressed by the recycle ratio) the inlet temperature and pressure, catalyst volume and activity, and converter configuration (see Sect. 6.4.3).

This means that for a given situation neither the purge rate nor the inert level can be changed without affecting operating pressure. It also means that if the operating pressure or the catalyst activity (by change of catalyst) is changed, then the other parameters will change until a new steady state is obtained.

6.4.2 Design of Ammonia Synthesis Reactors

6.4.2.1 General

Ideally, the design of a chemical reactor is a strictly logical process, in which *the* optimum solution to the problem is determined from experimental and other data by well-defined procedures. However, the real world does not work in this way. This is clearly shown, if not by other evidence, then by the multitude of different solutions which have been found in reactor designs for the same problem. All the solutions are normally technically and economically very reasonable – and none of them can claim to be clearly superior in every case.

The term "reactor design" is often interpreted as meaning "calculation of necessary reactor (catalyst) volume". There are, however, [457] several other factors to consider in reactor design – at least when this is interpreted as all the information required as an instruction to a workshop to manufacture a reactor for a specific practical application. Some of the more important points are listed below (from [459]).

Chemistry and thermodynamics

- the reaction including possible side reactions
- thermodynamics
- equilibrium
- heat of reaction
- properties of reactants and products

Reaction kinetics

- intrinsic + diffusion parameters
- pellet kinetics

Properties of catalyst

- thermal stability
- particle size and shape
- pore system
- mechanical strength
- change in properties during activation and operation

Process optimization

- energy recovery
- pressure drop
- limitations on volumetric flow

Process control

- temperature control
- flow distribution
- safety aspects, start-up/shut-down

Mechanical design

- materials of construction
- workshop manufacture
- sea and land transport
- erection

The basic data – the chemistry and the thermodynamics for the process – will define limits for a broad range of conditions at which the reaction is possible. The kinetics of the reaction and the properties of the catalyst, especially its thermal stability, will further narrow the range of possible reaction conditions and define a "window" of possible operating parameters. Process optimization, energy efficiency, and safety aspects will then determine at what conditions within the "window" the reactor should operate to give the optimum result. And *then* mathematical models are used to determine how big the reactor must be to obtain the performance (conversion and pressure drop) determined by the process optimization.

6.4.2.2 Reactor Calculations

In mathematical models for design, simulation, and/or optimization of ammonia synthesis reactors, the normal principles for reactor calculations are, of course, applied. Reference is made to [463–465]. A general purpose computer program for reactor calculations is described in [466, 467]. Application of the program for analysis of converter performance is described in [462, 468]. General discussions about modelling of ammonia synthesis reactors and comparison of different converter types may be found in [469–472]. Models of multibed quench converters and optimization studies for such reactors are reported in [473–479]. In [473] is described a study on conversion of tubular converters to quench cooled converters inside the same pressure shell. In [474] is discussed the effect of variation of operating parameters such as feed temperature, pressure, inert level, feed rate, and hydrogen to nitrogen ratio, on conversion and converter stability. Similar studies are reported in [476, 477].In [475] is compared the performance of a quench cooled and a TVA type converter with emphasis on analysis of the quench converter performance. Models for tubular converters are discussed in [480–486]. In [480] is discussed the effect of variations in operating parameters such as feed rate, feed gas concentration of inerts and ammonia, catalyst activity, and heat conductivity on conversion and converter stability. In [481] is analyzed the transient behaviour of the converter. In [483] is contained a critical review of the work reported in [455, 456, 482]. A proposal to optimize converter performance by optimizing the heat transfer coefficient between the catalyst bed and cooling tubes is presented in [484]. In [485] are reported results from the use of a model to control an existing ammonia plant, and in [486] is analyzed the effect of catalyst deactivation on converter performance.

In addition to the above, the effect of operating conditions on conversion in synthesis reactors is discussed in [487–489]. The general results are:

- *Pressure*: Increasing pressure increases conversion due to the increasing reaction rate and the increasing equilibrium conversion.
- *Inlet Temperature*: Has a dual effect since increasing temperature increases the reaction rate, but decreases the adiabatic equilibrium concentration.
- *Space velocity*: Increasing space velocity will normally decrease outlet ammonia concentration, but increase total ammonia production.
- *Inlet Ammonia Concentration*: Increasing inlet ammonia concentration decreases conversion since reaction rate decreases and adiabatic equilibrium conversion decreases.
- *Inert Level*: Increasing inert level decreases conversion since both the reaction rate and the adiabatic equilibrium concentration decrease due to the lower partial pressure of hydrogen and nitrogen.
- *Hydrogen/Nitrogen ratio*: Exhibits a true maximum since reaction rate has a maximum at a certain hydrogen to nitrogen ratio. The position of the maximum depends on space velocity, but will generally be between 2.0–3.0. It also may be noted that the equilibrium conversion of hydrogen-nitrogen mixtures to ammonia is maximum at a hydrogen to nitrogen ratio below 3 (see Chapter 1).
- *Catalyst Particle Size*: Smaller catalyst particles give a higher conversion due to lower diffusion restrictions (higher efficiency factor, see [459]).

6.4.2.3 Principal Reactor Types

In ammonia synthesis, the desirable conditions around the reactor system as described above are such that it is normally not possible to go from the inlet to the outlet conditions in the reactor in one adiabatic step. Some type of cooling is required. In principle the cooling can be applied in three different ways:

- internal cooling with cooling tubes in the catalyst bed or with catalyst in tubes surrounded by a cooling medium. The internal cooling can be effected either with gas flow in the same direction in the catalyst bed and cooling channels (cocurrent flow), with gas flow in opposite directions in the catalyst bed and cooling channels (countercurrent flow), or with gas flow in the catalyst bed perpendicular to the flow in cooling channels (cross flow). The cooling medium can be either synthesis gas or some other medium, for example boiling water.
- quench cooling by injection of cold gas. The injection of quench gas can be either between adiabatic beds or into a catalyst bed at different locations. Flow in the catalyst beds can be either axial or radial in vertical converters or downwards in horizontal converters.
- external cooling by heat exchange between catalyst beds. The cooling medium can be either synthesis gas or some other medium, for example boiling water. Flow in the catalyst beds can be either axial or radial in vertical converters or downwards in horizontal converters.

It is of course possible to combine several cooling methods in the same converter system. Furthermore, the catalyst beds and/or heat exchangers including feed effluent heat exchanger may be arranged in one pressure shell or in individual pressure shells. It is clear that this leads to a very significant number of possible converter configurations.

The various converter types may be characterized by the temperature profile through the catalyst bed(s) or by the temperature/concentration profile (plots of temperature vs ammonia concentration for the gas passing the converter) (see Fig. 6.6a–d below). Such profiles are often compared to maximum reaction rate profiles, see Fig. 6.5 (from [460]). It is seen from this figure that when the temperature is increased (at otherwise constant conditions, including constant ammonia concentration), then the reaction rate will increase up to a maximum value; when the temperature is further increased, the rate decreases until it becomes zero at the equilibrium temperature. The temperature/concentration points where maximum rate is achieved describe a curve, the maximum rate curve, which will normally be roughly parallel to the equilibrium curve, but at 30–50 °C lower temperature. It is clear that the minimum catalyst volume would be obtained in a converter where this maximum rate curve were followed. In the early days of ammonia production, available technology limited the obtainable size of the converter pressure shell, and the physical dimensions of the converter

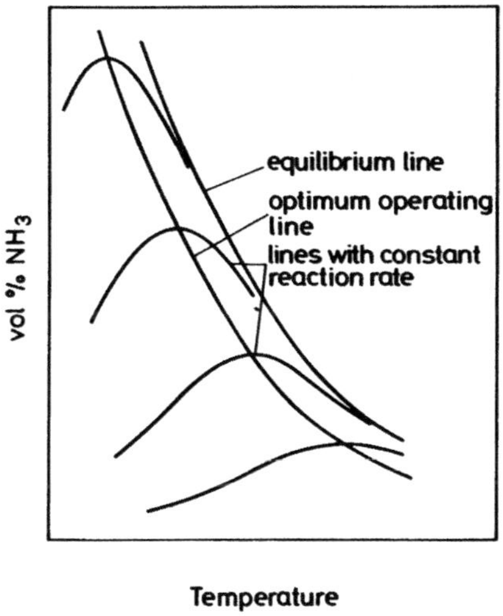

Fig. 6.5. Optimum reaction path or maximum rate curve. The difference between the lines with constant reaction rate is a factor of 10. Lowest reaction rate at highest NH_3 concentration (from [460])

thus limited the achievable production capacity. Great emphasis was therefore given to the ammonia production per unit converter volume. Quite complicated mechanical constructions were used to maximize the production capacity of a given volume, and the converters were compared to the "ideal" converter where the temperature/concentration plot follows the maximum rate curve (see e.g. 469)). As technology developed, other considerations such as optimum heat recovery, reliable mechanical construction, easy catalyst loading and unloading, etc. became more important than maximum production per unit volume, and different – mechanically simpler – converter configurations have, therefore, become more popular.

In connection with "revamps", i.e. modification of existing plants for increased capacity and/or increased energy efficiency (see Sect. 6.4.3.4), where the pressure shell of existing converters is reused, it has, however, again become a major consideration to maximize the production capacity of the volume available inside the pressure shell.

Ammonia converters most often consist of two separate parts, an outer pressure vessel and an internal "basket" containing the catalyst bed(s), internal piping for gas distribution, heat exchangers (when applicable) for control of catalyst temperatures, and in some cases a feed-effluent heat exchanger, so that all high temperatures are contained inside the pressure shell.

Early types of ammonia synthesis converters are described in [4, 50, 385]. More recent developments are discussed in [23, 490–492]. Good overall reviews of different converter designs may be found in [493, 494].

A discussion of the types of heaters used for start-up of ammonia synthesis converters (fired vs. electrical heaters) is given in [910].

6a)

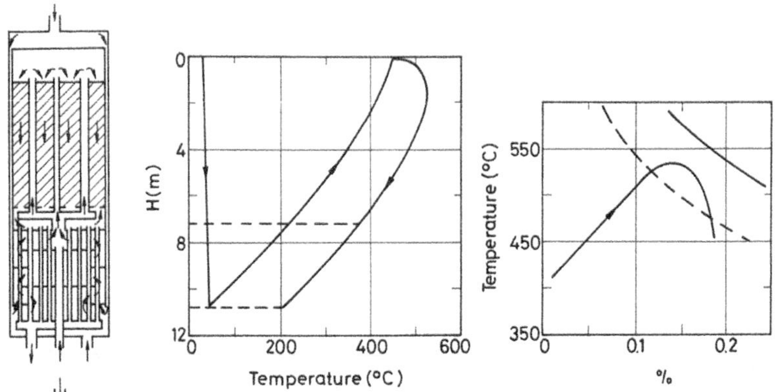
H(m)
0
4
8
12
0
200
400
600
Temperature (°C)
Temperature (°C)
550
450
350
0
0.1
0.2
%

6b)

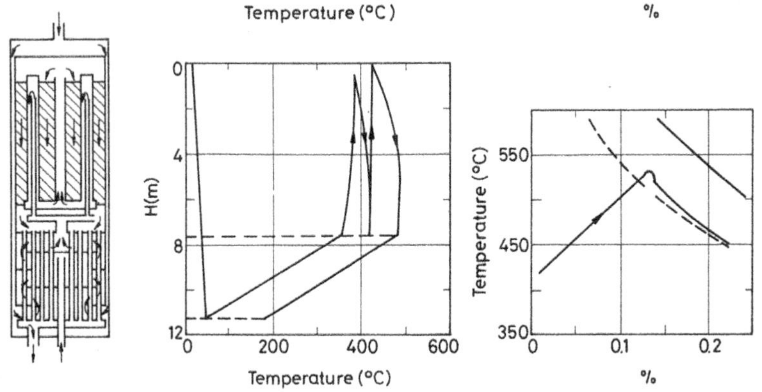
H(m)
0
4
8
12
0
200
400
600
Temperature (°C)
Temperature (°C)
550
450
350
0
0.1
0.2
%

6c)

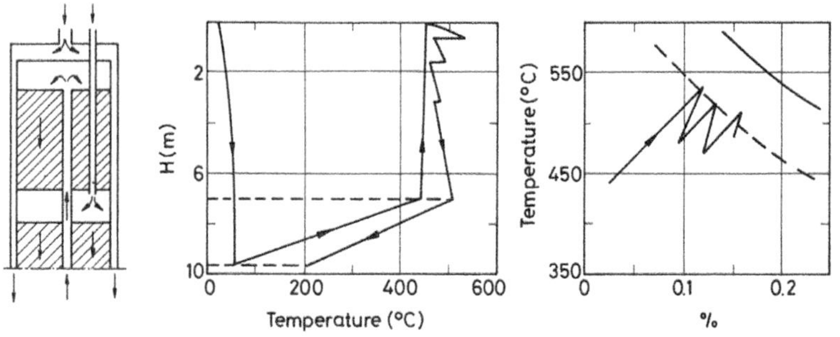
H(m)
2
6
10
0
200
400
600
Temperature (°C)
Temperature (°C)
550
450
350
0
0.1
0.2
%

6d)

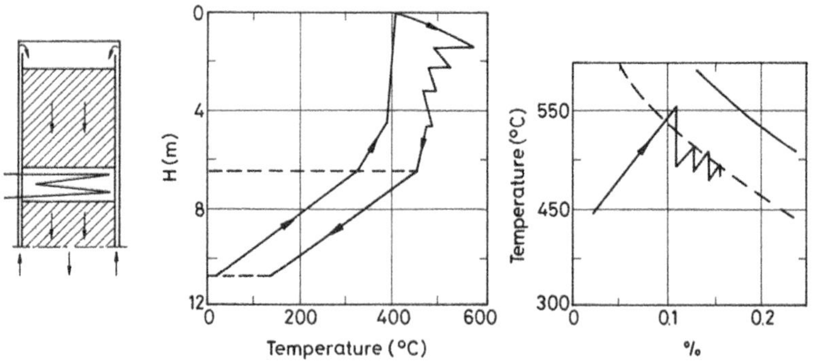
H(m)
0
4
8
12
0
200
400
600
Temperature (°C)
Temperature (°C)
550
450
300
0
0.1
0.2
%

6.4.3 Commercial Ammonia Synthesis Converters

A very significant number of different ammonia synthesis converter designs have been used in industrial practice. In the following survey the different designs have been characterized, first by the cooling principle applied, and thereafter mainly by flow direction through the catalyst bed(s).

6.4.3.1 Converters with Internal Cooling in Catalyst Beds

In converters with internal cooling the required cooling is supplied by a cooling medium which flows in most cases in cooling tubes in the catalyst bed or, in a few designs, around tubes containing the catalyst. The flow in the cooling tubes may be countercurrent or cocurrent (in the opposite direction or the same direction as the flow in the catalyst bed).

6.4.3.1.1 Countercurrent Flow in Cooling Tubes

The most important converter type using internal cooling with countercurrent flow in cooling tubes is *the TVA-converter*. A schematic drawing of the converter and the corresponding temperature- and temperature/conversion profile is shown in Fig. 6.6a. Feed gas enters at the top and passes in the annulus between the pressure shell and the basket to the bottom of the converter. In this way the pressure vessel is cooled (the feed gas is used as "shell cooling gas") so that a lower design temperature is applicable for the expensive pressure vessel.

The feed gas passes through the feed/effluent heat exchanger, which is contained in the same pressure shell as the catalyst bed (when installed in this position the heat exchanger is often referred to as "the lower heat exchanger"), on the shell side and further up through the cooling tubes in the catalyst bed. After passing in up flow through the cooling tubes the gas turns and flows in down flow through the catalyst bed and further through the tube side of the lower heat exchanger to the exit. A bypass steam (often called "cold shot") is added for temperature control to the feed gas between the lower heat exchanger and the catalyst bed cooling tubes. It is seen from the temperature profile that the gas is at relatively low temperature at converter outlet. This means that the waste heat can be used only for preheating boiler feed water or producing low pressure steam.

Fig. 6.6. Schematic drawing, typical temperature profile, and operating curve (temperature/ammonia concentration plot) for four important converter types. (from [471])
a Internal cooling, countercurrent flow (TVA-converter)
b Internal cooling, cocurrent flow (NEC-converter)
c Quench cooling
d Indirect cooling (heat exchange)

If a design is used where the feed/effluent heat exchanger is located in a separate pressure shell, then the converter outlet temperature becomes equal to the catalyst bed outlet temperature, and the converter outlet system must be designed for this higher temperature. In such systems the waste heat may be recovered at higher temperature so that it can be used, for example, for production of high pressure steam.

The temperature/concentration profile illustrates how the gas reacts almost adiabatically in the first part of the catalyst bed. As the temperature difference between the cooling gas and the reacting gas increases, and the reaction rate decreases due to the approach to equilibrium, the temperature is decreased so that the approach to the equilibrium line increases, and the profile moves close to the maximum reaction rate line. In the bottom of the catalyst bed the reaction rate becomes low, and the temperature drops below the maximum rate line. It is evident that in order to obtain the best performance of this converter type, it is important that the cooling must match the heat evolution. Too efficient cooling will lead to a low reaction rate or even to the loss of reaction, whereas too little cooling will lead to a too close approach to equilibrium and therefore to inefficient use of the catalyst. A TVA-type converter of special design is described in [495]. In this design, core rods are installed in the cooling tubes to enhance heat transfer, thereby reducing the required heat transfer area.

Stability problems and performance optimization of TVA-converters are discussed in [480, 481, 484, 485]. A classical account of operating problems in a synthesis unit using TVA-converters is given in [496] which also gives mechanical details on the construction of a TVA-converter. Drawings showing mechanical details of TVA-converters may also be found in [23, 385, 497, 498]. TVA type converters have been used extensively, and may are in operation today. This converter type has been suggested quite recently for installation in new, relatively small plants (up to about 300 MTPD of ammonia) by process licensors such as Tosøe [499] and ICI [500].

Another converter type using countercurrent flow in cooling tubes is the SBA-converter [501]. In this converter the feed gas enters at the bottom of the pressure shell and passes upwards in the annulus between the pressure shell and the basket, flows down through tubes which serve both as heat exchanger tubes in a gas/gas heat exchanger located in the top of the converter and as cooling tubes in the catalyst bed. The preheated gas passes in up flow through the catalyst bed and through the gas/gas exchanger on the shell side before it leaves the converter at the top. Converters of this type have recently been revamped into Tube Cooled Radial Flow converters [502].

Old converter types with countercurrent flow in cooling tubes are the Mont Cenis reactor [385, 503], the original Haber-Bosch converter [504], the Claude converter [505], and the "old" Fauser converter [506, 507]. These early converter types were all used in relatively small plants; they are not used in modern processes.

Combinations of countercurrent cooled catalyst beds with adiabatic beds have been suggested by ICI [508], Uhde [509] and in [510, 511].

6.4.3.1.2 Cocurrent Flow in Cooling Tubes

The most important converter type using internal cooling with cocurrent flow in cooling tubes in the NEC converter or the Chemico converter. A schematic drawing of the converter and the corresponding temperature and temperature/conversion profile is shown in Fig. 6.6b. Process gas enters the converter at the top, passes downwards in the annulus between the catalyst basket and the pressure shell as shell cooling gas to the shell side of the lower heat exchanger and therefrom to cooling tubes in the catalyst bed. The tubes are bayonet tubes, and the gas flows up through the inner tube, down in the annular space between the two concentric tubes, up through a center pipe and finally down through the catalyst bed and through the tube side of the lower heat exchanger to the exit.

The cocurrent flow makes it possible to obtain a temperature profile which is close to the optimum; the gas is heated adiabatically to a temperature close to the maximum reaction rate curve, and the temperature/conversion profile then follows closely the maximum rate curve for the test of the catalyst bed length.

The NEC- or Chemico converter with cocurrent flow in the cooling tubes has – like the TVA-converter – been used since the early days of the ammonia industry, and many modifications have been suggested. Descriptions of various versions may be found in [23, 385, 492, 512–516]. A converter with cocurrent cooling in flat cooling "fins" in the catalyst bed is described in [517]. A converter type which combines adiabatic catalyst beds with a catalyst bed using cooling tubes with cocurrent flow has been developed by the Japan Consulting Institute [518, 519]. The flow pattern in this converter is as follows: Feed gas enters at the top and passes as shell cooling gas to the bottom of the converter. It passes then on the shell side of a lower heat exchanger and further in an annulus between the tubular cooled catalyst bed and the outer wall of the basket to the top of the tubular cooled bed, in down flow through cooling tubes in the catalyst bed, up through a central pipe to the top of the converter, down through two adiabatic catalyst beds in series with quench cooling between the beds and finally through the tubular cooled catalyst bed and through the tube side of the lower heat exchanger to the exit. Cold gas is added for temperature control after the first passage of the lower heat exchanger and (as quench gas) after the passage of the first adiabatic bed.

6.4.3.1.3 Cross Flow

Ammonia synthesis converters with radial flow in tubular cooled catalyst beds have been suggested by Toyo Engineering Corp. [520] and in [502]. The Toyo concept has, so far, not been used industrially, while the concept described in [502] has, as mentioned above, been demonstrated in revamps of converters originally designed by SBA. It is claimed that the cross flow makes it possible – through proper design of the cooling tube bundles – to optimize the temperature profile so that it follows very closely the maximum reaction rate curve. It is furthermore reported that the heat transfer coefficients obtained in practice in the cross flow configuration are significantly higher than in axial flow. A converter with radial flow and with two zones in each of a series of catalyst beds, an adiabatic zone and a zone with cooling tubes, is described in [521].

6.4.3.2 Converters with Quench Cooling

In quench cooled synthesis converters the required cooling is supplied by injection of cold, unconverted feed gas either between separate catalyst beds or through proper distributors directly into a catalyst bed at different levels.

A schematic drawing of a quench cooled converter and of the typical temperature and temperature/conversion profile is shown in Fig. 6.6c. Only part of the feed gas passes through the first catalyst bed where the temperature increases adiabatically from the inlet temperature to a temperature normally quite close to the equilibrium temperature. After the first bed an amount of cold feed gas is added so that the temperature is decreased to the desired inlet temperature to the second bed. At the same time the ammonia concentration decreases because the exit gas from the first bed is diluted with unconverted gas. The quench cooling concept is adaptable to large capacity converters and was almost exclusively used in the early generation of large single stream units constructed in the 1960s and 1970s.

Even today quench converters are extensively used although they are increasingly being replaced by indirectly cooled converters, both for applications in new plants and by replacement or modification of converters in existing plants in revamp projects (see Sect. 6.4.3.4).

Quench converters appear to have been used first in plants constructed in the USA during the second world war with M. W. Kellogg as the engineering contractor [522]. The early Kellogg quench converter is also described in [385, 523].

The converter is described as a modified NEC converter in which the catalyst cooling tubes have been replaced by cold gas injections between layers. 10–12 catalyst beds were used in this early generation of quench converters. Other early designs of quench converters were the Uhde converter [524, 525] and the so called “modified Haber-Bosch” reactor [385, 526] or the BASF reactor [435, 490].

As mentioned above, quench cooled converters were typical for the first generation of the large single stream units constructed in 1960s and 1970s. The most important quench type converter designs used in these plants may be characterized by the arrangement of the quench cooling and by the flow direction in the catalyst bed.

6.4.3.2.1 Axial Flow, Quench Cooling Between Separate Catalyst Beds

The most important example of this converter type is the M. W. Kellogg 3- or 4-bed Quench Converter which was used in a large number of plants. The converter is described in [490, 491, 527, 528]. A simplified drawing of a 3-bed converter is shown in Fig. 6.7 (from [491]). The feed gas enters the converter at the bottom and passes as shell cooling gas in the annulus between catalyst basket and pressure shell to the feed-effluent heat exchanger mounted in the top of the converter.

The feed gas passes on the shell side of the exchanger to the first catalyst bed. Between the exchanger and the first catalyst bed and between the following catalyst beds cold gas is added via quench gas distributors for temperature control. The converted gas passes from the last catalyst bed through a center pipe to the tube side of the feed-effluent heat exchanger and leaves the converter at the top.

This converter type has given excellent service in industry. It is, however, due to the inherent weaknesses of the quench cooling system, not very efficient, and the performance has in many cases been improved by revamping, see Sect. 6.4.3.4.

Designs similar to the Kellogg design have been used by others, e.g. Grande Paroisse [385, 492, 529], Kubec [530, 531] and Casale [490]. A design in which quench cooling is applied between catalyst beds contained in separate vessels has been used by Pritchard [437].

A later development is the M. W. Kellogg horizontal quench converter, see Fig. 6.8 (from [492]). This converter design is described in [533–536]. The catalyst beds are arranged in a basket which fits into a horizontal pressure shell;

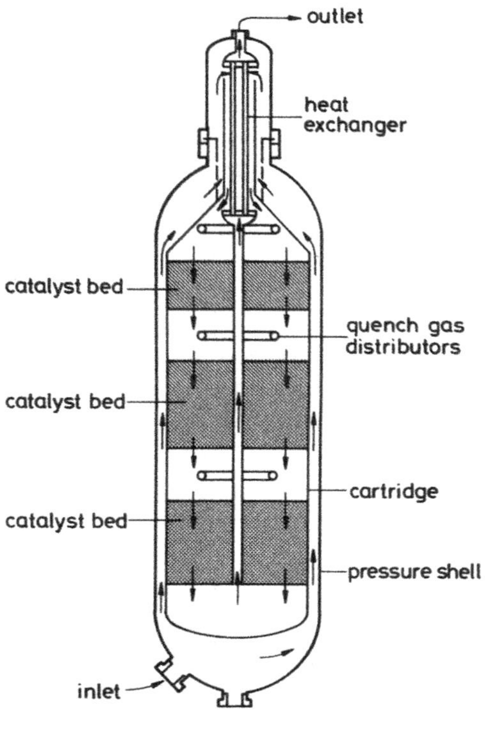

Fig. 6.7. Three-bed Quench Cooled Axial Flow Converter. M. W. Kellogg design. (from [491])

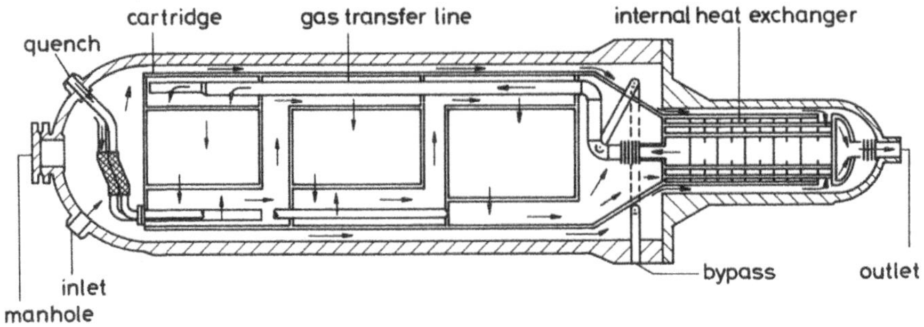

Fig. 6.8. Quench cooled horizontal converter. M. W. Kellogg design (from [492])

in this way catalyst loading and unloading are facilitated and there is no need for overhead structure or crane. The basket may be removed from the pressure shell simply by drawing it out on tracks.

In operation, gas flow is downwards through the catalyst beds. Because of the large area, shallow beds, a low pressure drop is obtained even with small catalyst particles. This means that the horizontal design offers some of the same advantages offered by radial flow converters (see Sect. 6.4.3.2.3.).

6.4.3.2.2 Axial Flow, Quench Cooling by Injection of Gas into the Catalyst Bed

This converter type has mainly been used by ICI; it is described in [23, 490, 491]. A schematic drawing is shown in Fig. 6.9 (from [491]). Synthesis gas enters at the bottom of the converter, flows as shell cooling gas in the annulus between the catalyst basket and pressure shell to the top of the converter, down on the shell side of a centrally mounted feed-effluent heat exchanger, upwards in an annulus between heat exchanger and catalyst bed, down through the catalyst bed, and up on the tube side of the feed effluent exchanger to the outlet in the top of the converter. Quench gas is added at various levels in the catalyst bed through gas distributors. A special version of this converter type – the opposed flow converter – was developed for very large capacities [537, 538]. In this design the converted gas is collected in the middle of the catalyst bed, and there is down flow in the upper half and up flow in the lower half of the bed. The special feature of the ICI design – one uninterrupted catalyst bed which can be emptied through an opening in the bottom of the converter – is maintained in the opposed-flow converter.

A design similar to the ICI-design has been suggested by Chemico [539].

6.4.3.2.3 Radial Flow or Axial/Radial Flow

A common problem for all converters with axial flow is that when the capacity increases, the pressure drop increases due to the increased depth of the catalyst bed. It is of course possible to some extent to compensate by increasing the bed

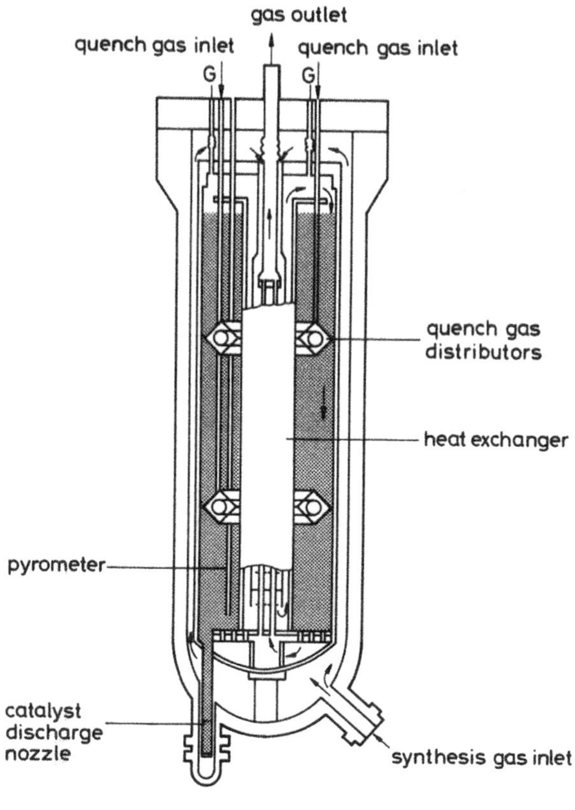

Fig. 6.9. Quench cooled converter with injection of gas into the catalyst bed. ICI design. G is the cold gas inlet for temperature control (from [491])

diameter, but above a certain size this becomes technically and economically unacceptable. In order to partially overcome the pressure drop problem it is normal in axial flow converters to use relatively large catalyst particles.

However, these have lower activity than small particles due to diffusion restrictions, and a larger catalyst volume is therefore required [459, 540].

In converters with radial flow the above mentioned disadvantages do not exist. Radial flow converters can be designed for very large capacities without excessive reactor diameter, and a low pressure drop can be maintained even with very small catalyst particles. The advantages of radial flow reactors for ammonia synthesis is discussed in [459, 502, 541, 542]. The pressure drop as function of catalyst particle size is discussed in [488]. Radial flow has been used in converters with cooling tubes [520], in quench cooled converters (see below) and in converters with indirect cooling (see Sect. 6.4.3.3.2).

The most widely used quench cooled radial flow converter has been the Topsøe designed S-100 converter, which is a two-bed quench cooled reactor. A schematic drawing is shown in Fig. 6.10 (from [544]). The converter is described in [23, 427, 490, 491, 534, 535, 543–547]. The main part of the synthesis gas enters the converter at the top and passes downwards as shell cooling gas in the annulus between the pressure shell and the catalyst basket. It then passes

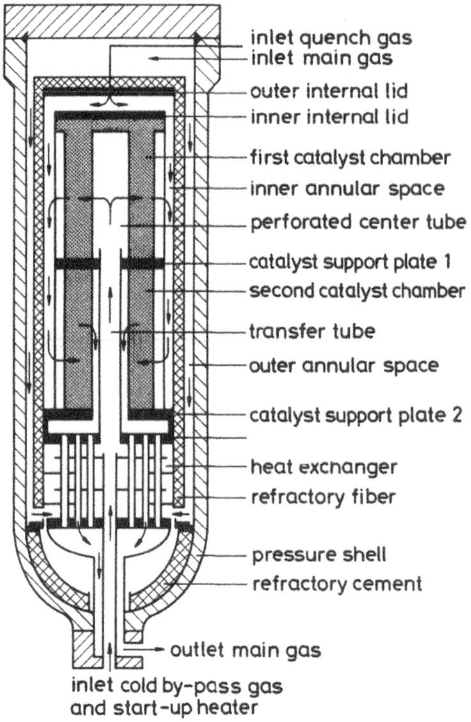

Fig. 6.10. Two bed radial flow quench cooled converter. Haldor Topsøe S-100 Converter (from [543])

through the feed/effluent heat exchanger ("lower heat exchanger") on the shell side and up through a central pipe to the first, upper catalyst bed. After passing the catalyst bed in radial flow in the outwards direction, the gas is cooled by quenching with fresh synthesis gas before it passes through the second, lower catalyst bed in inward direction, downwards in an annulus between the catalyst bed and the center pipe to the tube side of the lower heat exchanger and out through the bottom of the converter. Cold gas is added through the bottom and mixed with the inlet gas to the first catalyst bed for temperature control.

Radial flow quench converters have also been used by Chemoproject [548], Österreichische Stickstoffwerke [549], and Lummus [550]. In these converters, as well as in the Topøe converter, the flow direction is restricted to be essentially radial.

An axial-radial flow converter has been introduced by Ammonia Casale [490, 492, 551–553]. The special feature of this design is that gas can enter each catalyst bed both from the top (in axial direction) and from the side through perforations (in radial direction), see Fig. 6.11 (from [492]). The gas leaves the catalyst bed through perforations in the inner wall. There are no perforations in an upper part of the inner wall, and the gas entering at the top of the bed is therefore forced to flow through part of the catalyst in partially axial flow before it can leave the catalyst bed. This flow principle can be used in quench cooled

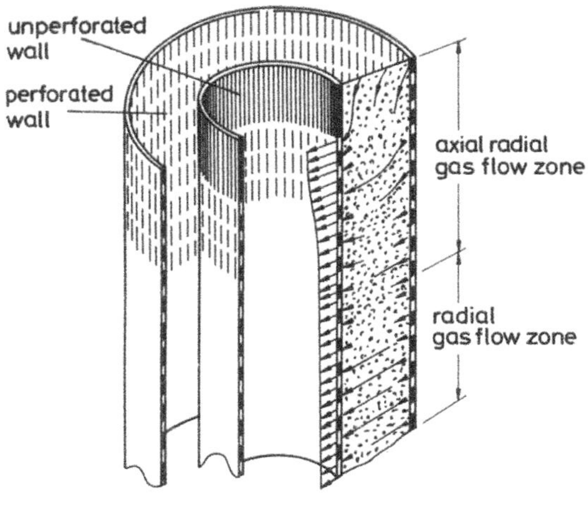

Fig. 6.11. Principle of axial-radial flow converter. Ammonia Casale design (from [492])

converters and in indirectly cooled converters (see Sect. 6.4.3.3.2). The design has found widespread use in modification of existing converters, see Sect. 6.4.3.4.

6.4.3.3 Converters with Indirect Cooling

In converters with indirect cooling, the cooling required to reach a desired high conversion in the converter is provided between the catalyst beds by some cooling medium, which most often is synthesis gas or boiler feed water (in steam boilers). The heat exchange most often takes place in exchangers installed together with the catalyst beds inside one pressure shell, although heat exchangers installed between separate, adiabatic converters have also been used.

Typical temperature and temperature/conversion profiles for an indirectly cooled converter are shown in Fig. 6.6d. In contrast to the quench cooled converter (see Fig. 6.6c) all gas passes through all catalyst beds, and there is no dilution of the partially converted gas between catalyst beds. This means that at otherwise identical conditions, a higher conversion can be obtained in the same number of catalyst beds in the indirectly cooled converter than in the quench cooled converter.

Converters with indirect cooling are used in almost all large, new ammonia plants constructed today. The cooling principle is also introduced in many existing plants by converter revamps, whereby quench cooled converters are modified into converters with indirect cooling by reconstruction of the converter internals (see Sect. 6.4.3.4).

6.4.3.3.1 Axial Flow

Converters with indirect cooling have been used since the early days of ammonia synthesis, for example by Fauser-Montecatini [23, 385, 490, 492, 506, 554–556]. In this converter, cooling is provided by circulating water which in turn delivers the heat to an external high pressure steam boiler.

In the converter design developed by Österreichische Stickstoffwerke (ÖSW) the cooling between the catalyst beds is provided by the converter feed gas. The ÖSW converter – also known as the Uhde-Chemie Linz converter – is described in various versions in [23, 490–492, 557–568]. A schematic drawing is shown in Fig. 6.12 (from [568]). The feed gas enters at the top of the converter, flows as shell cooling gas in the annulus between the pressure shell and catalyst basket, up through the shell side of the lower (feed/effluent-) heat exchanger and – via a center pipe – through the interbed heat exchangers to the top of the converter. The gas then passes downwards in axial flow through the catalyst beds and on the tube side of the interbed heat exchangers and the lower heat exchanger to the exit at the bottom of the converter. Cold synthesis gas ("cold shot") is added through the bottom of the converter and mixed into the feed gas to the individual catalyst beds for temperature control. The ÖSW/Uhde-design with interbed heat exchange was further developed into a horizontal converter [563] and into several Uhde designs with radial flow (see below).

A horizontal converter design with indirect cooling also was proposed in [569] and has been used by M. W. Kellogg in their low energy process [492, 570, 911, 571]. A schematic drawing of this converter is shown in Fig. 6.13 (from [492]). Gas enters at one end of the horizontal pressure shell, passes as shell cooling gas between the pressure shell and the catalyst basket to the shell side of the interbed heat exchanger which is installed at the other end. After preheating

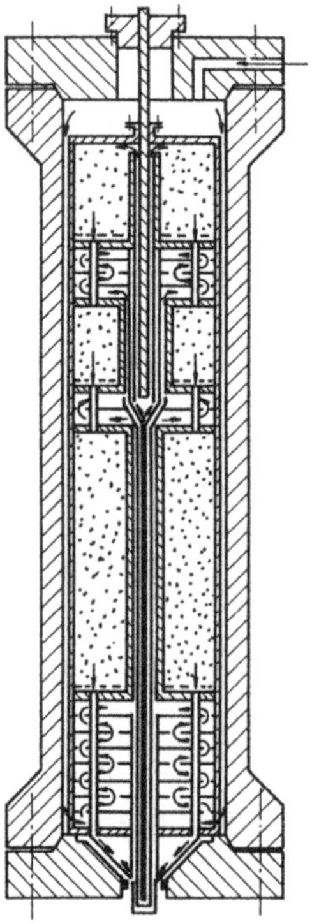

Fig. 6.12. Converter with indirect cooling, axial flow. ÖSW design (from [568])

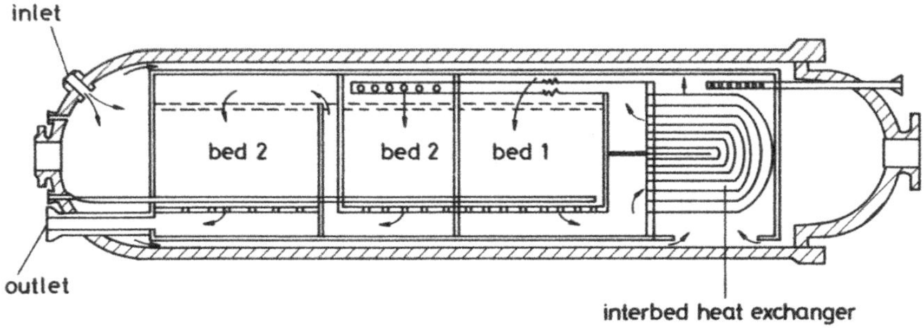

Fig. 6.13. Horizontal converter with indirect cooling between catalyst beds. M. W. Kellogg design (from [492])

in the heat exchanger and mixing with cold gas ("cold shot") for temperature control, the gas passes downwards through the first catalysed bed, through the interbed heat exchanger on the tube side, downwards through the second catalyst bed (which is divided into two sections) and out in the same end of the pressure shell as the main inlet. This design offers the same advantages as a horizontal quench cooled converter – a low pressure drop with small catalyst particles and the easy removal of converter internals without requiring a crane or overhead structure.

A design with indirect cooling between catalyst beds is also used by C. F. Braun. In this case the catalyst beds and the heat exchangers are each in their separate pressure shell. In an early design [401, 411, 572, 573] two adiabatic converters were used, and the cooling between the converters was by heat exchange between feed and product gas to/from the first converter. The arrangement and the converter design is shown in Fig. 6.14, and Fig. 6.15 (from [572]). In a later design [573–576, 952] three reactors are used, and the cooling between the reactors is provided partly by gas/gas heat exchange, partly by external cooling, for example by raising or superheating steam. The C. F. Braun converter (see Fig. 6.15) contains one adiabatic catalyst bed with axial flow. Gas enters at the bottom, flows as shell cooling gas to the top, down through the catalyst bed and out at the bottom. The design of the bottom part of the converter is critical due to the relatively high temperatures prevailing there.

ICI has proposed a split flow converter with indirect cooling between beds [577]. Feed gas is introduced in the top of the converter and flows to the shell side of an interbed heat exchanger, where it is preheated to the inlet temperature of the first conversion stage. It is then passed to at least two parallel axial flow beds installed above the heat exchanger, where the first conversion takes place. After these "upstream" beds the gas is cooled by passing through the interbed heat exchanger on the tube side. It then reacts further in at least two parallel axial flow beds installed below the heat exchanger. A somewhat similar system has been proposed by M. W. Kellogg [578]. In this case the beds are arranged in separate vessels, the heat exchanger is external, and the beds are with opposed flow.

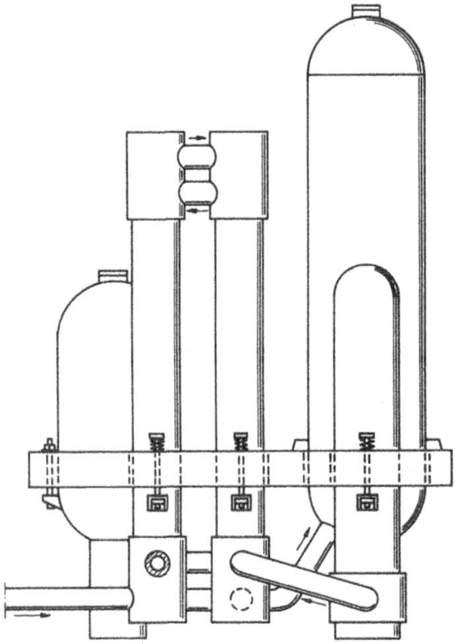

Fig. 6.14. Separate converters with axial flow and indirect cooling between the converters. C. F. Braun design (from [572])

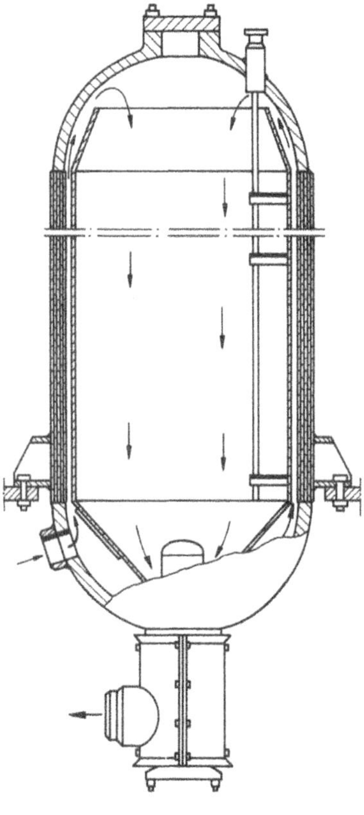

Fig. 6.15. C. F. Braun converter design (from [572])

6.4.3.3.2 Radial Flow or Axial/Radial Flow

The radial flow principle (see Sect. 6.4.3.2.3) has also been used in indirectly cooled reactors.

The Topsøe S-200 reactor [459–461, 492, 579–581] is illustrated in Fig. 6.16 (from [492]). Two versions are shown, with and without a lower heat exchanger. In both cases the converter consists of a pressure shell and a basket. In the converter without a lower heat exchanger, the feed gas enters at the bottom and flows as shell cooling gas to the top of the converter and through the interbed heat exchanger, which is installed centrally in the first bed. After the passage through the interbed heat exchanger the gas is mixed with cold gas ("cold shot") for temperature control before it passes through the first bed radially from outside in, through the tube side of the interbed heat exchanger, to and through the second bed, also from outside in, and finally out at the converter bottom. In the converter type with lower heat exchanger the gas flow is modified to allow the feed gas and the exit gas to pass through this heat exchanger.

The S-200 converter may be considered a further development of the S-100 converter, and it has largely replaced the latter in industry. The S-200 converter design has also been used in the revamping of existing converter installations, see Sect. 6.4.3.4. In a similar way the axial/radial flow principle of Casale has been used in indirectly cooled converters [582, 583]. This design is also being offered both for new plants and for revamps. Other radial converter designs have been suggested by Uhde [584–586], by M. W. Kellogg [587, 588] and by Snamprogetti [589]. In the last case the cooling is provided by a boiler centrally installed in the first radial flow catalyst bed.

6.4.3.4 Revamp of Ammonia Synthesis Converters

In many cases it is of interest to an ammonia producer to improve the production economics of an existing plant by the addition of new equipment or modification of existing equipment, by optimization of the operation, for example by installation of new and more efficient catalysts, or in some cases by a change of feedstock, etc. An important element in such a modernization of an ammonia plant, which is most often referred to as a *revamp* of the plant, is in many cases a modernization of the converter system.

Three obvious options exist for revamp of ammonia synthesis converters (see [590]):

- Replace the old converter by a new, larger and/or more efficient type.
- Add an extra new converter in parallel or in series with the existing converter.
- Replace or modify the internals of the existing converter with new internals to obtain higher conversion.

The first option is mainly of interest when the existing converter is in poor condition so that complete replacement is indicated. The two other options require at least that the pressure shell is in good condition.

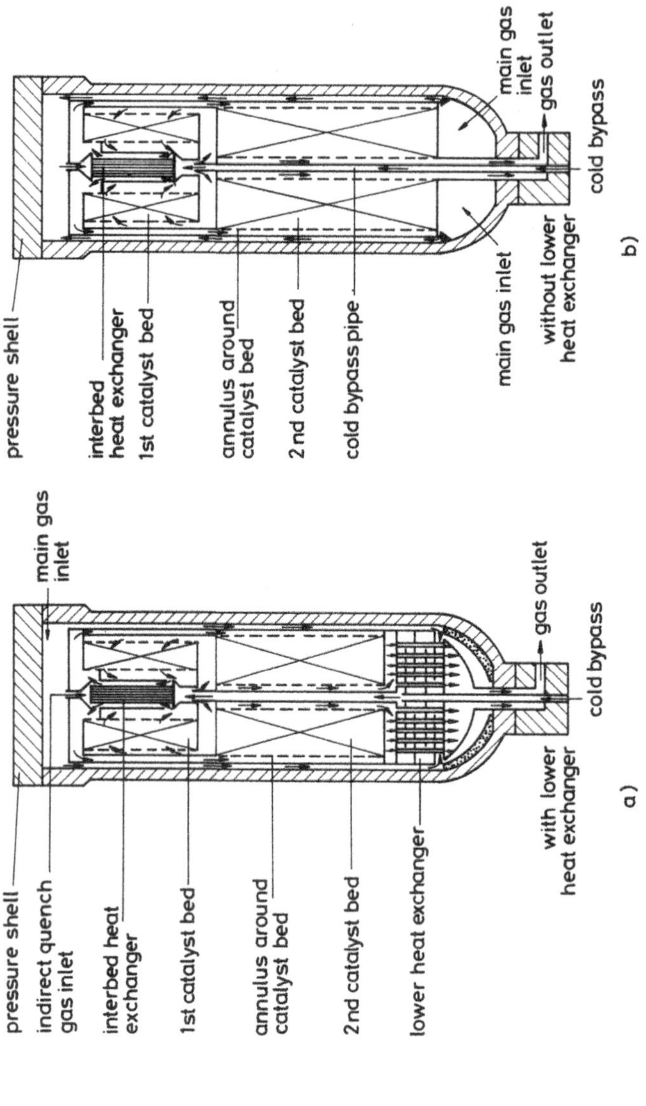

Fig. 6.16 Two bed radial flow converter with indirect cooling between catalyst beds. Haldor Topsøe S-200 converter. **a** with lower heat exchanger. **b** without lower heat exchanger (from [492])

A survey of converter revamp technologies is given in [590, 591]. The largest number of converter revamps have been modifications of the M. W. Kellogg 3- or 4-bed quench converters – simply because a very large number of plants using these converter types were constructed in the 1960s and 1970s. But also Topsøe S-100 quench type radial flow converters, ICI and Uhde quench converters, as well as other types of converters have been modified.

Three options for revamp of the M. W. Kellogg quench converters by modification of the converter internals (the so-called "in-situ revamp") are illustrated in Figs 6.17, 6.18, 6.19 (from [591]). M. W. Kellogg has proposed a solution whereby the four bed quench cooled reactor configuration is changed into a kind of two bed configuration with split flow through parallel beds in the second conversion step and with indirect cooling between the conversion steps. Haldor Topsøe proposes to change the internals into a "S-200 configuration", a two-bed radial flow converter with indirect cooling between the beds. Ammonia Casale proposes to maintain a four bed quench cooled configuration, but with axial/radial flow. More detailed descriptions of these converter modification technologies as applied to M. W. Kellogg quench converters and to other converter types may be found for the M. W. Kellogg technology in [592–594], for the Haldor Topsøe technology in [595, 596], and for the Ammonia Casale technology in [597–602], [913]. Other converter revamp technologies are the Uhde technology [586], which changes axial flow converters partly or completely into radial flow converters and in some cases also changes quench cooling into indirect cooling, and the technology described in [502] where a kind of radial flow principle is introduced into tubular cooled reactors. The

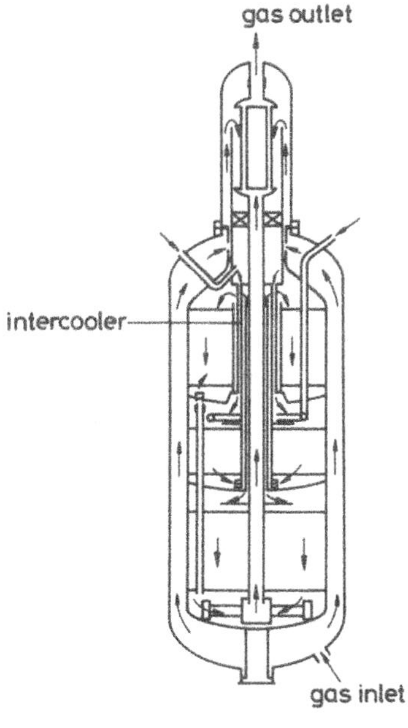

Fig. 6.17. Four bed Kellogg quench cooled converter after modification to Kellogg's split flow concept (from [591])

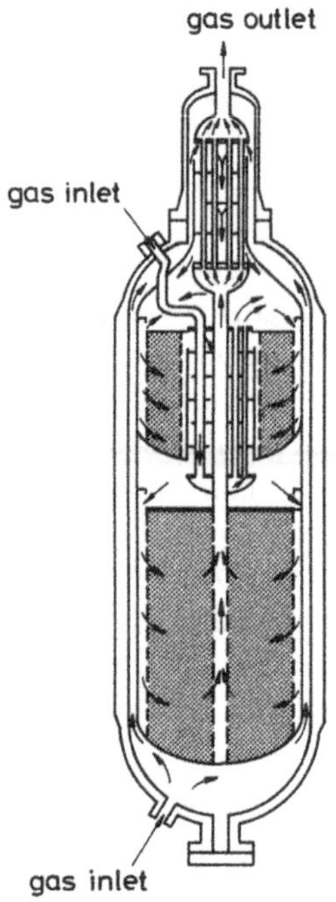

Fig. 6.18. Four bed Kellogg quench cooled converter after modification to Haldor Topsøe's S-200 two bed radial flow concept (from [591])

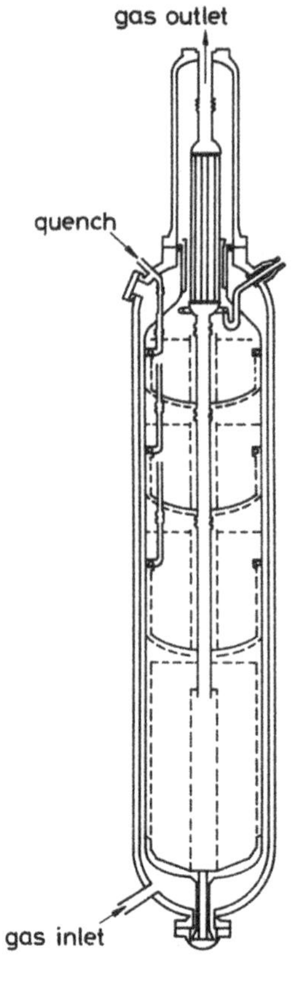

Fig. 6.19. Four bed Kellogg quench cooled converter after modification to Ammonia Casale's four bed quench cooled axial-radical flow concept (from [591])

M. W. Kellogg and the Uhde revamp technologies have only been applied to converters originally designed by M. W. Kellogg and Uhde respectively, whereas the Haldor Topsøe and the Casale technologies have been used to modify a number of different converter designs (see [591]).

Revamps consisting of the replacement of old converters with completely new converters are described in [409], (C. F. Braun: in one case a complete new synthesis loop replaced 5 identical old loops; studies are reported on the addition of a third reactor to an existing two reactor configuration) and in [603] (Haldor Topsøe: an S-200 converter replaced an existing M. W. Kellogg 4-bed quench converter). Revamp by addition of an extra converter in series with the existing converter has been proposed by Topsøe [955], and by M. W. Kellogg using their new KAAP technology (see Sect. 6.5.3.2.3) [949].

6.4.4 Waste Heat Recovery in Ammonia Synthesis Loops

The heat of reaction in ammonia synthesis is 46.22 kJ/mol ammonia at standard temperature and pressure or, in the units most often used in the industry, 1168 Btu/lb or 649 kcal/kg of ammonia. The efficient use of this heat, preferably for the production of high pressure steam, is of major importance in modern ammonia production. In almost all cases, some heat must be removed from the converter to ensure sufficient conversion, and this heat is used either to preheat the feed or for steam production as outlined in Sect. 6.4.3.

In early plants the effluent from the ammonia converter was often cooled with cooling water without any recovery of heat (see [604]). However, it soon became evident that the economics of ammonia production could be improved by better heat recovery, and steam production and/or boiler feed water preheating directly by the reactor effluent was introduced. The temperature of the reactor effluent depends on the reactor inlet temperature for each type of reactor. In the early 1960's it was suggested to increase the reactor inlet temperature by heat exchange between the outlet gas from the waste heat recovery and the reactor feed [605–607]. This corresponds to moving the lower heat exchanger (which in many converter types exchanges heat between the feed to the first catalyst bed and the effluent from the last bed) partly or completely to a position outside the reactor and downstream of the waste heat recovery. In this way the temperature at which the heat can be recovered is increased, ultimately to a point where the inlet temperature to the waste heat recovery is equal to the outlet temperature from the last catalyst bed (converters without lower heat exchangers). The effect is to allow steam production at a higher pressure. In modern large ammonia plants, steam is most often produced at 100–120 bar; the waste heat downstream from the synthesis converter is typically available between about 450 °C and about 260 °C with the exact value depending on the layout and the conversion within the converter system. This means that the heat must be recovered partly in a boiler (down to about 350 °C), and partly in a boiler feed water preheater. In low energy plants the tendency is to increase the conversion by increasing the outlet concentration of ammonia from the last catalyst bed, which requires a reduced outlet temperature (at constant pressure). At the same time, maximum high pressure steam production in the synthesis loop is required in order to satisfy the overall energy balance in the plant. In such cases it may be necessary to recover part of the heat before the reaction in the converter system has been completed. This may be accomplished [425, 460, 461, 585] in a converter system with two or more converters by cooling by steam production, not only after the last converter, but also between the converters. It may, of course, also be done by indirect cooling by steam production inside the converter pressure shell [554, 589], but this solution is not preferred in modern plants because of the safety risk involved.

Synthesis loop heat exchangers, and especially the waste heat boilers, are critical items in ammonia plants. The ammonia content and the high partial pressure of hydrogen in the reactor effluent makes proper boiler design and selection of construction materials essential. Several designs are available. Reviews and examples may be found in [558, 559, 608–615, 919].

It has been proposed [616, 617] to recover heat, not by converting water to steam, but by evaporating liquid ammonia at high pressure. This could take place inside the ammonia converter, since it does not involve the safety risks involved in the evaporation of water in such equipment. The hot, high pressure ammonia would be used for power production by expansion in a turbine in a way similar to the production of power from high pressure steam. The proposed system has, however, not been used in industrial practice.

6.4.5 Construction Materials in Ammonia Synthesis Loops

The corrosion phenomena causing the most problems in ammonia synthesis loop equipment are hydrogen attack, nitriding, and stress corrosion. A survey of the types of corrosion is given in [618, 619]. Steel for service at different hydrogen partial pressures and temperature may be chosen on the basis of the so-called Nelson-Curves [620, 621]. A survey of materials problems in ammonia plants is given in [622].

Hydrogen attacks steel by dissociation and diffusion of atomic hydrogen into the metal. The hydrogen atoms react with carbon in the steel causing the formation of methane. Methane cannot diffuse in steel, and very high pressure may build up in "pockets" in the steel, ultimately causing destruction of the material. The reaction can be counteracted by alloying the steel with carbide-forming elements such as Cr and Mo. Hydrogen is to some extent soluble in all construction materials. Too rapid cooling may trap the dissolved hydrogen, whereby the material becomes brittle. Slow cooling will allow the hydrogen to escape, and the material stays ductile. Examples of hydrogen attack are given in [623–625].

Nitriding is caused by ammonia reacting with the surface of the steel forming a hard and brittle surface layer of nitride [626, 627]. It has been found by both experience and experiments that low alloy steel can be safely used in gas containing ammonia at temperatures below approximately 400 °C, whereas stainless steel, Inconel and Incoloy can be used at any temperature relevant for ammonia synthesis loop equipment. Formulas for prediction of the rate of nitriding of stainless steels in ammonia synthesis loops are given in [628].

Stress corrosion cracking may be caused by a number of reasons including presence of trace compounds, especially chlorine, poor heat treatment after welding, etc. Some examples of stress corrosion on loop equipment are given in [629–633]. Stress corrosion cracking of ammonia storage tanks is dealt with in Sect. 6.5.6.1.

The above indicates that low alloy steel, for example 2 1/2 Cr 1/2 Mo, can be used for most loop equipment including the converter pressure shell. Critical parts are the reactor internals (the "basket"), which are most often made from stainless stell (SS 304), and the inlet part of the waste heat boiler which must either be made from or cladded by corrosion resistant material, most often Incoloy.

6.4.6 Product Recovery from Ammonia Synthesis Loops

Ammonia is without exception recovered as a product from synthesis units by cooling at synthesis pressure to condensation followed by separation of the liquid product from the gas, which is recirculated to the converter. The arrangement of the ammonia separator(s), the recirculator, and the addition of make-up gas and extraction of purge gas is discussed in Sect. 6.4.1.1. In high pressure synthesis loops the cooling can be done by water cooling and/or air cooling [634]. In modern plants, which operate at moderate pressures, this is in most cases supplemented by refrigeration. The refrigeration is typically supplied from a mechanical ammonia refrigeration cycle with one, two or several refrigeration levels. Refrigeration down to − 25 °C has been used which corresponds to cooling by evaporation of ammonia at about atmospheric pressure.

The liquid ammonia from the separator contains a small amount of dissolved gases. These are partly released by pressure reduction in a "let-down vessel" normally to about 20 bar. The gases released are normally used as fuel. Recycle to the synthesis gas compressor has been proposed [635]. Flashing at an intermediate pressure with recycle of the released gas to the process feed has been used, e.g. in revamp projects [452].

After the let-down vessel the ammonia is further flashed to cool to the temperature required for product storage. The ammonia vapour is removed from the let-down gas by washing with water, and the gas is used for fuel, while the ammonia water is distilled together with ammonia water obtained from washing of the purge gas. The ammonia vapour from the last flash is normally sent to the refrigeration compressor.

Other fluids than ammonia have been used in the refrigeration circuit (see [636]), and absorption refrigeration (normally based on ammonia/water) [637, 638] has been used instead of mechanical refrigeration. It has been proposed to remove the ammonia from the partly cooled reactor effluent by absorption in water. This procedure makes it necessary to dry the recycle gas before it reenters the converter. The drying may be done with molecular sieves or by injection of liquid ammonia into the gas stream and separation of the resulting ammonia – water mixture [32, 639].

Ammonia recovery in part by adsorption of ammonia vapour on a solid adsorbent has also been proposed [640–642]. The warm gas from heat recovery passes through a cold adsorption vessel in which the adsorbent is saturated with ammonia. The ammonia is desorbed and goes with the product gas to further refrigeration cooling and product separation. The cold gas from the separator passes through another adsorption vessel which has been heated and depleted from ammonia by warm converter effluent. This vessel is now cooled and ammonia is adsorbed from the separator effluent. In this way the adsorption beds act as both heat and mass exchangers, transferring heat from the reactor effluent to the reactor feed and at the same time ammonia from reactor feed to reactor effluent. Both energy savings and increased production capacity are claimed.

Separation of ammonia through semipermeable membranes has been suggested. Use of a polyethylene membrane was proposed in 1954 [643], whereas a more recent development based on anion-exchange membranes is described in [644, 645].

6.4.7 Purge Gas Recovery

The synthesis gas produced by any of the methods described in Sect. 6.3. will contain certain concentrations of "inerts", i.e. compounds which are not consumed by the ammonia synthesis reaction and do not interfere with the catalyst performance. The inerts are typically methane, argon, and traces of other rare gases. In some cases significant amounts of helium may be present originating from He-containing natural gas.

In plants using a liquid nitrogen wash for final purification (see Sect. 6.3.5.2) the concentration of inerts may be so low – about 100 vol ppm or less – that they are dissolved in the liquid ammonia leaving the separator. In all other cases it is necessary to purge a gas stream from the synthesis loop in order to prevent excessive build up of the inerts.

The purge gas is ideally extracted at a point, where the inerts concentration is highest and the ammonia concentration lowest, i.e. after product separation and before make-up gas addition. However, for various reasons this is not always possible, as discussed in Sect. 6.4.1.1. The purge gas will, wherever it is taken from the loop, contain a few percent ammonia, 10–30% inerts, mainly methane and argon, and 60–85% of a 3 to 1 mixture of hydrogen and nitrogen. Ammonia is normally recovered by condensation after cooling to a very low temperature in a purge gas chiller and/or by washing with water. The remaining gas may be used as fuel in the reformer or in auxilliary installations.

The hydrogen and nitrogen content of the purge gas represents, however, potential raw materials for ammonia production, and energy has been invested in the production of especially the hydrogen and in bringing the gases to the synthesis pressure. As energy cost increased, it became increasingly attractive to separate at least the hydrogen from the purge gas and return it to the synthesis loop. In this way the yield of ammonia from hydrogen in the make-up gas in a typical synthesis loop may be increased from about 94% to above 98%. Recovery of nitrogen is less attractive, especially in plants based on steam reforming, because preferential recovery of hydrogen will shift the required hydrogen/nitrogen ratio in the synthesis gas to lower values, from 3.0 to in some cases as low as 2.8. This makes it possible to transfer some of the reforming duty from the primary to the secondary reformer (see Sect. 6.3.1 and [646, 647]), which means a saving in the capital cost of new plants and possibilities for a capacity increase in existing plants. The addition of purge gas recovery is an attractive revamp option, partly because it can result in both a reduction in the specific energy consumption and in an increased capacity, partly because in most cases it can be done as a separate undertaking with minimum interference to other parts of the plant.

Several types of processes have been developed for purge gas recovery. The most important are discussed in the following. Reviews and comparisons may be found in [648–650].

6.4.7.1 Cryogenic Separation

The inerts may be separated from the purge gas in a cryogenic unit similar to the process used in some cases for synthesis gas purification (see Sect. 6.3.5.2). The purge gas is washed with water and then dried over molecular sieves to prevent freezing of water and ammonia on the heat exchanger surfaces. It is thereafter cooled to partial condensation in a heat exchanger. Liquid and gas are separated, and the hydrogen-rich gas is reheated by heat exchange with the incoming gas. The liquid, which contains most of the inerts, is depressurized, and the resulting gas is reheated by heat exchange with the incoming gas. It is further used for regeneration of the molecular sieves in the drier, before it is used as fuel. Refrigeration is supplied by the expansion of the off gas and, if required, by additional chilling from the ammonia refrigeration system.

The allowable operating pressure in cryogenic purge gas recovery units depends on the heat exchanger type. If plate heat exchangers are used, the maximum pressure is about 90 bar. With other heat exchanger types, a higher pressure is possible. The pressure drop on the hydrogen rich stream is marginal, whereas the off gas becomes available at low pressure; typical recoveries (the fraction of a compound in the incoming gas which is found in the hydrogen rich gas) are about 90–95% for hydrogen, 25% for nitrogen, 25% for argon, and 4% for methane. Cryogenic purge gas recovery is discussed in [650–659]. Results obtained in industrial plants are described in [660–662]. Argon recovery in a cryogenic purge gas recovery unit is discussed in [663].

6.4.7.2 Separation Through Membranes

A very rapid development has taken place in the use of membranes for gas separation. Reviews are given in [664–666]. The most popular process for purge gas recovery using membranes is based on the use of hollow fibres, although other types such as spiral wound membranes and stacks of flat elements have also been used.

In the processes using hollow membranes, the feed gas, which has been chilled and normally washed to remove ammonia to a low level, is passed to the "shell side" of an assembly which contains a large number of hollow fibres. The fibres pass through a "tube sheet" at one end while they are sealed at the other end. Hydrogen will – together with smaller amounts of other compounds – pass through the membrane and pass out on the "tube side". Hydrogen recovery and purity depend on the membrane material, on the pressure differential, and on space velocity, and typical values are difficult to give. Feed pressure can be high and is often equal to synthesis pressure. It is of fundamental disadvantage of the membrane processes that the pressure drop is on the hydrogen rich gas, while the off gas is obtained essentially without pressure drop. In some cases two stage processes are used so that two hydrogen rich streams are obtained, one at intermediate pressure and one at lower pressure. The process is, however, simple and reliable, and it has become popular in the ammonia industry both for new plants and for revamps. Descriptions of membrane processes may be found in [667–670]. An example of industrial applications is given in [671].

6.4.7.3 Separation by Pressure Swing Adsorption

Pressure swing adsorption (PSA) is in many respects a rather attractive process for hydrogen recovery in ammonia plants. High recovery of hydrogen (about 90%) can be obtained with 100% rejection of inerts, and the pressure loss through the PSA unit is on the off gas, while the hydrogen is available approximately at feed gas pressure.

However, the feed gas pressure is limited to about 40 bar with today's technology so that the overall pressure loss of the hydrogen is high, and this has, together with investment considerations, limited the utilization of the technology in the ammonia industry. Descriptions of the technology may be found in [358–361]. A comparison between membrane separation and separation by pressure swing adsorption is given in [672]. Cryogenic separation and PSA are compared in [673].

6.4.7.4 Use of Metal Hydrides

It has also been considered to use metal hydrides for hydrogen recovery from ammonia synthesis purge gas. Hydrogen may react with certain metals with the formation of hydrides, and the reaction can be reversed by increasing the temperature and/or reducing the pressure. The process has not yet found acceptance in the ammonia industry. Description of the technology may be found in [674, 675].

6.5 Complete Ammonia Production Processes

6.5.1 General

In the foregoing sections the individual process steps involved in the production of ammonia from various feedstocks have been described. However, it is very important how these "building blocks" are combined with each other, and with the steam and power systems, to form a complete facility for the production of ammonia. The way this is accomplished has a major impact on plant efficiency and reliability, and much of the difference between the several ammonia processes and much of the development in ammonia production technology may today be found in these areas. It may be said that while "Ammonia Technology" was in the early days of the industry most often understood as "Ammonia Synthesis Technology" or even "Ammonia Converter and Catalyst Technology", it is today interpreted as the complete technology involved in transformation of the primary feedstock to the final product ammonia.

The most important factor for the choice of process layout is, of course, the type of feedstock. Before the Second World War coke dominated (see Sect. 6.3.6). During the war several plants based on natural gas were constructed in the USA [522], and natural gas has since then been the preferred feedstock in the USA as well as in other parts of the world. There has, however, also been a significant production based on partial oxidation of heavy fuel oil or gasification of coal, especially in Europe and in countries like India and China; also naphtha has been a preferred feedstock in some areas. During the 1970s there was, due to the oil crises, a renewed interest, especially in the USA, in coal as feedstock for ammonia production, but an expected major change to coal-based production of ammonia did not materialize. Comparisons of the economics of ammonia production from different feedstocks may be found in [160, 676–691].

Another important factor is plant capacity. Until 1960 the capacity of individual installations had increased to a maximum of about 400 MTPD. The units were often multi-train units meaning that several parallel trains were installed in synthesis gas preparation and the synthesis loop – not necessarily the same number of trains in each part of the plant. In the early 1960s technological developments made it possible to construct large capacity single train plants, [692–695], and since then this has been the dominating concept. In special situations small plants may, of course, be of interest [696–700]; but in "normal" conditions plants with capacities below about 1000 MTPD (metric tons per day) are rarely considered [958]. On the other hand, it seems that logistics and the dwindling economy of scale have limited the maximum capacity to below 2000 MTPD. Today there are no technical limitations which prevent the construction of single train ammonia plants with a larger capacity.

In the following section, energy balances in ammonia production will be discussed. The difficulties related to the comparison of energy consumption in different process schemes will be dealt with, and finally a description will be given of the main concepts in the integration of ammonia production units and steam and power systems.

6.5.2 Energy Balances

6.5.2.1 Thermodynamic Analysis

In an energy analysis a reference value for energy must be selected. Normal practice in the industry is to select the lower (or "net") heating value, which is generally used as a basis for pricing of fuels. (The lower heating value (LHV) of a hydrocarbon is the enthalpy released when the compound is converted by reaction with air to carbon dioxide and water vapour at atmospheric pressure and 25 °C). If a production process consumes liquid water, the energy balance can only be fulfilled if the energy for evaporation of the water is taken into consideration. In this case, therefore, it is more correct to use the higher heating value of the gas, where the product of the combustion is liquid water. The use of the higher heating value (HHV) is, however, a concept which is not generally accepted in the gas processing industry. If steam and/or power is imported or exported to or from the production process, it should be converted to the equivalent amounts of fuel by using standard conversion factors. Examples of energy balances based on lower heating value may be found, for example, in [16, 246, 426]. In [246] balances based on higher heating value are also given. The difficulties associated with balances based on lower heating value are briefly discussed in [701].

The balances based on a heating value are based on the first law of thermodynamics. A more stringent analysis should, however, also take the second law of thermodynamics into consideration. The first law simply states that the energy is conserved, and an energy analysis using this law can, therefore, only be used to estimate where and how the energy is converted. It cannot be used to estimate where energy is degraded. This can, however, be shown by a second law analysis which shows where and to what extent energy is wasted through irreversibilities in the process. The combination of the first and second laws of thermodynamics leads to a concept called ideal work, chemical availability or *exergy*. The latter notation will be used in the following. For literature on Second Law Analysis and Exergy, see [702–708]. Second Law Analysis of ammonia production is given in [703, 709–711] and for CO_2 removal systems in [291].

The first law of thermodynamics is a simple conservation balance for energy. In a steady-state flow system it is

$$\Delta H = Q - W_s \tag{9}$$

where ΔH is the enthalpy difference between the outlet and inlet streams, Q the heat added to the system, and W_s the shaft work delivered by the system.

In any process requiring or delivering work, there is a limiting value called the ideal work or maximum work, which results when all changes in the system are carried out in a completely reversible process. This implies that all heat transfer between the system and the surroundings is carried out reversibly and at the temperature of the surroundings, T_e. The second law of thermodynamics then requires:

$$Q_{rev} = T_e \Delta S \tag{10}$$

where ΔS is the entropy change in the system.

Insertion of (10) in (9) then gives the equation for the ideal work:

$$W_{ideal} = -(\Delta H - T_e \Delta S) \tag{11}$$

The exergy is now defined as the ideal work which can be obtained from a stream when it is converted to a standard state, an equilibrium state of the surroundings where no more work can be extracted from it, i.e.:

$$Ex_1 = -(\Delta H_{1->e} - T_e \Delta S_{1->e}) \tag{12}$$

where 1 is the actual state, e the equilibrium state of the surroundings, and Ex is the exergy. The exergy is therefore also the theoretical minimum work that must be used to produce the material from the surroundings in a steady-state flow process.

The change in exergy from a state 1 to a state 2 is:

$$\Delta Ex = Ex_2 - Ex_1 = \Delta H - T_e \Delta S \tag{13}$$

The change is, therefore, independent of the selected equilibrium standard state except for the temperature T_e.

It is also seen that the exergy change is equal to the change in Gibbs free energy when the process is carried out isothermally at T_e.

As an example, the differential exergy change for a gas which is heated with no change in composition and pressure can be seen from the following equation:

$$\begin{aligned} dEx &= d(\Delta H) - T_e d(\Delta S) \\ &= C_p dT - T_e^* C_p/T dT \\ &= d(\Delta H)^* (1 - T_e/T) \end{aligned} \tag{14}$$

In this case the exergy change is consequently equal to the change in enthalpy multiplied with the ideal Carnot efficiency of a heat engine. The exergy change is, therefore, small when all heat is transferred to or from the surroundings at a temperature close to T_e.

The maximum work a system can perform is found by combining Eqs. (11) and (13).

$$W_{ideal} = -\Delta Ex \tag{15}$$

i.e. the maximum work is equal to the negative value of the exergy change, which is the sum of the exergies of the leaving streams minus the sum of the exergies of the entering streams. The *lost work* is then found as the difference between the ideal work the system can perform and the actual work it delivers. In the actual work should also be included work which could be extracted from useful heat transferred to or from the system by using an ideal Carnot engine. The lost work is then:

$$W_{lost} = -\Delta Ex - W_s + (1 - T_e/T_s)^* Q_s \tag{16}$$

where T_s is the temperature of the source delivering the heat Q_s to the system.

In processes with chemical reactions it is necessary, if one is interested in absolute values of the exergy, to define a convenient standard state where, in principle, the compound can do no more work. Different standard states have been defined in the literature depending on how many 'spheres' (atmosphere, hydrosphere, lithosphere) are included in the analysis.

In gas processing systems such as the production of ammonia from natural gas, liquid water and air at 25 °C and atmospheric pressure are normally selected as standard states for the environment. This implies that these raw materials are assigned zero exergies. This may not always be realistic, in particular in places where clean water is expensive to provide.

There is also the problem of carbon dioxide which is only present in a very small amount in the atmosphere. It can, therefore, be argued whether it is correct to assign a value of the exergy of carbon dioxide equal to the theoretical minimum work to extract it from the atmosphere.

For this reason, some authors prefer to define the standard state for each pure component at atmospheric pressure, and some prefer to assign a partial pressure to carbon dioxide as it exists in a typical flue gas. The difference between different standard states is mainly the value assigned to air. For other gas processing streams the difference in exergy is less than 1–2%.

In the following, the standard state selected for each component in air is the partial pressure in dry air at atmospheric pressure. The standard state for water is liquid at 1 atmosphere. The temperature of the surroundings is 25 °C. The exergy of a stream of arbitrary composition can now be calculated by use of Eq. (12).

This is done in two steps. The first contribution Ex_{25} is the Gibbs energy of formation of the mixture from the compounds in the surroundings and considering the work required to transfer the individual reactants and products to 1 atm partial pressure, which is the standard state for the Gibbs energy.

The second contribution Ex is the transfer of the compounds in the mixture from 1 atm partial pressure at 25 °C to the actual temperature and the actual partial pressures in the mixture. This is done by use of Eq. (13), where the enthalpy and entropy differences are calculated by standard methods. An example of results obtained by exergy analysis of an ammonia production process is given in Sect. 6.5.3.1.2.

6.5.2.2 Comparison of Energy Consumption for Different Cases

Ammonia can be produced from many different feed materials and with different process design. In order to compare and study all these cases, it is necessary to have a common basis for comparing the energy consumption.

6.5.2.2.1 Theoretical Minimum Energy Consumption

In all processes there is a theoretical minimum energy consumption which is equal to the chemical 'value' of the pure ammonia.

The simplest raw materials are air and liquid water, which are both present in sufficient quantities, and the 'value' of the ammonia can then be expressed as the ideal chemical work that can be obtained when ammonia is converted to air and water. This work is the exergy of ammonia.

For gaseous ammonia at 25 °C and 1 atm, and for liquid ammonia at − 33 °C, 1 atm, the exergies in Gcal/MT (MT = metric tons) NH_3 are shown in Table 6.2 together with the higher and lower heating values, which are the first law "values" of ammonia.

It is interesting to note that whereas the heating values both drop when ammonia is condensed, this is not the case for the exergy. The reason is that the exergy also considers the entropy change. It is also seen that the exergy lies between the higher and lower heating values; the reason being that the amount of work obtained by condensation of water is much smaller than the corresponding enthalpy change.

Table 6.2. Energy value of ammonia, Gcal/MT

	Exergy	HHV	LHV
Gaseous 25 °C, 1 atm	4.73	5.37	4.44
Liquid − 33 °C, 1 atm	4.81	5.01	4.08

6.5.2.2.2 Influence of External Conditions

Even when the thermodynamic basis has been defined, it may be quite difficult to compare the energy consumption for different process schemes on the basis of information available in the literature. The only safe way is to prepare, on the same basis and using the same calculation procedures, complete mass and energy balances for the different schemes within well defined battery limits. This will obviously not always be possible – due to lack of information about the processes or due to lack of resources to do the calculations – and it may therefore not be possible to compare figures gives in the literature for different cases. There are, however, a number of external factors which should in any case be considered, and which may have a significant influence on the calculated energy consumption. Important among these are the following.

Cooling Water Temperature. The temperature of the cooling water has an influence on the power consumption for air compression and make-up gas compression and more importantly on the power required for the refrigeration compressor because a larger amount of the ammonia is condensed in the water cooler when the cooling water temperature is low. The overall effect in a natural gas based plant is shown in Fig. 6.20 (from [701]). It is seen that the effect is quite significant, especially at low cooling water temperature. However, it should be noted that a significant part of the effect is due to a reduced temperature and, therefore, a reduced pressure in the turbine condensers. This effect may or may not be obtained in the specific case depending on the choice of type of turbines and turbine condensers.

Feed Composition. It is evident that the type of feedstock has a major influence on energy consumption in an ammonia plant. But even for the same type of feedstock, differences in composition may have an effect. As an example, consider natural gas, which can be characterized by its content of "bulk" impurities (most often carbon dioxide and/or nitrogen), "trace" impurities

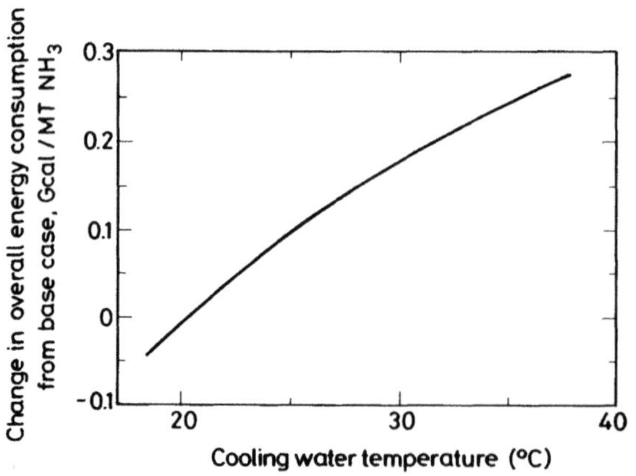

Fig. 6.20. Influence of cooling water temperature on overall energy consumption. Basis Haldor Topsøe's low energy ammonia process (from [701])

(sulfur compounds, etc.), and by the mix of hydrocarbons which can be expressed as the hydrogen to carbon ratio of the gas.

- Trace impurities will normally have no influence on the energy consumption. They may be catalyst poisons in which case they must be removed carefully before the gas is introduced in the process.
- Carbon dioxide may be present in certain natural gases in significant concentrations. If the concentration is very high, it will normally be reduced in a separate purification step before the ammonia process proper. But natural gas with up to 15–20% carbon dioxide can be used without problems as feed gas for an ammonia plant. The presence of such amounts of carbon dioxide will have an effect on the operation in several sectors of the plant, but the effect on overall energy consumption is modest, especially when a physical carbon dioxide removal process is used. A typical increase in energy consumption caused by the presence of 10% carbon dioxide is about 0.05 Gcal/MT ammonia.
- Nitrogen could, in contrast to carbon dioxide, be regarded as a desirable impurity since it is a component required for the synthesis of ammonia. It has, however, only a very limited influence on energy consumption. A concentration of 10 vol % nitrogen in methane will cause a decrease in the overall energy consumption of about 0.02 Gcal/MT ammonia and a slight change in the operating conditions in the reformer due to the decrease in the amount of process air.
- The hydrogen to carbon ratio of the gas has an influence on the overall stoichiometry of the theoretical synthesis reaction. When the hydrogen to carbon ratio decreases, more water is consumed in the reaction and more carbon dioxide is produced together with the ammonia (see Section 6.5.4.1). The effect on the energy consumption of the process is modest. For pure n-butane as a feed, the consumption expressed as the lower heating value will increase by about 0.1 Gcal/MT of ammonia. If the higher heating value is considered, the same consumption corresponds to a decrease in energy consumption of about 0.07 Gcal/MT of ammonia compared to a case with a pure methane feed. The hydrogen to carbon ratio of the feed will also have an influence on the acceptable feed preheat temperature due to the risk of steam cracking in the preheat coil when higher hydrocarbons are present.

Ammonia Supply Temperature. A part of the energy consumption in ammonia plants is used to condense the product ammonia. If the ammonia could be supplied as a gas, this energy could be saved. On the other hand, extra energy must be used if the product must be supplied to an atmospheric storage as liquid at − 33 °C. The differences compared to supply of liquid ammonia at ambient temperature are:

- Ammonia supplied as gas at 3 atm, savings: 0.14 Gcal/MT
- Ammonia supplied as liquid at − 33 °C, extra consumption: depends on cooling water temperature; 0.07 Gcal/MT with cooling water at 30 °C.

6.5.2.3 Steam and Power System

Fundamentally an ammonia production process is a combination of endothermic and exothermic processes. Heat is required at high temperature for the initial conversion of feedstock, e.g. by steam reforming; this heat is supplied by external combustion of fuel in the primary reformer and by internal combustion in the secondary reformer (or by internal combustion alone in partial oxidation or gasification processes).

Heat is also required for the generation of process steam, for the preheat of process streams, and in most cases for the regeneration of solvent in a carbon dioxide removal unit.

Furthermore, motive power is required for compression of feed streams and make-up gas for the synthesis loop, for recovery and refrigeration of product ammonia, and for certain other purposes, mainly pumps and blowers.

Heat is generated at various temperature levels by exothermic processes, notably in gasifiers or secondary reformers, by shift conversion, and by the ammonia synthesis reaction. Heat at high temperature is also available in the flue gas from the primary reformer.

In the design of an energy efficient ammonia production process a number of considerations are relevant:

- Efficient use of the feedstock must be ensured. In modern natural gas based plants with recovery of hydrogen from the loop purge gas, ammonia yields above 95% of the theoretically possible (corresponding to stoichiometric conversion of feedstock with air and steam to ammonia, see Eq. (17), Sect. 6.5.3.1.1) may be achieved.
- Non-converted feedstock must be collected and used as fuel with the highest possible efficiency.
- Excessive firing in primary reformers and/or preheaters must be avoided by efficient equipment design and by maximum use of heat recovery for preheating of feed streams and, in some cases, combustion air.
- Consumption of power for compressors, pumps etc. must be minimized by the choice of optimized operating conditions and by the choice of efficient machines including drivers. In all large modern plants this involves use of centrifugal compressors driven by stream turbines or gas turbines.
- Heat available in various process streams must be recovered to the maximum possible extent for useful purposes. In all large modern plants this involves recovery for the production of high pressure steam.
- The release of waste heat must be adjusted to various temperature levels as required by the steam generation, so that matching amounts of heat are available for preheating boiler feed water, steam generation, and steam superheating.

The above points indicate that the steam and power balance of an ammonia plant is of major importance for the overall efficiency.

The concept of recovering heat for the production of high pressure steam and using the steam for driving compressors via steam turbines in plants based on steam reforming was developed during the 1960s [426, 712–714]. Descriptions of systems for heat recovery by high pressure steam production may be found in [709, 715–718]. See also Sect. 6.4.4. A discussion of the energy and money value of waste heat and steam is given in [719].

An example of a steam generation system in a large, modern, natural gas based ammonia plant is shown in Fig. 21 and 22 (from (720)). Fig. 6.21 shows how heat is recovered for preheating boiler feed water and for generating and superheating high pressure (110 bar, 510 °C) steam. The total heat recovery in the steam production system is 3.00 Gcal/MT of ammonia corresponding to 4.52 tons of high pressure superheated steam per ton ammonia. It could be possible by adjusting the design – for example by introducing combustion air preheat – to reduce the heat available for steam production to about 2.4 Gcal corresponding to about 3.6 tons of steam per ton ammonia. (It is of course possible by extra firing to increase the steam production to any desired level).

Fig 6.22 shows how low grade heat is recovered in the same plant; as much as possible – in this case slightly above 0.4 Gcal/MT of ammonia – is used to preheat demineralized water before deaeration. Other uses of low grade heat include regeneration of solvent in the carbon dioxide removal unit and production of low pressure steam which is used in deaeration and in the carbon dioxide removal unit. It may also be noted that condensate stripping is in this plant integrated with deaeration and carbon dioxide removal. General descriptions of process condensate treatment are given in [721, 722, 920].

Fig 6.23 shows schematically how steam is used in the same plant. Superheated high pressure steam from the process units generated as shown in Fig. 6.21 is passed through the back pressure part of the turbine driving the synthesis gas compressor; part of the steam is condensed in the condensing part to provide sufficient power, while the main part is extracted as medium pressure steam. The medium pressure steam is partly used for process steam, partly for other turbines driving the air compressor, refrigeration compressor, and flue gas fan. Low pressure steam to balance the needs in carbon dioxide removal, boiler feed water deaeration, and other uses is generated in the back pressure turbine driving the flue gas fan and extracted from the turbine driving the air compressor. Total power production in the four large turbines is about 23 MW in this 1000 MTPD ammonia plant.

Figs. 6.21, 6.22, and 6.23 show only one possible layout of the steam system in large ammonia plants. Other similar systems are described in [31, 615, 718, 723, 724, 949]. Other types of drivers for ammonia plant compressors than steam turbines are discussed in [398] (electrical motors) and [399, 400] (gas turbines). Steam and power production for ammonia plants in cogeneration units is described in [725–727, 921]. Steam and power balances in ammonia plants and ammonia-urea complexes are discussed in [959] with special reference to the use of gas turbines.

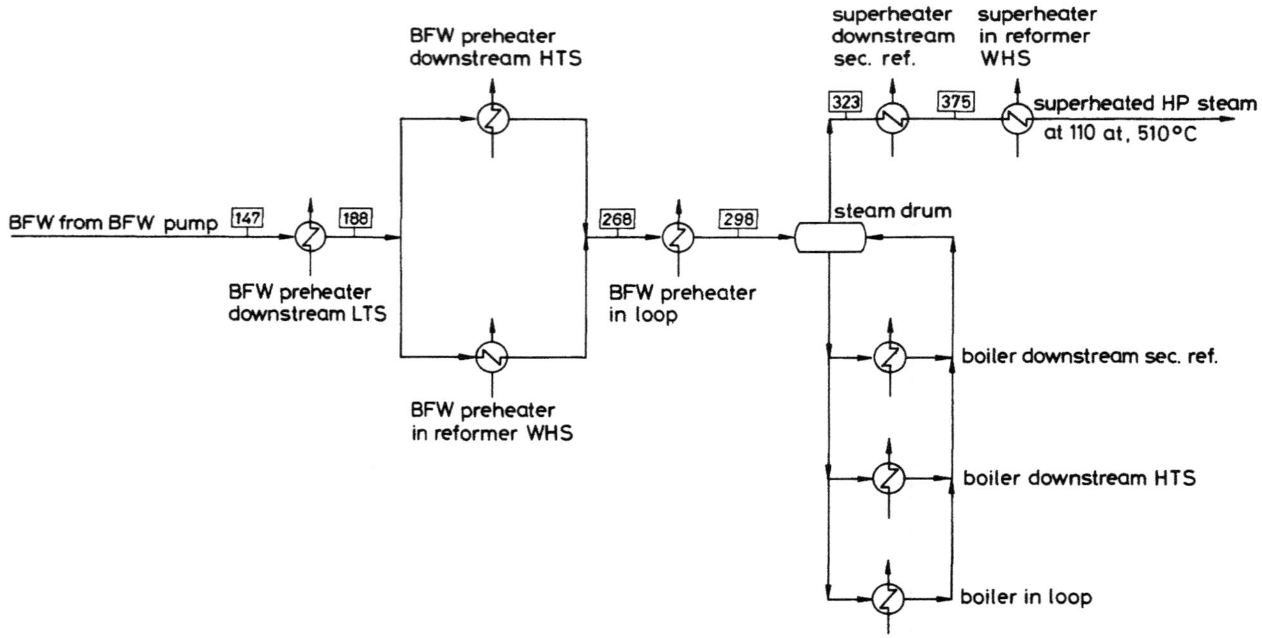

Fig. 6.21. Steam production in a large, modern natural gas based ammonia plant (from [720])

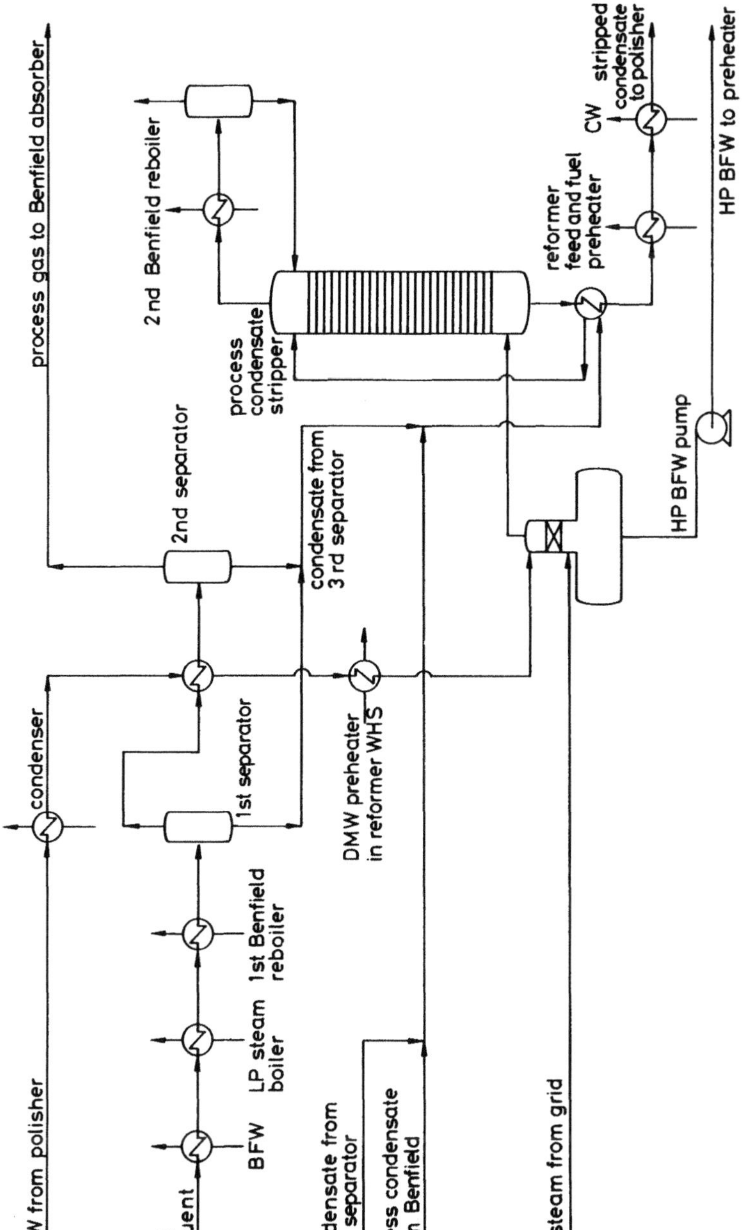

Fig. 6.22. Recovery and use of low grade heat in a large, modern natural gas based ammonia plant (same plant as Fig. 21) (from [720])

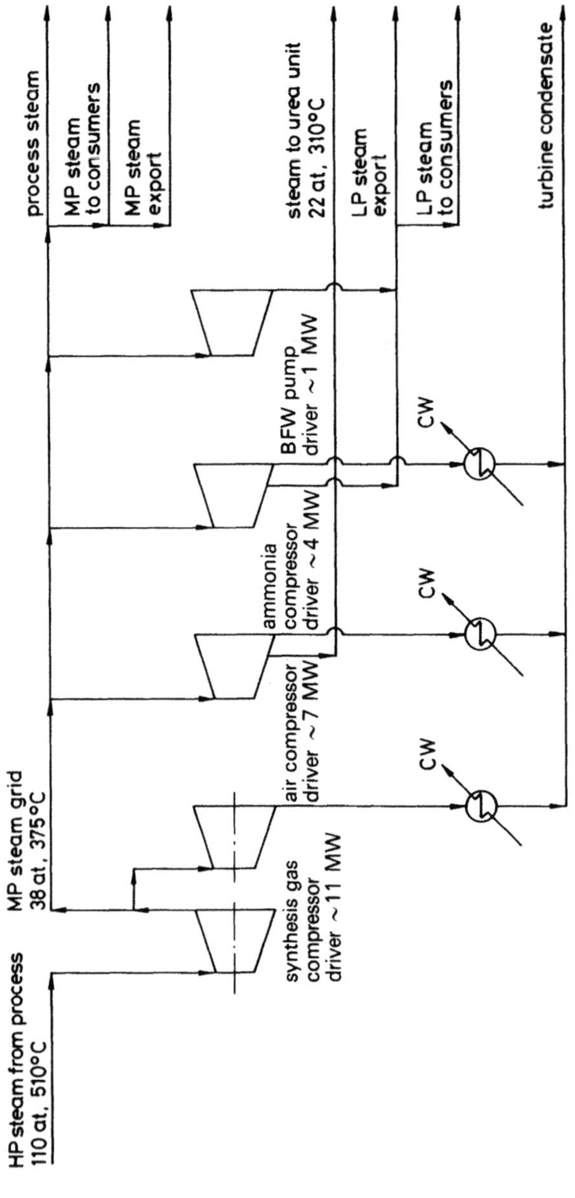

Fig. 6.23. Use of high pressure steam in a large modern natural gas based ammonia plant (same plant as Figs. 21 and 22)

6.5.3 Processes Based on Steam Reforming of Light Hydrocarbons

6.5.3.1 Theoretical Considerations

6.5.3.1.1 Stoichiometric Production of Ammonia from Methane

Consider the production of pure liquid ammonia and pure carbon dioxide from pure methane, pure liquid water, and dry air. By pure in the context is meant that the component is present in its pure form at 1 atm partial pressure.

The reaction scheme may be written as:

$$\begin{aligned} &0.4422\,CH_4 + 0.6155\,H_2O(\ell) + 0.6407\,(0.2099\,O_2 + 0.7804\,N_2 \\ &+ 0.0003\,CO_2 + 0.0094\,Ar) = NH_3(\ell,\ -33\,°C) \\ &+ 0.4424\,CO_2 + 0.0060\,Ar \end{aligned} \tag{17}$$

The thermodynamic properties of the components are given in Table 6.3.

It can be seen from the table that the exergy change of the reaction is much smaller than the corresponding enthalpy change. This is mainly due to the difference caused by condensation of ammonia. It also appears from the table that it is necessary to assign a negative heating value to the liquid water if the heat balance is to be fulfilled in the case using the lower heating value.

6.5.3.1.2 Energy Analysis of a Low Energy Ammonia Process

A *first law* analysis based on both lower and higher heating values was reported in [246].

This analysis was repeated in a slightly updated version for one of the cases, i.e. for a low energy process with a net energy consumption (lower heating value) of 7.01 Gcal/MT of ammonia, and an exergy analysis was also made. The process concept is an "optimized conventional" scheme with the following features:

- Reforming at a steam to carbon ratio of 2.5 with combustion air preheat.
- Shift conversion in two stages with copper based catalysts.
- Carbon dioxide removal with the Selexol process.
- Final purification by methanation.

Table 6.3. Thermodynamic values of components in ammonia production (Gcal per ton Ammonia).

	CH_4	$H_2O(\ell)$	Air	$NH_3(\ell, -33)$	CO_2	Ar	ΔReact
Exergy	5.15	0	0	4.81	0.12	0	−0.22
LHV	4.98	−0.38	0	4.08	0	0	−0.52
HHV	5.53	0	0	5.01	0	0	−0.52

- Synthesis at 140 kg/cm^{2}g using a two bed radial flow converter with indirect cooling.
- Use of centrifugal compressors driven by steam turbines.

The energy balance was based on the following assumptions:

- Feed and fuel: 100% methane available at the required pressure.
- Cooling water available at 30 °C at the required pressure.
- Ambient conditions: 30 °C, 65% relative humidity, 1 atm.
- Product ammonia supplied at − 33 °C to atmospheric storage.
- No import or export of steam or power across ammonia plant battery limits.

The results of the analysis are summarized in Table 4.

Losses are calculated as the energy content (heating value or exergy) of the ingoing streams minus the content in the outgoing streams including in- or outgoing work. The total loss, therefore, includes the heat loss to the surroundings and the heat loss to the cooling water. For the exergy, these two losses are in most cases much smaller than the losses due to the irreversibilities

Table 6.4. Energy analysis, low energy ammonia plant (Gcal/MT Ammonia)

	HHV	LHV	Exergy
Natural Gas Consumption			
Reformer Feed	5.89	5.32	5.56
– Fuel	1.79	1.62	1.69
Fuel, Aux. Boiler	0.08	0.07	0.08
Total Consumption	7.76	7.01	7.33
Energy to HP Steam System			
Primary Reformer	0.20	0.20	0.11
Secondary Reformer	1.06	1.06	0.53
Shift – CO_2-removal	0.73	0.73	0.32
Synthesis	0.61	0.61	0.29
Aux. Boiler-Misc. Recovery	0.17	0.17	0.07
Total HP Steam Production	2.77	2.77	1.32
Steam Use			
Steam for Turbines	1.56	1.56	0.84
Process Steam	0.97	0.97	0.35
Various Consumers	0.24	0.24	0.13
Total Steam Use	2.77	2.77	1.32
Losses			
Reforming	0.09	0.09	1.18
Steam Generation	0.08	0.08	0.57
Shift + CO_2-removal + Methanator	0.31	0.31	0.16*
Synthesis	0.41	0.41	0.37
Turbines + Compressor	1.56	1.56	0.13
Misc. Losses	0.08	0.08	0.05
Stack	0.23	0.08	0.06
Total Losses	2.76	2.61	2.52
Energy in NH_3 Product	5.01	4.09	4.81

* Mainly CO2 Product

produced in the process. The reason is, of course, that the heat loss is multiplied by the Carnot efficiency.

The largest transfer of heat to the cooling water occurs in the steam system, where the cooling needed to condense the exit streams from the turbines and for cooling in compressor interstage coolers is equal to 1.56 Gcal/MT. But since most of the heat is transferred at about 50 °C, the Carnot efficiency is only 8%, and the exergy loss to the cooling water is only 0.13 Gcal/MT.

The heat loss from the primary reformer occurs at a high temperature, hence giving a high Carnot efficiency. The exergy value is best evaluated by considering the extra fuel necessary to compensate for the heat loss. This is equivalent to about 5% of the total exergy loss from the reforming section or about 3% of the heating value of the total fuel to the reformer.

Almost 70% of the total exergy loss is lost in reforming section and the associated steam generation, mainly due to irreversibilities present in the combustion.

This result can be compared with an ordinary energy or first law analysis, which shows that almost all energy is transferred to the cooling water.

The exergy analysis can be used to calculate a thermodynamic efficiency for production of ammonia. It is simply the ratio between the exergy of product and the total feed which is:

$$\text{Efficiency} = \text{exergy ratio} = 4.81/7.33 = 0.66 \text{ or } 66\%$$

6.5.3.2 Process Schemes

6.5.3.2.1 Early Developments

The overall process scheme for production of ammonia from light hydrocarbons, i.e. natural gas, LPG, or naphtha, has changed little since the early days of the industry. In most cases the plants have comprised the following process steps:

- feed purification
- primary and secondary reforming
- shift conversion
- CO_2-removal
- final purification of synthesis gas
- compression and ammonia synthesis
- ammonia recovery and refrigeration

but there has been, of course, a significant development within each of the process steps, and in addition a dramatic increase has been seen in the capacity of the units and in the integration of the process units with the steam and power system.

The development in design of natural gas plants from the Second World War up to 1954 is summarized in [25] and in [141]. The developments are summarized in Fig. 6.24 and Table 6.5 (adapted from [25]).

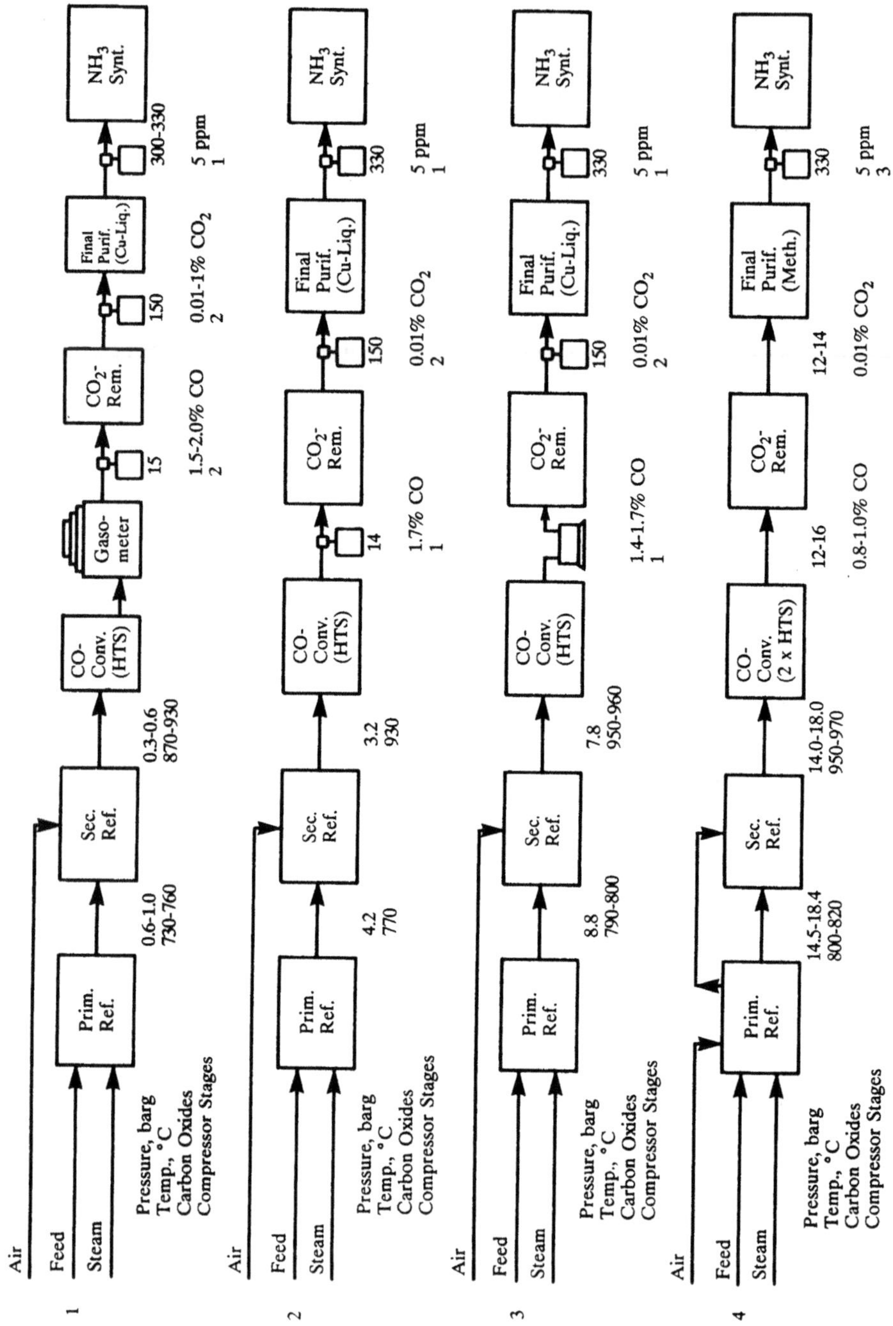
1
Air
Feed
Steam
Prim. Ref.
Sec. Ref.
CO-Conv. (HTS)
Gaso-meter
CO_2-Rem.
Final Purif. (Cu-Liq.)
NH_3 Synt.
Pressure, barg
Temp., °C
Carbon Oxides
Compressor Stages
0.6-1.0
730-760
0.3-0.6
870-930
15
1.5-2.0% CO
2
150
0.01-1% CO_2
2
300-330
5 ppm
1
2
Air
Feed
Steam
Prim. Ref.
Sec. Ref.
CO-Conv. (HTS)
CO_2-Rem.
Final Purif. (Cu-Liq.)
NH_3 Synt.
Pressure, barg
Temp., °C
Carbon Oxides
Compressor Stages
4.2
770
3.2
930
14
1.7% CO
1
150
0.01% CO_2
2
330
5 ppm
1
3
Air
Feed
Steam
Prim. Ref.
Sec. Ref.
CO-Conv. (HTS)
CO_2-Rem.
Final Purif. (Cu-Liq.)
NH_3 Synt.
Pressure, barg
Temp., °C
Carbon Oxides
Compressor Stages
8.8
790-800
7.8
950-960
1.4-1.7% CO
1
150
0.01% CO_2
2
330
5 ppm
1
4
Air
Feed
Steam
Prim. Ref.
Sec. Ref.
CO-Conv. (2 x HTS)
CO_2-Rem.
Final Purif. (Meth.)
NH_3 Synt.
Pressure, barg
Temp., °C
Carbon Oxides
Compressor Stages
14.5-18.4
800-820
14.0-18.0
950-970
12-16
0.8-1.0% CO
12-14
0.01% CO_2
330
5 ppm
3

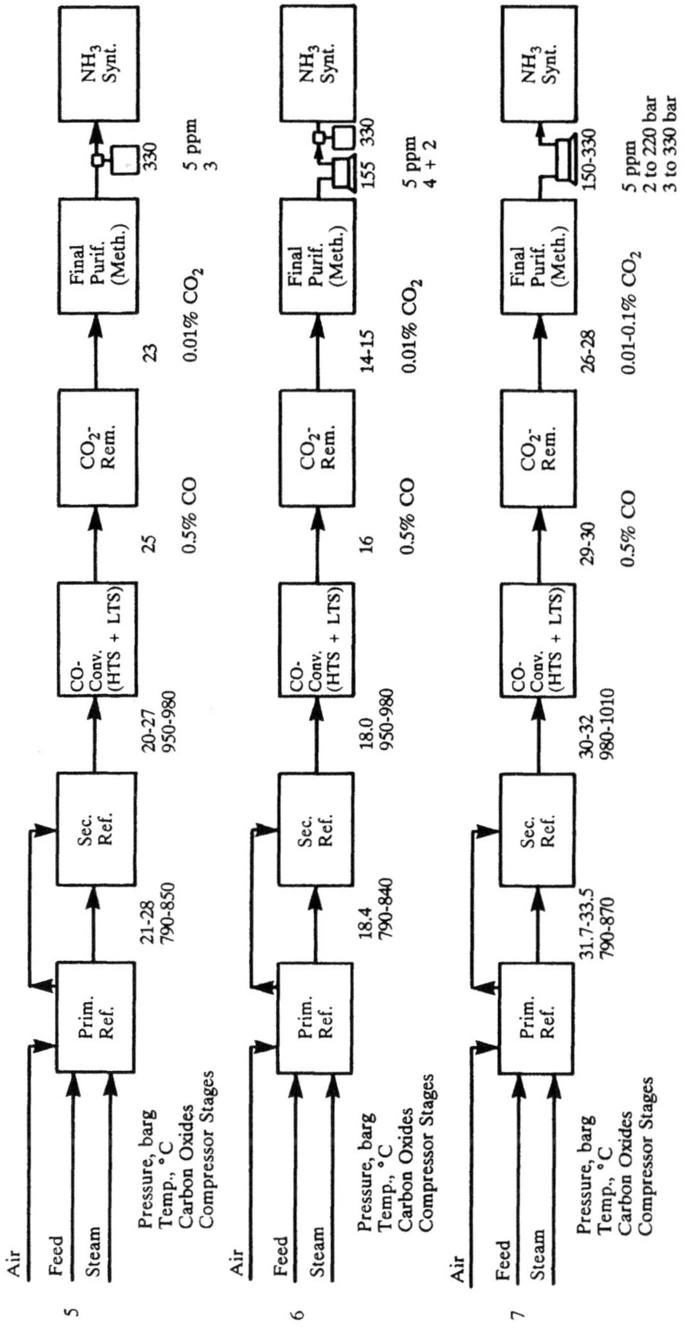

Fig. 6.24. Some steps in the development of ammonia process layout between 1940 and 1972 (cfr. also Table 5) (adapted from [25]).

Table 6.5. Comments to Fig. 6.24

No.	Year	Capacity short tons/day	CO_2-Removal	No. of NH_3 Converters	Process Features and Improvements
1	1940–52	150	Water wash 20% MEA	3	– low pressure steam reforming – poor heat recovery – high energy consumption
2	1953	162	20% MEA	2	– first increase of reforming pressure – significant reduction of power consumption – deletion of gasometer and associated equipment
3	1955	300	20% MEA hot potassium carb. vetrocoke	1	– further increase of reforming pressure – reduction of raw material and power consumption – first introduction of turbocompressor for compression of raw synthesis gas – first single stream 300 tons/day unit – use of raw materials such as refinery off gas and naphtha – use of hot potassium carbonate wash for CO_2 removal
4	1960–62	360	20% MEA	1	– further increase of reforming pressure, improved tube materials – Cu liquor wash replaced by methanation – deletion of compression between gas preparation and purification – preheat of process air – modification of reformer for reduction of heat loss – improved heat recovery at all catalytic process steps

5	1963	320	20% MEA	1	– increase of reforming pressure – high heat recovery – reduced fuel consumption – improved LT shift catalyst – reduction of purge from synthesis loop
6	1964	600	20% MEA	1	– first 600 tons/day single stream unit – first use of turbocompressor for synthesis gas compression to 150 atm – about 80% of power for synthesis gas compression by turbocompressor
7	1965–72	600–1700	20% MEA Vetrocoke Catacarb Carsol TEA/MEA Sulfinol	1	– further increase of reforming pressure combined with low pressure synthesis – all synthesis gas compression up to 200 atm by one two-stage turbocompressor, to 300 atm by three-stage machine – introduction of steam turbines for compressor and pump drivers. Production of high pressure steam – reduction of electric power consumption (in some cases to practically zero) – low (for the time) raw material and operation costs – high preheat temperature of process air in spite of high reforming pressure – first single stream 1700 tons/day unit – improved converter construction gives lower pressure drop in synthesis loop

Descriptions of plants designed in the various periods may be found in the following references:

1940–1952: [139, 522, 728–730]
1952–1960: [143, 433, 731–734]
1960–1965: [144, 145, 735]
1965–1972: [426, 430–432, 736–738]

The "end product" of the development described above (and in more detail in [25]) was, as mentioned, the large capacity single stream ammonia plant.

Fig. 6.25 from [430] shows a simplified process flow diagram for an ammonia plant of this type. This process design – by M. W. Kellogg – was used in a major part of the very rapid expansion of the ammonia industry in the late 1960s and early 1970s. A detailed description of the process including composition of major process streams is given in [739].

6.5.3.2.2 Recent Developments

A major concern of process licensors, engineering contractors, and plant owners, especially since the changes in energy price levels in 1973 and thereafter, has been to develop process schemes with better energy efficiency than the above mentioned classical process.

The effort has been quite successful; the overall energy consumption has been reduced from about 10 Gcal/MT of ammonia in the earliest large capacity single train units to less than 7 Gcal/MT in the most modern plants. Energy saving developments are discussed e.g. in [30–32, 35, 38, 40, 740, 741, 954].

The most important improvements and modifications which have been introduced in the individual process steps as compared to the classical process scheme are:

- In feedstock purification – mainly desulfurization – the development was from adsorption of sulfur-compounds on active carbon to catalytic hydrogenation of sulfur-compounds followed by adsorption of hydrogen sulfide on hot zinc oxide. This concept has been of major importance for the development of processes based on reforming at severe conditions including reforming of naphtha. See also Sect. 6.3.1.
  Feedstock purification as such has no influence on the energy consumption. But efficient purification is a prerequisite for other energy saving measures, especially in reforming and shift conversion.
- In the reforming section, which is the most critical section of the synthesis gas preparation (the "front end"), the most important developments have been the increase in operating pressure, decreased steam addition, use of more sophisticated alloys for reformer tubes, increasing exit temperatures from the primary and secondary reformer, increasing preheat temperatures for feed, process air, and combustion air, and the use of catalysts with higher activity and improved resistance to poisoning and coking. See also Sect. 6.3.1.

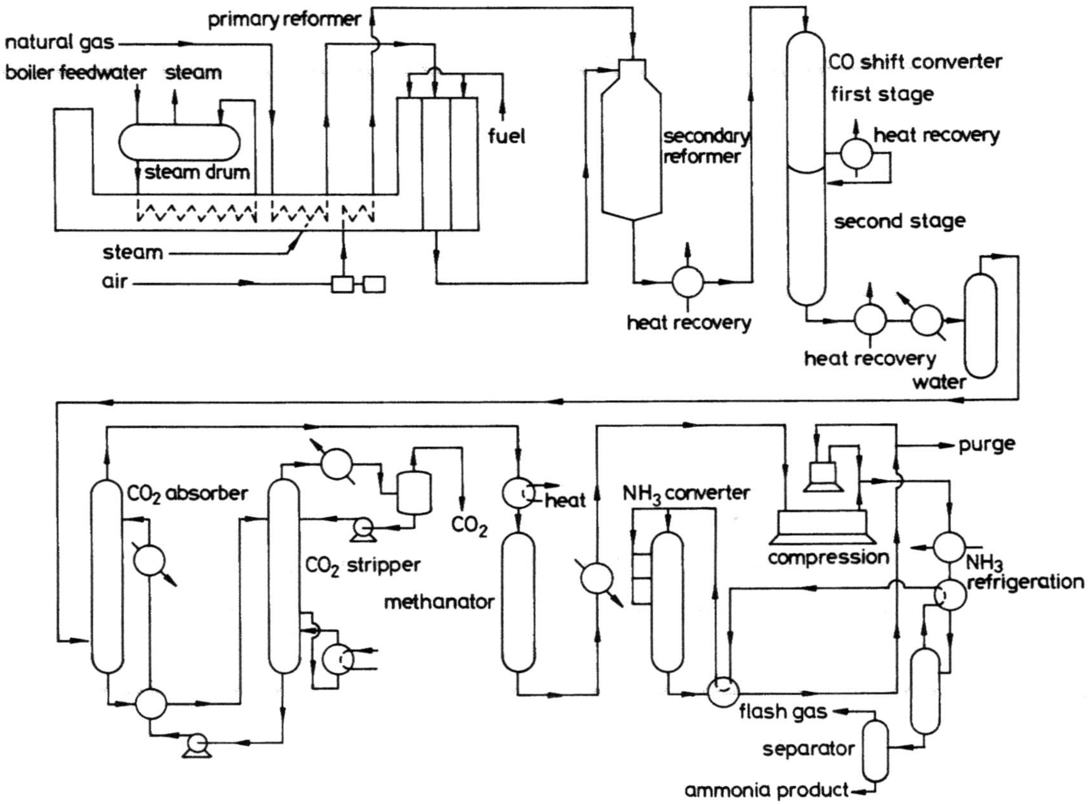

Fig. 6.25. The classical single-train M. W. Kellogg ammonia process. Simplified process flow diagram (from [430])

Reforming at higher pressure saves energy due to reduced power requirements for synthesis gas compression. Reduction of the steam to carbon ratio also saves energy due to savings in process steam consumption. In both cases, more severe conditions are required in the reformer to obtain the required high conversion of the feedstock. This has led to the requirement for better catalysts and better materials for the reformer tubes.
High preheat temperatures for feed and combustion air lead to reduced fuel consumption and thereby to savings, especially when the steam production in the plant must be minimized. Reduction of the flue gas stack temperature reduces the heat loss to the atmosphere. A similar effect has, perhaps more importantly, been obtained by improved insulation in the reformer.
In some processes excess air is added to the secondary reformer. This decreases the duty and the temperature level in the primary reformer, so that firing is reduced. But the energy required to compress the process air is increased, and an extra process step becomes necessary to remove excess nitrogen from the synthesis gas.

- In shift conversion, only one stage–high temperature shift conversion–was used in early plants. In the early 1960s two stage shift units were introduced, originally with two stages of high temperature shift.
  Later low temperature shift catalysts were developed. Since then more efficient catalysts have facilitated operation at the reduced steam content and the correspondingly reduced temperature required to obtain high conversion at the modified conditions. The risk of byproduct formation at a low steam to dry gas ratio has led to the development of new types of catalysts which avoid these problems. See also Sect. 6.3.3.

- In carbon dioxide removal, the water wash was very early replaced with more efficient systems using MEA or hot potassium carbonate as the absorbent. Later, the efficiency of these processes was gradually improved, and also new processes based on physical or combined physical and chemical absorption were developed. See also Sect. 6.3.4.
  The energy requirements have been reduced in part by the introduction of better solvents which reduces both the power requirement for solvent recirculation and the heat requirement for regeneration. Savings have also been obtained by process modifications which have improved the heat recovery from the process streams. The effect of this has been to reduce the requirement for low grade heat which has mainly been supplied by condensation of excess process steam. Therefore, the improvements in carbon dioxide removal have further added to the advantages obtained by reforming at reduced steam to carbon ratio.

- Final purification was in the early plants most often done with a copper liquor wash. With the advent of the low temperature shift, it became feasible to do the final purification by methanation, and since then this has been the preferred method. See also Sect. 6.3.5.

In cases where excess air is used in the secondary reformer, it is necessary to adjust the composition of the synthesis gas by the removal of excess nitrogen. This has been done in a cryogenic purification unit or in a Pressure Swing Adsorption unit. See also Sect. 6.3.5.

- In the ammonia synthesis, improved reactor design, increased converter volume, and to some extent improved catalysts have improved the conversion efficiency. Improved heat recovery has been introduced with the production of high pressure steam instead of preheating of boiler feed water, and single train capacity has been dramatically increased. Drying of make-up synthesis gas and recovery of hydrogen from the purge gas have been introduced to improve performance. See also Sects. 6.4.3, 6.4.4 and 6.4.7, where the improvements in loop and converter design have been extensively reviewed.
- In the steam and power system, developments in compressor and turbine technology have led to significant energy savings. The use of gas turbines as direct drivers for compressors or as part of cogeneration units producing steam and power has also been introduced. Heat recovery has been improved by increasing the steam pressure, by introducing high pressure steam production in the synthesis loop, and by combining steam production and steam superheating downstream of the secondary reformer. Production of process steam from low level heat by feed gas saturation has been used in some cases, and stripping of process condensate with low or medium pressure steam for reuse as boiler feed water has been introduced. See also Sects. 6.3.7, 6.4.4, and 6.5.2.3.
- Developments in instrumentation and computer science have led to the increased use of advanced control systems and computerized optimization of plant operations. Advanced control systems may be used to keep key operating parameters (e.g. steam to carbon ratio at inlet reformer, reformer exit temperature, hydrogen to nitrogen ratio of synthesis gas, synthesis loop pressure, etc.) constant in spite of variations in feedstock properties or other external changes [742–750, 922, 965]. The control systems can be in open loop, where the computer system evaluates data collected by the plant instrumentation and gives new set points to the plant operator who then manually changes the controls. The system can also be in closed loop where the set points are directly controlled and changed by the computer to optimum values calculated in a computer model on the basis of operating data and external (economic) data.

Computer simulation of plant operation may also be made off line real time using computer models of the complete installation, which are capable of simulating dynamic plant response to changes in operating parameters, plant upsets, etc. Such systems may be used for off line optimization studies and for operator training in handling emegencies, start-up- and shut-down situations, etc. Without risk to plant or personnel. Simulators are described in [751–758, 923].

6.5.3.2.3 Commercial Processes

A modern ammonia plant consists, as described in the previous sections, of a combination of several more or less independent sections each with their own options for choosing basic technology and for optimizing process layout and operating parameters. The value of the plant and of the product is very significant and certainly large enough to justify individual design and optimization in each case. This has lead to a situation where all plants are different, and where it may be difficult to give a general description of what is called an "ammonia process".

A limited number of companies possess the basic knowledge and the experience which makes them able to prepare the basic engineering information–process flow diagrams with heat and mass balances, equipment specifications, specification of instrumentation and process control including safety precautions, operating manuals etc. –which is required for the design of an ammonia plant. These companies have also developed their own proprietary design of critical equipment, most often the ammonia synthesis converter and the primary reformer.

The sum of the basic engineering information and the detailed design of proprietary equipment may then be said to represent an "Ammonia Technology", which is made available to others under a license agreement. The basic information is then further developed to a detailed engineering specification by an engineering contractor, which may be the process licensor or another company. A discussion of the contribution of the engineering contractor to the plant design may be found in [36].

In the past, many different companies have offered ammonia technology to the market. The competitive situation has, however, been such that today (1992) only five licensors of ammonia technology are active, namely Braun (original name: C. F. Braun, later Santa Fe Braun, and most recently Brown & Root Braun), ICI, M. W. Kellogg, Haldor Topsøe A/S, and Uhde GmbH.

In the following, each of these commercially important technologies will be discussed in some detail. In addition, a somewhat more limited description will be given of other process schemes which have less commercial importance today. The discussion will be limited to "stand alone ammonia processes", i.e. processes producing ammonia without special considerations for production of other products. Simultaneous production of ammonia and other products, especially the simultaneous production of ammonia and a stoichiometric amount of carbon dioxide to convert all of the ammonia to urea, is treated in Sect. 6.5.4.

The Braun Purifier process [28, 32, 33, 37, 324–330, 409–412, 759, 952] has been available for more than 20 years. A simplified process flow scheme of the updated version is shown in Fig. 6.26 (from [330]). Key features of the process are:

- Reforming with a large ($\approx$ 50%) excess of air. Steam to carbon ratio at inlet of the primary reformer approximately 2.8. Relatively high methane leakage ($\approx$ 1% dry) from the secondary reformer.

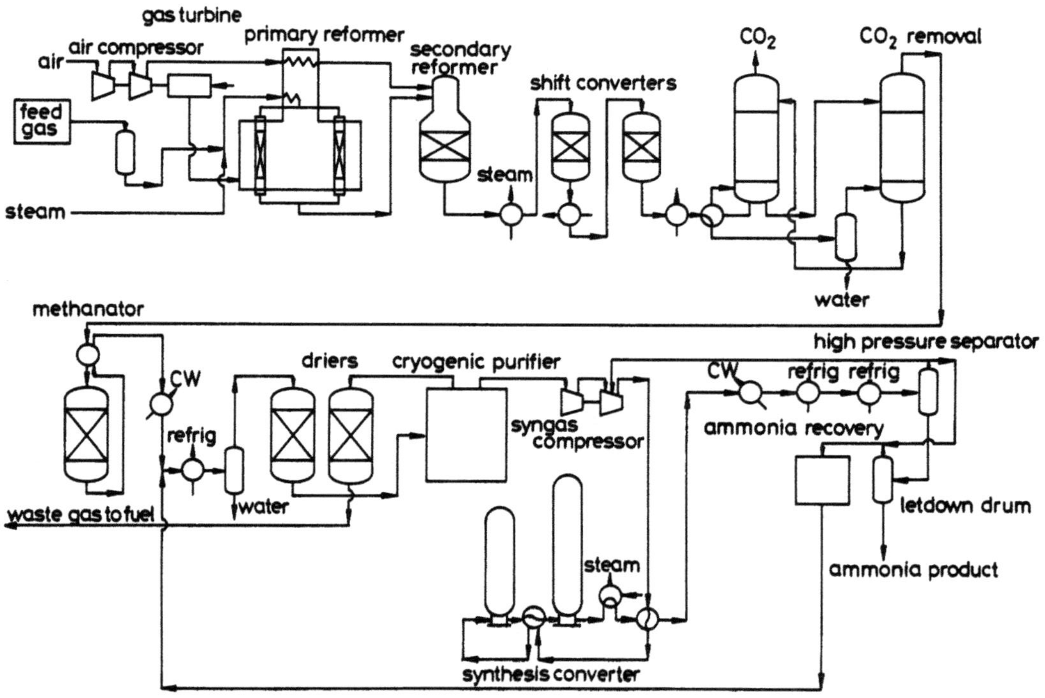

Fig. 6.26. The Braun purifier process, simplified process flow diagram (from [330])

- Air compressor driven directly by gas turbine. Exhaust from the gas turbine is used as combustion air in the primary reformer.
- Conventional high temperature and low temperature shift conversion.
- Carbon dioxide removal, normally by BASF's MDEA process.
- Final purification by methanation followed by adjustment of synthesis gas composition (removal of excess nitrogen and part of the inerts) in a cryogenic unit (referred to as the Braun purifier). Gas drying upstream of the purifier.
- Synthesis of ammonia in a "dry" synthesis loop (with addition of make-up gas after the ammonia separator) at approximately 180 kg/cm^2g with the Braun converter system consisting of two or three single-bed, adiabatic, axial flow converters. Cooling between the converters is by synthesis gas preheating and/or by steam production.

Proprietary items are the purifier and the converters. The process scheme as such is unique and is not used by other licensors. The use of excess air in the secondary reformer and the high methane leakage both contribute to a significant reduction in the size of the primary reformer compared to more conventional process schemes and thereby to savings in fuel consumption. On the other hand, the power consumption in the air compressor is higher than in conventional plants; the whole front end up to the purifier must handle more gas, and the purifier adds pressure drop. Energy consumption in the synthesis loop is low due to the dry and relatively pure synthesis gas. The use of a gas turbine as the driver for the air compressor adds extra firing and thereby increases the amount of waste heat generated in the plant. The extra heat can be recovered for steam production, but the plants will by necessity generate a surplus of steam for export.

About fifteen plants have been designed using the Braun process. Experience from operating plants using the Braun technology is reported in [329, 330, 332, 333, 760–763, 952]. A net energy consumption of 6.7 Gcal/MT of ammonia has been reported [330] for a specific case, operating in a cold climate with partial production of gaseous ammonia and with full credit for a large export of 18 kg/cm^2g steam. In [952] energy consumptions of 7.18 and 6.92 Gcal/MT of ammonia are reported for two plants, also in a cold climate.

ICI has been active in the design and operation of ammonia plants since before the Second World War. In recent years, they have commercialized two processes namely the AMV process, and the LCA process.

The AMV process[37, 413–418, 764] was introduced in the early 1980s. A simplified process flowsheet is shown in Fig. 6.27 (from [416]). Key features of the process are:

- Reforming with an excess of process air. Steam to carbon ratio at inlet reformer approximately 2.8. Relatively high methane leakage ($\approx 1\%$ dry) from the secondary reformer. Addition of part of the process steam via a saturator. Combustion air preheat.
- Process air compressor driven by a steam turbine (all steam produced in the

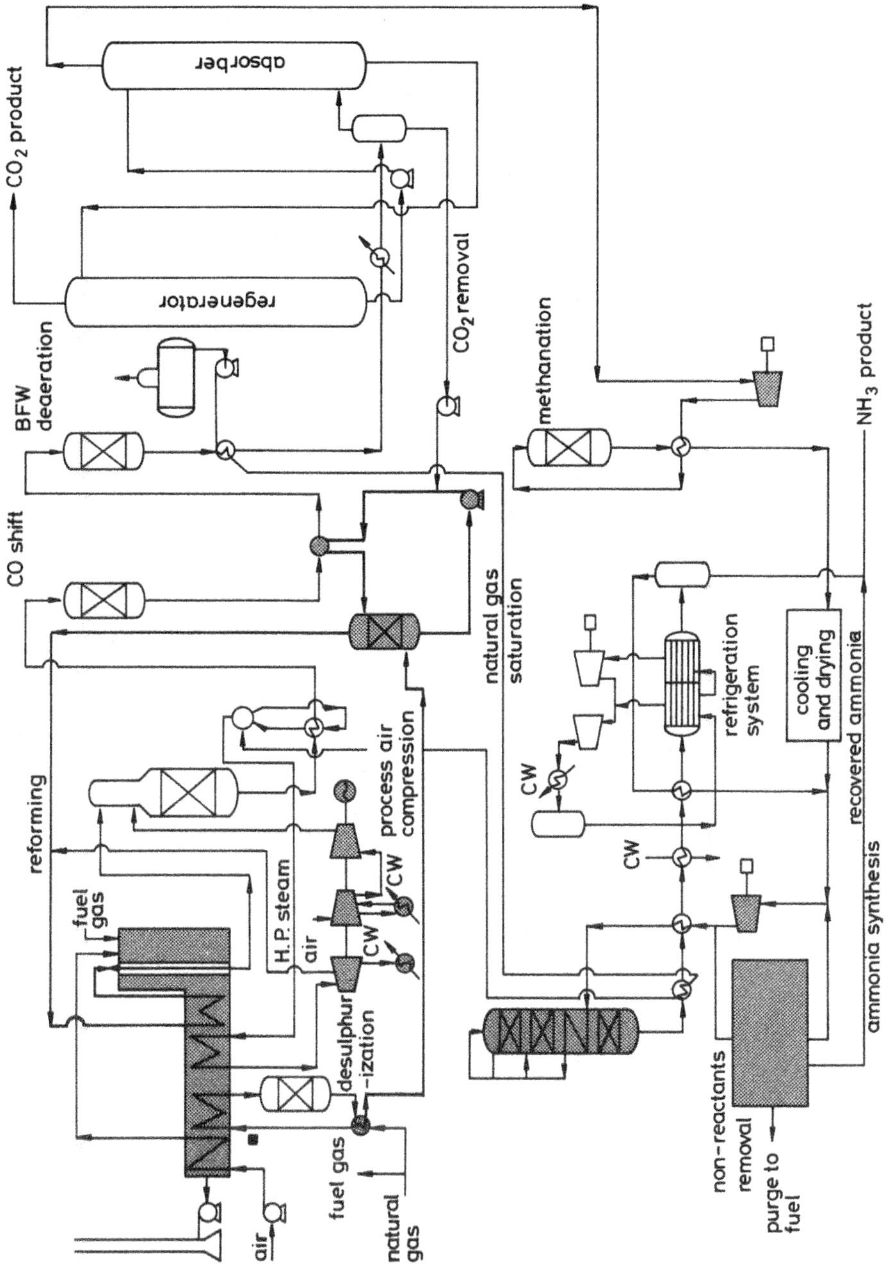

Fig. 6.27. The ICI AMV process. Simplified process flow diagram (from [416])

plant goes to this turbine) with a turbo-alternator on the same shaft. All other power consumers, including synthesis gas compressor, use electric motors.

- Conventional high temperature and low temperature shift conversion.
- Carbon dioxide removal by the Selexol process; compression to synthesis pressure before final purification.
- Final purification by methanation and drying at synthesis pressure. After drying, the gas is mixed with recycle gas and off-gas from a cryogenic purge gas recovery unit. The mixed gas goes via a recirculation compressor and a gas-gas heat exchanger, the hot exchanger, to the synthesis converter. Between the recirculator and the exchanger, a part of the gas is taken to a cryogenic purge gas recovery, where inerts are removed and the gas composition adjusted by removal of excess nitrogen.
- Synthesis at about 90 kg/cm^2g in a three-bed converter with quench between the two first beds and indirect cooling (converter feed preheat) between the second and third bed.

As in the Braun process, the size of the primary reformer is reduced and the size of the air compressor increased in the AMV process compared to more conventional process schemes. This is due to the operation with excess process air and with high methane leakage. Power consumption in the synthetic gas compressor is low because of the low synthesis pressure and the low suction temperature (gas direct from the low temperature Selexol CO_2 removal unit), but this is compensated by increased power consumption for the compression of excess nitrogen and high power consumption in the refrigeration section.

The AMV process scheme may be considered as a variation of the classical scheme with some unique features. It has been used as described in the design of one plant and– with some modifications to accommodate integration with a urea plant–in another plant. Experience from operation of a plant using the process has been reported in [417]. The expected energy consumption has been given as less than 7.0 Gcal/MT of ammonia (for a plant located in a cold climate), but actual data are not available.

The LCA process was announced by ICI in the late 1980s [350–357, 924–927, 951]. A simplified flowsheet is shown in Fig. 6.28 (from [350]). Key features of this process are:

- Reforming in gas-heated reformer (unfired, heat supplied by heat exchange with product gas from the secondary reformer) at a steam to carbon ratio of 2.5. Process steam supplied via a saturator. Large excess of process air; high methane leakage (1%). Process air preheat in fired heater.
- One-stage shift conversion in a cooled reactor; exit temperature approximately 265 °C.
- Gas purification: PSA with adjustment of the hydrogen to nitrogen ratio. Carbon dioxide removal (optional) in a low pressure MDEA wash on the off-gas from PSA unit. Methanation upstream of the synthesis gas compressor.
- Synthesis in a low pressure loop (approx. 80 kg/cm^2g) with a tubular converter (TVA type).

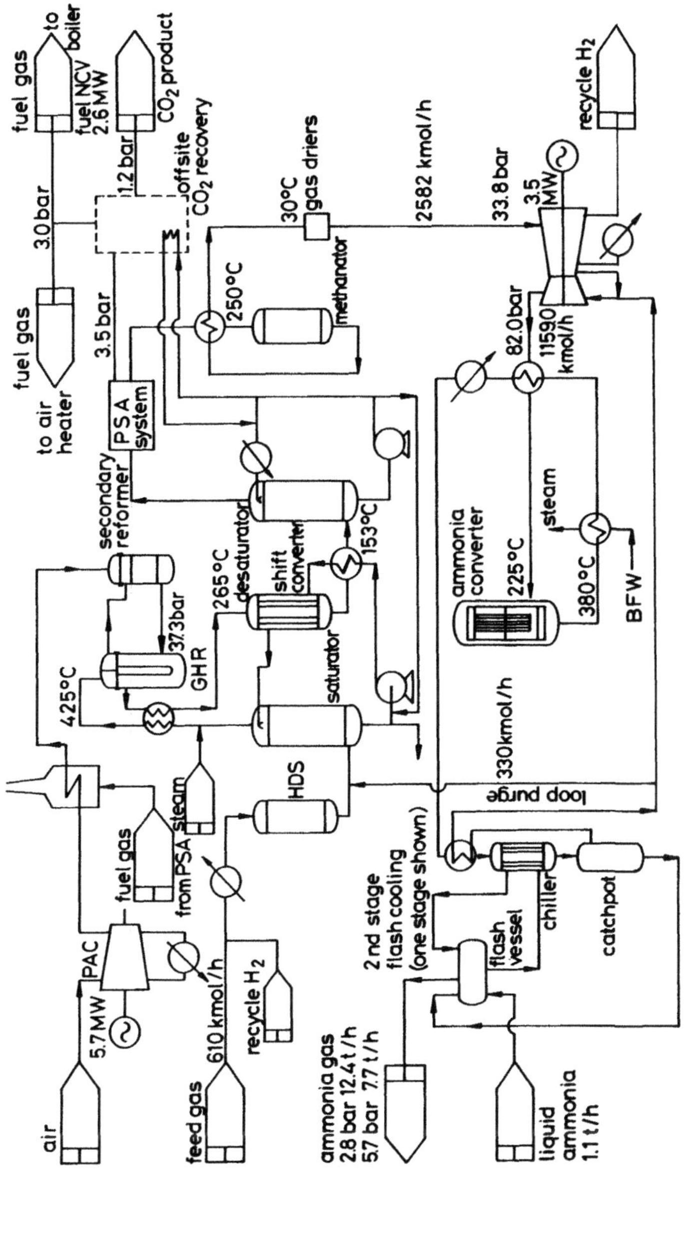

Fig. 6.28. The ICI LCA process. Simplified process flow diagram (from [350])

Proprietary items are the gas-heated reformer [117, 889], the PSA unit [765] and the synthesis converter. The process scheme as such is unique. It is, according to literature, suitable only for small-capacity plants. The process has so far only been used in one installation comprising two units, each with a capacity of 450 MTPD.

The amount of process air is adjusted so that sufficient heat is available in the effluent from the secondary reformer to supply all heat required for the primary reformer; the excess of process air is approximately 75% compared to the conventional process. The extra power for compression of this excess air and the loss of hydrogen in the PSA unit increase the energy consumption. On the other hand, firing in the plant is greatly reduced, and the natural gas consumption is low, only 6.44 Gcal/MT ammonia. The process produces very little steam (the only boiler is a 60 kg/cm^{2}g boiler in the synthesis loop), and extra power must be supplied from outside battery limits for compressor drivers etc. It is suggested to raise the required power in a gas-fired, combined cycle power plant with high efficiency. If this system–which is not incorporated in the operating plants–is considered, a net energy consumption of 6.95 Gcal/MT ammonia is claimed. The energy consumption in the actual plants is reported to be 7.6 Gcal/MT [952].

M. W. Kellogg has played a major role in the development of the modern ammonia industry (see Sect. 6.5.3.2.1). Their most recent Low Energy Process [37, 419, 766–768, 960] represents a further development of the classical process maintaining basically the same process scheme, but introducing updated technology and energy saving features in the various process steps. A simplified process flowsheet is shown in Fig. 6.29 from ([419]). Key features of the process are:

- Reforming with the stoichiometric amount of process air (i.e. the amount of air required to give a hydrogen to nitrogen ratio of 3.0 at the ammonia synthesis converter inlet without dedicated installations for adjustment of the gas composition). Steam to carbon ratio about 3.3; low methane leakage (approximately 0.3% dry).
- Air compressor driven by a steam turbine; preheat of process air to a very high temperature in the reformer waste heat section.
- Gas purification by conventional shift; carbon dioxide removal normally by Selexol or Benfield; methanation; drying of synthesis gas with molecular sieves.
- Synthesis at 140 kg/cm^{2}g or 180 kg/cm^{2}g (depending on plant capacity) in a "dry" synthesis loop (make-up gas added after ammonia separator) using a two-bed horizontal converter with indirect cooling by gas/gas exchange. Refrigeration in so-called unitized heat exchanger. Purge gas recovery.

Proprietary items are the primary reformer, the synthesis converter, and the unitized heat exchanger. About 6 plants have been designed using various versions of this technology. (The total number of ammonia plants designed by M. W. Kellogg is much higher). Operating experience from a plant using the

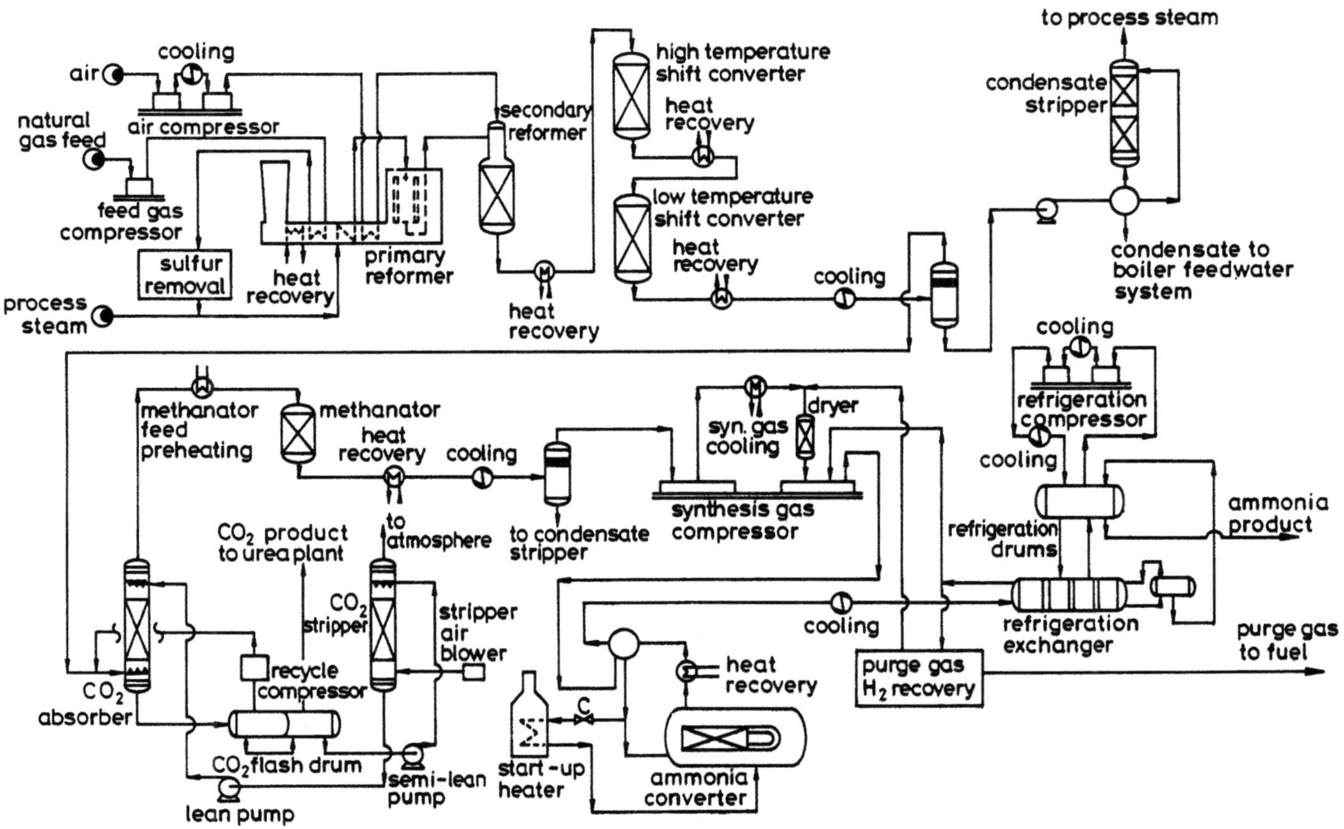

Fig. 6.29. M. W. Kellogg's low energy process. Simplified process flow diagram (from [419])

Selexol process for carbon dioxide removal and located in a cold climate is reported in (766). Energy consumption is reported to be below 29 GJ (LHV)/MT of ammonia.

M. W. Kellogg has also suggested a process (the KAAP process) based on a new catalyst [928, 949]. This technology has been used in a revamp project where a new reactor was installed downstream of the existing reactor in an ammonia synthesis loop (see Sect. 6.4.3.4). In addition, more radically new process schemes deviating from the traditional route have been described. An example is a scheme based on so-called parallel reforming [30, 32, 120, 121] and a low pressure loop with ammonia recovery by water absorption [769]. None of these new developments have been implemented in practice.

Haldor Topsøe A/S has supplied technology for ammonia production for more than 30 years. Unlike the other process licensors, Topsøe is also a catalyst supplier, and the development of catalysts and process has gone hand in hand. The most modern, the Topsøe low energy ammonia process is, as is the M.W. Kellogg low energy process, based on the traditional process layout with new energy saving developments introduced in the individual process steps. Descriptions are given in [28, 29, 32, 33, 37, 246, 595, 770, 929]. A simplified flow sheet is shown in Fig. 6.30 (from [246]). The process is available in two main versions, the main difference being the steam to carbon ratio at the reformer inlet. The first version (with a steam to carbon ratio about 3.3) has been used in about 15 industrial plants. The second version (with a steam to carbon ratio about 2.5 and

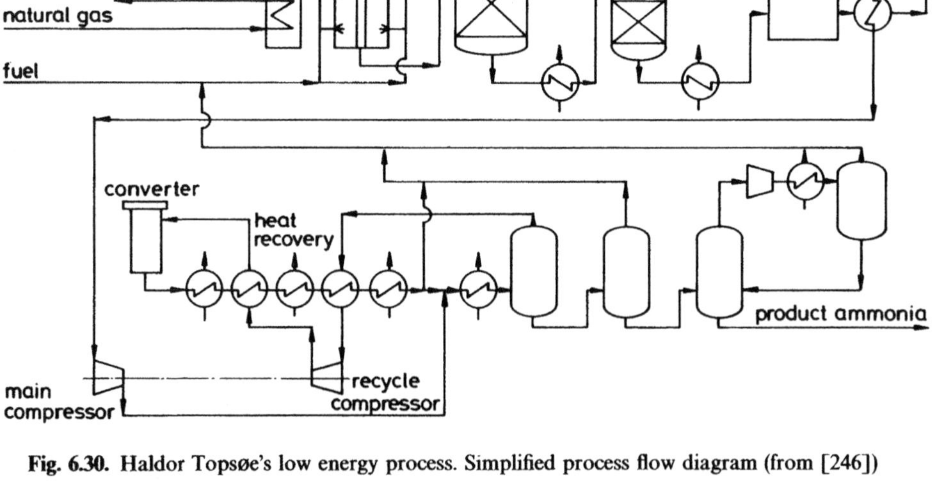

Fig. 6.30. Haldor Topsøe's low energy process. Simplified process flow diagram (from [246])

other energy saving features) is not yet proven in any industrial plant. But each process step is individually proven, and two plans based on the concept are under construction.

Key points in the two versions of the Topsøe process are:

First version:

- Reforming with stoichiometric air and low methane leakage. Steam to carbon ratio about 3.3.
- Process air compressor driven by a stream turbine (or gas turbine).
- Gas purification: conventional shift; carbon dioxide removal by Benfield or Vetrocoke; methanation.
- Synthesis at 220 or 140 kg/cm^{2}g with Topsøe 2-bed radial flow converter (S-200).

Second version:

- Reforming with stoichiometric air and low methane leakage. Steam to carbon ratio 2.5.
- Process air compressor driven by a steam turbine (or gas turbine).
- Gas purification: shift conversion with medium and low temperature catalysts, both copper-based; carbon dioxide removal by MDEA or Selexol; methanation.
- Synthesis at 140 kg/cm^{2}g with Topsøe 2-bed and 1-bed radial flow converters (S-250).

Proprietary items are the primary reformer, the ammonia synthesis converter, and certain catalysts. Operating experience in plants designed by Topsøe has been described in [720, 771–774, 956, 966, 967]. In [967] a net energy consumption of 6.97 Gcal/MT ammonia was reported for a plant located in a warm climate. For the most energy efficient process concept, which has so far not been demonstrated in an industrial installation, a net consumption of 6.67 Gcal/MT ammonia is claimed for a stand alone plant [920].

Uhde GmbH is one of the pioneers in the ammonia industry. Even before the Second World War they constructed ammonia plants (based on the Mont Cenis process). Later they built plants in cooperation with many other companies, such as ÖSW, Topsøe, and ICI.

Uhde has recently commercialized a new low energy process scheme [41, 585, 775, 930]. A simplified process flow is shown in Fig. 6.31 (from [41]). The process scheme is, like the Topsøe and Kellogg process schemes, a further development of classical process schemes. Key features of the process are:

- Reforming at a slightly increased pressure with stoichiometric air and high methane leakage ($\approx$ 0.6%). Steam to carbon ratio about 3.0.
- Process air compressor driven by a steam turbine (or gas turbine).
- Gas purification: conventional shift; carbon dioxide removal by MDEA; methanation.
- Synthesis at approximately 180 kg/cm^{2}g in two converters, a two-bed and a one-bed radial flow converter with steam production between the beds.

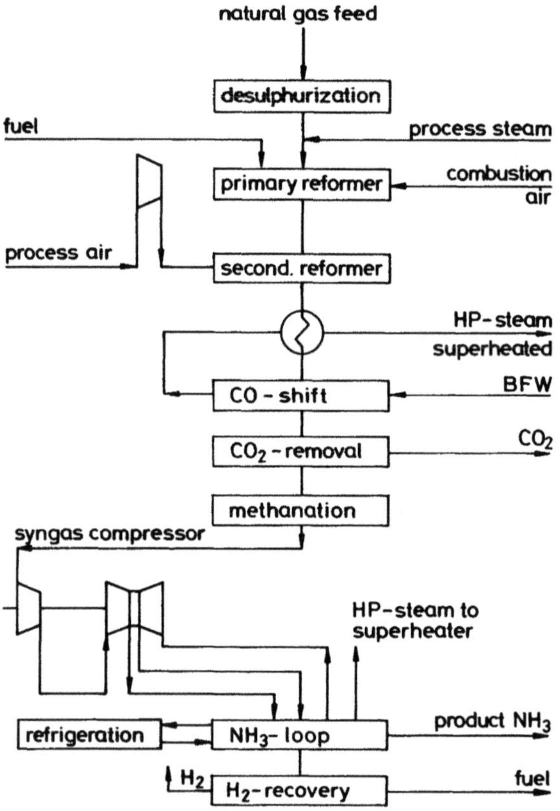

Fig. 6.31. Uhde's low energy process. Block diagram (from [585])

Proprietary items are the primary reformer and the synthesis converters. Two plants are operating with a design according to this process scheme, while a third is under construction. An expected net energy consumption of 6.67–7.20 Gcal/MT ammonia is claimed, depending on the conditions.

As mentioned above, several other companies have proposed low energy process schemes, some based on the conventional process route, some representing a more innovative approach. In the following the most important of these process schemes will be briefly discussed.

The Exxon Chemical Low Energy Ammonia Process has been used in one plant. Key features of the process are:

- Reforming with stoichiometric air and low methane leakage. Steam to carbon ratio about 3.3.
- Air compressor driven by a gas turbine. Exhaust from the gas turbine to the reformer as hot combustion air.
- Conventional shift conversion (with low temperature shift guard vessel); carbon dioxide removal with the Catacarb process; methanation.
- Drying of synthesis gas with molecular sieves installed between the first and second compressor stages.

- Ammonia synthesis at 140 bar with a two-bed radial flow converter with indirect cooling. Make-up gas addition after the separator.
- Cryogenic purge gas recovery.

The energy consumption of the process is given as 29 GJ/MT (6.93 Gcal/MT) ammonia.

The Exxon Chemical process is a good example of processes which are designed for a specific case in cooperation between the plant owner (Exxon Chemical), the contractor (Bechtel), and process licensors (Haldor Topsøe for the synthesis loop and converter). The process and experience from operation of the plant is described in [776, 777].

The *Fluor Ammonia Process* [28, 30, 32, 298, 778] is based on a conventional process scheme. The main features are:

- Use of Fluor's proprietary polypropylene carbonate process (see Sect. 6.3.4) for carbon dioxide removal. It is suggested [778] that carbon dioxide removal and methanation be installed after synthesis gas compression.
- Use of heat downstream of the low temperature shift converter for absorption refrigeration which is used in the ammonia recovery section. In conventional plants this low level heat is used for reboiling in the carbon dioxide removal process.
- Air compressor driven by a gas turbine; exhaust from the gas turbine to the reformer as hot combustion air.
- Primary reforming in a special combination of adiabatic reactors and reheating elements [125, 126].

It is claimed that the above modifications will reduce the energy consumption from 32.0×19^9 Btu/short ton (8.89 Gcal/MT) to 29.5×10^9 Btu/short ton (8.19 Gcal/MT) of ammonia [778].

Foster Wheeler's AM 2 Process [779–781] belongs to the group of processes using an excess of process air to the secondary reformer, so that a specific process step is required to separate excess nitrogen from the synthesis gas. Key features of the process are:

- Only 20–50% of the total hydrocarbon feedstock is treated in the primary reformer which operates at high pressure and a relatively low temperature.
- The remaining feedstock is, together with product gas from the primary reformer, reacted in the secondary reformer with a large excess of process air. Methane leakage from the secondary reformer is high, 2.5 vol%.
- Gas purification is by conventional shift conversion followed by a physical carbon dioxide removal (e.g. Selexol) and final purification in a cryogenic unit. The hydrogen to nitrogen ratio of the synthesis gas at the cryogenic unit inlet is 1.0–1.8.
- Synthesis loop conditions have not been specified.

It is claimed that the energy consumption in the process may be as low as 25.1 $\times 10^9$ Btu/short ton of ammonia (7.0 Gcal/MT).

Foster Wheeler has also suggested a process where the excess of process air is so high that the primary reformer is deleted. This process (see Sect. 6.5.3.3) can operate on all types of hydrocarbon feedstock including feedstocks which can not be used as feed for a primary reformer.

Humphreys & Glasgow have suggested a number of process schemes i.e. the MDF-process, the BYAS-process, and the LEAD-process.

The MDF-process [28, 30, 32, 341–343, 782] consists basically of a hydrogen plant with PSA purification, an air separation plant for production of nitrogen, and a synthesis loop converting the hydrogen and nitrogen to ammonia. Key features of the process are:

- Steam reforming (primary reforming only) at a pressure of about 25 bar.
- Purification of hydrogen by high temperature shift conversion and Pressure Swing Absorption. Optional carbon dioxide-removal located upstream of the PSA unit in case carbon dioxide is required.
- Mixing of the pure hydrogen with pure nitrogen which is imported or produced in an air separation plant.
- Compression and synthesis in an inert-free, dry synthesis loop.

Energy consumption including energy for air separation is given as 32.8 GJ/MT of ammonia (7.84 Gcal/MT). Significant savings in capital costs are claimed [782].

The BYAS-process [783–785] differs from the conventional process scheme mainly in the reformer section and in the final gas purification. Key features of the process are:

- A significant part of the process feed is sent direct to the secondary reformer bypassing the primary reformer.
- An excess of air is used in the secondary reformer; the amount of excess depends on the accepted methane leakage, which may be relatively high.
- Conventional shift conversion and carbon dioxide removal.
- Final purification and adjustment of the hydrogen to nitrogen ratio in a cryogenic unit.
- Synthesis in an inert-free, dry synthesis loop.

Energy consumption is claimed to be as low as 6.85 Gcal/MT of ammonia. The process is well suited for increasing the capacity of existing plants. An example of a revamp case is given in [786].

The LEAD Process [32, 420] is basically a variation of the classical process scheme. Key features are:

- Reforming with a stoichiometric amount of process air using high preheat temperatures for process feed, process air, and combustion air. Low methane leakage. As a special feature, natural gas feed is preheated downstream of the high temperature shift converter, and reformer feed is preheated downstream of the process gas boiler after the secondary reformer.

- Conventional high temperature and low temperature shift; carbon dioxide removal by the Selexol process; methanation; and synthesis gas drying.
- Synthesis at about 125 kg/cm^{2}g in a dry synthesis loop using two converters, the first a two-bed quench-cooled converter, the second a one-bed converter. Absorption refrigeration for ammonia recovery. Cryogenic pure gas recovery.

Energy consumption for the process is claimed to be 7.0 Gcal/MT of ammonia (32).

The PARC Process by KTI [32, 345–349, 785] is very similar to the MDF-process by Humhpreys & Glasgow as described above. Key features of the process are:

- Reforming of natural gas (primary reformer only) at 2.9 MPa. Enriched air from the nitrogen production unit (air separation plant) is used as combustion air in the reformer.
- High temperature shift conversion.
- Production of power in a Rankine cycle with trichlorofluoromethane using heat available downstream of the shift converter.
- Optional carbon dioxide removal before final purification.
- Final purification of feedstock and nitrogen addition in a specially designed Pressure Swing Absorption unit.
- Synthesis in an inert-free, dry synthesis loop at 350 kg/cm^{2}g without refrigeration.

An energy consumption of 29.21×10^9 Btu/short ton of ammonia (8.11 Gcal/MT) is given for the PARC process [345–348]; in [349] a figure of 27.86 $\times 10^9$ Btu/short ton (7.74 Gcal/MT) is given. Experience from an operating plant is reported in [787].

Linde has also suggested a process scheme [344] similar to the MDF-process and the PARC-process. Key features are:

- Production of pure hydrogen by tubular reforming; high temperature shift conversion; and purification by proprietary Pressure Swing Absorption technology.
- Production of pure nitrogen by proprietary air separation technology.
- Mixing of hydrogen and nitrogen in the correct ratio and synthesis in a dry, inert-free synthesis loop.

The process is said to be especially suited for small plants. Energy consumption is claimed to be 7.60 Gcal/MT of ammonia.

The Lummus Process [28, 30, 32, 788] is an optimized conventional process scheme. Key features are:
- Reforming with stoichiometric air at a steam to carbon ratio of approximately 3.0. Combustion air preheat.
- Conventional high and low temperature shift conversion; carbon dioxide removal by a physical process; methanation.
- Synthesis at 231 kg/cm^{2}g using the Lummus quench cooled converter.

Energy consumption values from 8.0 Gcal/MT [32] to 7.06 Gcal/MT [28] are quoted.

Montecatini has suggested a process involving a so-called tertiary reformer [30, 123] and low pressure synthesis, in many ways similar to process schemes suggested by M. W. Kellogg [120, 121]. Key features of Montecatini's process scheme are:

- Reforming with stoichiometric air in a special reforming process involving split flow to two primary reformers, one heated by direct firing, the other heated by heat exchange with the effluent from the secondary reformer.
- Air compressor driven by a gas turbine; exhaust from the gas turbine is used as hot combustion air in the fired primary reformer.
- Conventional high temperature and low temperature shift; carbon dioxide removal using a chemical absorption process; methanation.
- Synthesis of ammonia in a low pressure loop (at 60–70 kg/cm^2g). Drying of the mixture of make-up gas and recycle gas by molecular sieves. Synthesis in a proprietary three-bed radial flow converter with indirect cooling between the beds.
- Ammonia recovery by absorption in water and distillation. Cryogenic purge gas recovery. The energy consumption is claimed to be 6.7 Gcal/MT of ammonia [123].

Projects and Development India Limited (PDIL) has proposed a scheme using enriched air as process air [789, 790]. The scheme is very similar to other modern low energy concepts with the exception of the use of enriched air. Key features are:

- Air separation to produce pure oxygen and nitrogen; nitrogen is used as plant inert gas or exported.
- Primary reforming at elevated pressure, about 45 kg/cm^2g, and low exit temperature, about 725 °C.
- Secondary reforming using enriched air with 24–30 vol% oxygen. Low methane leakage.
- Air compressor driven by a gas turbine; compressed air for oxygen production may be drawn from an inter mediate stage of the air compressor. Exhaust from the gas turbine used as hot combustion air for the primary reformer.
- Conventional high and low temperature shift conversion; carbon dioxide removal preferably by a physical absorption process; methanation.
- Compression and synthesis at 220 kg/cm^2g using a two-bed radial flow converter.

Energy consumption for the process scheme is given [789] as 7.40 Gcal/MT of ammonia.

PDIL has also proposed a process [791] where so much oxygen is added to the process air that the primary reformer duty is reduced to zero, i.e. the process uses an autothermal reformer only in the reforming step.

6.5.3.2.4 Revamp of Existing Plants

A large part of the world's ammonia demand is produced in plants which were designed in the late 1960s or early 1970s using the original Kellogg process (see Sect. 6.5.3.2.1) or other processes from the same generation. These plants will typically have an energy consumption of 9–11 Gcal/MT ammonia, and it has become attractive to modernize – "revamp" – such plants in order to increase the energy efficiency and/or production capacity.

Each revamp project must be tailor-made to enable optimization of production, consumption and capital investment based on the specific case. In the most comprehensive and most extensive revamp projects, the energy efficiency may be increased to a level comparable with that of the new low-energy ammonia plants. The required capital investment will, however, in most cases be too high to justify such extensive revamps.

Revamp of an ammonia plant will typically involve a combination of replacing or upgrading of equipment, replacing of catalysts and chemicals and changing the operating parameters. Replacing or upgrading of equipment will, in addition to more efficient operation, also result in an increased lifetime of the plant as well as reduced maintenance costs. A higher on-stream factor can be expected. The new or upgraded equipment is typically designed in such a way that future bottlenecks are eliminated in that part of the plant. Replacing catalysts and chemicals will improve efficiency of the plants and, in some cases, also the lifetime of the unit, while changing operating conditions will mainly improve plant efficiency.

When major improvements in the performance of a plant is desired, it is obviously not sufficient to consider only one part of the plant. A complete review of all sections is required in order to ensure that capacity of the individual units corresponds in all cases to the expected improved efficiency and/or higher capacity after revamp. Discussions on such "audits" may be found in [792, 793], and general discussions on the revamp of ammonia plants in [38, 595, 784, 794–803]. Descriptions of actual revamp projects are given in [452, 786, 804–806, 933, 934, 957].

Revamp Options–Front End.

The most important options for revamp of the front end of an ammonia plant–i.e. the part of the plant where the synthesis gas is generated–using steam reforming are:

- Reduction of the Steam to Carbon Ratio
- Introduction of Combustion Air Preheat
- Introduction of Feed Gas Saturation
- Installation of an Adiabatic Pre-reformer
- Modification of the Shift Unit
- Modification of the Carbon Dioxide Removal System

- Reduction of the steam to carbon ratio will reduce the pressure drop in the front-end of the plant. The reduced production of steam in the waste heat boiler downstream of the reformer is compensated for by the smaller amount of steam required for the process. The reduced flow inside the reformer tubes will result in a reduced firing required for the process. The methane leakage will increase slightly due to the reduced steam to carbon ratio, but the overall results is typically an energy saving of about 0.20 Gcal/MT ammonia when the steam to carbon ratio is reduced from 4 to 3.

Operation of the reformer at a reduced steam to carbon ratio requires modification of the downstream shift (if the steam to carbon ratio goes below the critical level for by-product formation) and carbon dioxide removal sections in order to fully obtain the potential energy savings. The reduced steam to carbon ratio will also require a highly active reforming catalyst in order to eliminate the risk of carbon formation and resulting hot bands on the reformer tubes.

- Introduction of combustion air preheat will reduce the amount of fuel required in the reformer by using the low-level energy in the fuel gas. When the stack temperature is reduced from about 200 °C to 100 °C, the corresponding energy saving is about 0.10 Gcal/MT ammonia. The lower temperature of the fuel gas may cause corrosion problems in the form of condensing sulfuric acid, if the sulfur content of the fuel is high. This problem may be eliminated by desulfurizing the fuel gas together with the process feed gas. The energy available in the preheated, pressurized fuel gas may be recovered in an expansion turbine, which will reduce the pressure to that required in the fuel gas header system.
- The feed gas saturator is used to transfer the low-level energy in the fuel gas into process steam. Hot boiler feed water or process condensate is contacted with cold natural gas in the saturator, whereby the natural gas is preheated, and, at the same time, saturated with water. The water leaves the saturator, make-up water or condensate is added and the water is then reheated in a coil in the reformer waste heat section. About 20% of the required process steam may be taken up by the natural gas in the saturator. The energy saving is typically around 0.10 Gcal/MT ammonia. The saturator also serves as a process condensate stripper whereby overall water consumption may be reduced, and a pollution problem may be solved. The choice between combustion air preheat and saturator depends on local conditions. Combustion air preheat saves fuel directly whereas the saturator saves steam.
- An adiabatic pre-reformer is an attractive revamp option, especially when the aim is to minimize steam export and/or to increase plant capacity in cases where the primary reformer is the bottleneck.

In the typical reformer waste heat section, the process feed is preheated to about 520 °C before entering the reformer tubes, where reforming takes place using the additional heat supplied by the burners. The burner fuel could be reduced, if the process feed were preheated to higher temperatures, but this may not be feasible due to the increased potential for carbon formation by the cracking of hydrocarbons. By passing the preheated process feed through an adiabatic pre-reformer, the higher hydrocarbons in the feed will react quantitatively with the steam causing a temperature drop. The partially reformed mixture can then be reheated safely in the waste heat section before entering the reformer tubes. The overall result is a reduction of reformer fuel consumption and a corresponding reduction in the waste heat available for steam production. The fuel consumption can, in certain cases, be reduced by as much as 15%.

- When the steam to carbon ratio in the reforming section is reduced, the conditions in the shift section must be carefully evaluated. If the steam to dry gas ratio becomes too low, severe problems may arise due to conversion of the iron oxide in the high temperature shift catalyst to iron carbide, which will promote formation of undesirable by-products (hydrocarbons and oxygenates) (see Sect. 6.3.3.1).
  When the steam to carbon ratio in the reforming section is not reduced below the critical limit for carbide formation, it is possible, by adding an extra shift catalyst bed after the conventional shift unit, to reduce the carbon monoxide leakage to around 0.05 dry vol%. This means that less hydrogen will be lost by reaction with carbon monoxide in the methanator, and that the synthesis loop will become more efficient due to the lower inert content in the make-up gas.

- Several revamp options are available for modification of the carbon dioxide removal section depending on the type of carbon dioxide removal process. The processes mostly used in ammonia plants are chemical absorption processes based on either hot potassium carbonate (HPC) such as Benfield, or Vetrocoke, or amine solutions such as MEA. The chemical carbon dioxide removal processes may be improved or replaced with a physical process in which the absorbent is regenerated by simply flashing off carbon dioxide. In this way the need for regeneration heat may be reduced or eliminated. A physical carbon dioxide removal system may result in energy savings of 0.01–0.35 Gcal/MT ammonia.
- In certain cases methanolation may be an attractive revamp option [907, 908]. In this process, the process gas after carbon dioxide removal is fed to a methanol synthesis reactor instead of a methanation reactor. Residual carbon oxides are converted to methanol, which can be recycled to the reformer on the process side, whereby hydrogen loss is avoided, and the inert content of the make-up gas is reduced. An overall reduction in reformer feed consumption of 0.15–0.30 Gcal/MT NH_3 can be obtained depending on plant layout. There will, however, be an increased consumption of reformer fuel to compensate for the reduced heating value of the loop purge gas caused by the reduced methane content. The total saving in energy consumption is therefore somewhat smaller, in the range of 0.05–0.10 Gcal/MT NH_3. As an alternative to recycling of methanol, it can be recovered as a product which is exported or converted to formaldehyde or ureaform, which can be used for coating of urea to prevent caking.

Revamp Options–Synthesis Loop.

The main revamp options of the ammonia synthesis loop are the installation of molecular sieves for drying of the make-up gas, installation of a purge gas hydrogen recovery system, and improvements in the ammonia synthesis converter system.

- The installation of molecular sieves to remove water and carbon dioxide from the make-up gas and the corresponding relocation of the make-up point to the inlet of the synthesis reactor is especially interesting in loops where the make-up gas addition point is upstream of the water cooler. In such cases, the modification will result in a reduced pressure drop as well as a slightly higher conversion. The energy savings will vary with the loop configuration and may be in the order of 0.10–0.15 Gcal/MT ammonia. In process such as the Topsøe and Uhde processes where the make-up gas is added just before the last chiller, the potential savings are much smaller, only about 0.03 Gcal/MT [461]. In such cases installation of make-up gas driers can not be justified.
- Hydrogen loss with the purge gas may be reduced substantially by installing a purge gas recovery unit such as a membrane unit, a PSA unit or a cryogenic unit. Energy savings of up to 0.15 Gcal/MT ammonia have been demonstrated (see Sect. 6.4.7).
- Converter Revamp options are discussed in detail in Sect. 6.4.3.4.

6.5.4 Processes based on Partial Oxidation of Hydrocarbons or Gasification of Solid Feedstocks

In certain economic circumstances it may be of interest to produce ammonia from feedstocks such as fuel oil or coal (see Sect. 6.5.1 and [676–691]). The technology for the production of synthesis gas from such feedstocks is partial oxidation or gasification (see Sect. 6.3.2). Partial oxidation of natural gas was used in a few early plants [137, 138, 807], and it was suggested in a few modern process schemes, for example by Foster Wheeler [808, 809], but it is of no importance today.

Plants based on partial oxidation of heavy fuel oil by the Shell process are described in [166–168] and by the Texaco process in [163–165, 169]. Coal based plants using the Koppers-Totzek process are described in [187, 188], the Winkler process in [202], and the Texaco process in [213–219]. Catalysts for coal based ammonia plants are discussed in [810], and problems in synthesis gas purification in [811, 812]. Complete process concepts are discussed in [179, 182, 813–818].

Two classical process schemes are illustrated in Fig. 6.32 (from [818]). The schemes as the illustrated can be used both with fuel oil or coal as the feedstock. In both cases the feedstock is converted to raw synthesis gas by reaction with oxygen and steam. In scheme A, which is used with gasification processes which require cooling of the raw gas to near ambient temperature for soot removal,

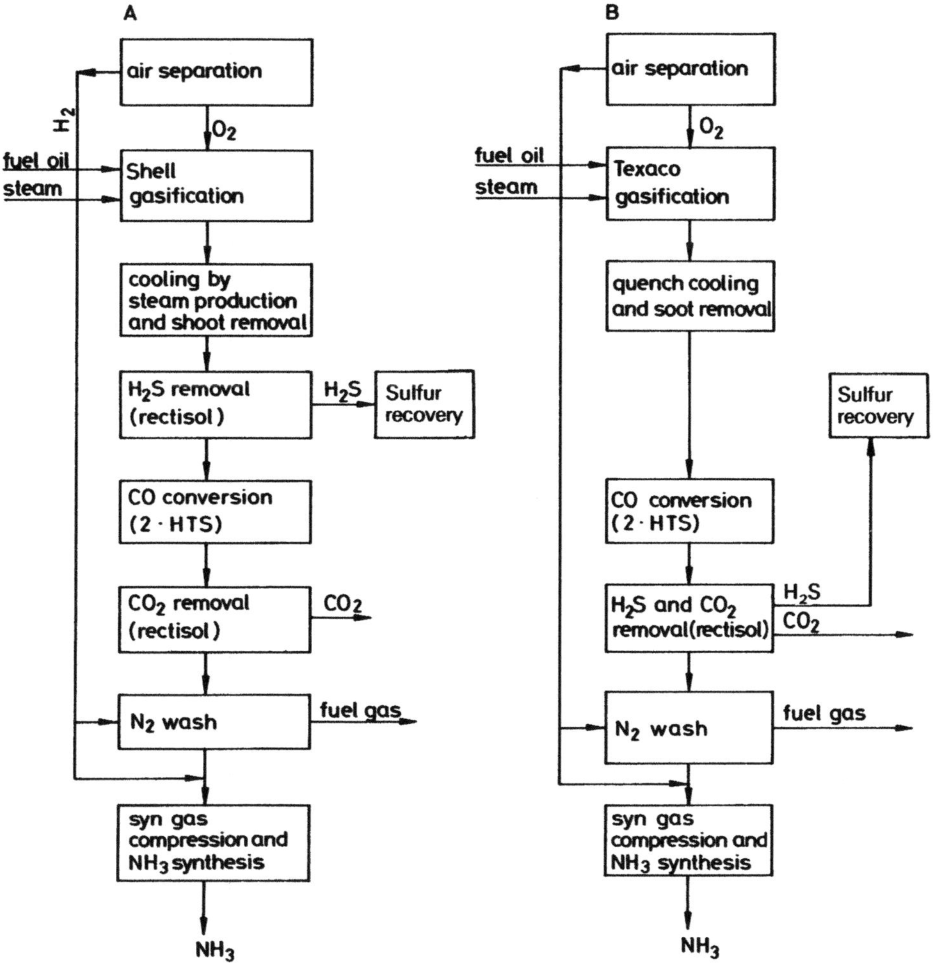

Fig. 6.32. Two classical process schemes for production of ammonia from fuel oil (or coal). (from [818])
a Scheme with Shell gasification (or other process with cooling of raw gas to near ambient temperature)
b Scheme with Texaco gasification (or other process with soot removal at elevated temperature)

hydrogen sulfide is removed upstream of the shift conversion, and conventional shift catalysts are used. In scheme B, which is used with gasification processes where soot is removed at elevated temperature, sulfur remains in the feed gas to the shift conversion section, and sulfur tolerant shift catalysts must be used. In both process schemes the final gas purification is done by acid gas removal and nitrogen wash, and ammonia synthesis is done in a dry, inert-free ammonia synthesis loop. In a special process proposed by Topsoe [818, 819], gasification is done with enriched air, the raw gas goes after cooling and soot removal direct to the shift conversion without sulfur removal. The shift conversion is done in two stages, using a special high activity sulfur tolerant shift catalyst. After the shift conversion, hydrogen sulfide and carbon dioxide are removed in an acid gas removal unit; the remaining traces of carbon oxides are removed by methanation; and ammonia is synthesized in a normal synthesis loop using the Topsoe S-200 two bed radial flow converter. A synthesis gas generation process similar to this scheme and using the same type of catalysts–but based on gasification with oxygen–is used in a large ammonia plant in Japan [213, 214].

6.5.5 Integrated Production of Ammonia and Other Products

In many cases ammonia is supplied direct–or via an intermediate storage–to another production unit where it is further processed to a final product, for example urea or ammonium salts (nitrate, phosphate, etc.). In other cases there may be on the same site production of other products, such as methanol, which do not require ammonia as raw material, but may be made from the same raw materials as ammonia. In both situations it may be advantageous to integrate the production units to obtain savings in investments and/or production costs. The integration may be at various levels, ranging from shared off-site and utility installations (cooling water installations, raw water treatment, auxiliary power production and distribution etc.) through integrated steam systems (with e.g. boiler feed water preheating in one unit for boilers located in other units) to complete integration on the process side. Only the last type of integration will be considered in the following. In addition, the special situations will be discussed which arise when ammonia and another product must be produced in a constant ratio because they are used in fixed amounts in downstream processes. A typical example is a urea complex without facilities for ammonia export. In such units there must be a fixed ratio between ammonia and carbon dioxide production in the ammonia plant and this will have an impact on the design of the production process.

6.5.5.1 Ammonia and Urea

Urea, $CO(NH_2)_2$, is formed by dehydration of ammonium carbamate, $NH_4CO_2NH_2$, which is again formed by the reaction between ammonia and carbon dioxide

$$2NH_3 + CO_2 \rightarrow NH_4CO_2NH_2 \tag{18}$$

It follows that one half mole of carbon dioxide is required for each mole of ammonia to obtain complete conversion. It is evident from Eq. (17) (p. 269) that when natural gas is the feedstock, and the highest possible conversion of feedstock to ammonia is achieved, there may be a deficit in carbon dioxide production.

The most important ways to overcome this problem and to obtain stoichiometric production of carbon dioxide for complete conversion of ammonia to urea can be illustrated by theoretical equations similar to Eq. (17) [701]:

$$\text{a}\quad 0.4422\,CH_4 + 0.6155\,H_2O + 0.6407\,\text{Air} \rightarrow$$
$$NH_3 + 0.4424\,CO_2 + 0.0060\,\text{Ar} \tag{19}$$
$$\text{b}\quad 0.4998\,CH_{3.0786} + 0.7306\,H_2O + 0.6407\,\text{Air} \rightarrow$$
$$NH_3 + 0.50\,CO_2 + 0.0060\,\text{Ar} \tag{20}$$
$$\text{c}\quad 0.4998\,CH_4 + 0.6956\,H_2O + 0.7421\,\text{Air} \rightarrow$$
$$NH_3 + 0.50\,CO_2 + 0.1952\,H_2 + 0.0651\,N_2 + 0.0068\,\text{Ar} \tag{21}$$
$$\text{d}\quad 0.4997\,CH_4 + 0.5006\,H_2O + 1.1896\,\text{Air} \rightarrow$$
$$NH_3 + 0.50\,CO_2 + 0.0112\,\text{Ar} + 0.4284\,N_2 \tag{22}$$
$$\text{e}\quad 0.4998\,CH_4 + 0.5004\,H_2O + 0.6407\,\text{Air} + 0.1153\,O_2 \rightarrow$$
$$NH_3 + 0.50\,CO_2 + 0.0060\,\text{Ar}$$
$$\text{Air} = 0.2099\,O_2 + 0.7804\,N_2 + 0.0003\,CO_2 + 0.0094\,\text{Ar}. \tag{23}$$

Case a above corresponds to Eq. (17). It is seen that in this case, which is based on the use of pure methane as feed, on 100% conversion of the feed and 100% yield of ammonia from hydrogen and nitrogen and of carbon dioxide from carbon and oxygen, there is a deficit of carbon dioxide of 13% compared to the amount required for full conversion of ammonia to urea.

Case b illustrates that if the feed contains heavy hydrocarbons corresponding to an overall hydrogen to carbon atomic ratio slightly above 3.0, then the stoichiometric production of carbon dioxide may automatically be achieved.

Case c shows a solution where the extra carbon dioxide is obtained by production of excess synthesis gas. The excess synthesis gas can be extracted as a side stream before the synthesis gas compressor and used as fuel in the reformer or exported as such or as hydrogen after purification. Alternatively, it can be compressed and passed through the synthesis loop which will then operate with a high purge rate and correspondingly low inert level. The actual solution will depend on the specific case, especially on the composition of the feedstock. Solutions of this type are used by the process schemes (see Sect. 6.5.3.2.3) using a stoichiometric amount of process air, such as the Kellogg, Topsoe and Uhde processes.

Case d shows that stoichiometric carbon dioxide production can also be achieved by using an excess of process air. However, in this case it becomes necessary to introduce an extra gas separation process step in order to remove the excess nitrogen which is introduced with the excess air. As explained in Sect. 6.5.3.2.3, several process schemes exist where excess process air is used, and the excess nitrogen is removed in dedicated units. Processes using excess process air are the Braun process, the ICI processes, and several process schemes of minor importance.

Case e illustrates that sufficient carbon dioxide can also be produced by adding pure oxygen, i.e. by using enriched air instead of normal air. In the theoretical situation illustrated in case e, the oxygen concentration in the enriched air becomes 33%. Less oxygen is required if the feedstock contains higher hydrocarbons. Use of enriched air has been suggested in several process schemes, see Sect. 6.5.3.2.3.

In addition to the above discussed influence of urea production on the design of ammonia plants due to the required carbon dioxide production, ammonia and urea plants may interact further on several levels. It is customary to integrate the steam systems, for example by driving the carbon dioxide compressor turbine with steam produced in the ammonia plant and extracting process steam for the urea plant from the same turbine or from one of the drivers in the ammonia plant. A discussion of further integration may be found in [820]. Integration of ammonia and urea plants on the process side has been suggested by several companies, for example Toyo Koatsu [821–823], M. W. Kellogg [824, 825] and Snamprogetti [826–828].

In all cases the main point of integration is the carbon dioxide removal, which is done by washing the process gas with an ammonium carbonate/ammonia solution. The gas from the shift conversion unit is cooled, compressed to 200–300 kg/cm^2g, and carbon dioxide is removed by contact with the solution. The remaining carbon oxides are removed by methanation, and the gas goes to the ammonia synthesis section. The Snamprogetti process, which has been described in considerable detail [826], features a falling film absorber for carbon dioxide absorption. In addition, a falling film absorber is used in the ammonia synthesis loop to recover the ammonia product by absorption in water. The aqueous ammonia is sent direct to the urea reactor. Traces of water are removed from the recycling gas in the ammonia synthesis loop by the injection of liquid ammonia [32, 639]. In [829] a process is described in which the raw ammonia synthesis gas containing carbon dioxide is introduced into the urea reactor where the carbon dioxide is removed by reaction with ammonia. Significant improvements in the urea production process are claimed.

6.5.5.2 Ammonia and Other Products

A review of integrated processes based on steam reforming of hydrocarbons is given in [825]. It is proposed to integrate production of ammonia and hydrogen by recovering hydrogen from the ammonia synthesis purge gas (see Sect. 6.4.7). The hydrogen to ammonia production ratio can, within limits, be adjusted by varying the purge rate from the synthesis loop. The suggested scheme for integrated production of ammonia and methanol features the installation of a methanol synthesis unit (with or without synthesis gas compression or synthesis gas recycle depending on required ammonia to methanol production ratio) between the secondary reformer and the shift conversion unit in the conventional ammonia process. Co-production of ammonia and methanol may also be achieved by installation of a methanolation unit (see Sect. 6.3.5.5) with or without partial by-pass of the shift unit and/or the carbon dioxide removal unit to adjust the ammonia to methanol production ratio. Conversion of existing ammonia plants to combined production of ammonia and methanol is described in [935]. Integration of ammonia production with the production of steel has also been described [380–384]. Production of ammonia from off gases from ethylene plants or methanol plants has also been used [378, 379]. Integrated production of ammonia and methanol by modifying the methanol process by incorporating an oxygen-fired secondary reformer and producing ammonia from the off-gas and nitrogen from the air separation unit is discussed in [968]. A plant producing both methanol and ammonia from synthesis gas produced by partial oxidation of heavy oil is described in [166–168].

6.6 Storage of Ammonia

Facilities for storage of ammonia are required at the production facilities either as buffer capacity to smooth out variations in production and demand between the ammonia plant and downstream units using ammonia as raw material or to hold product between shipments from plants exporting ammonia. Required storage capacity may range from less than one to a few days' production capacity for intermediate storage and up to one month's production capacity, or even more, for exporting plants located in remote areas. This means that ammonia storage tanks in ammonia plants may range in size from less than 100 t to above 40000 t.

In addition to this, very large ammonia storage facilities have in recent years been constructed at terminals served by ammonia pipelines. Pipeline transport of ammonia and related storage facilities are discussed in [830, 831]. In the construction and operation of ammonia storage facilities the special properties of ammonia–an inflammable and toxic compound–must be taken into account. The properties of ammonia and the hazards involved in handling and storage of this product are dealt with in chapter 7. General descriptions of the design of storage facilities are given in [832, 833]. Surveys of actual installations may be found in [834–840, 936]. Safety recommendations are given in [841–843] (see also chapter 7).

Depending on the capacity, ammonia may be stored by three different methods:

- In non-refrigerated pressure vessels–perhaps cooled by water spray and/or painted with reflecting paint. The storage pressure corresponds to ammonia vapour pressure at ambient temperature. The storage vessels, often referred to as "bullets", are usually cylinders; capacity is generally low, below about 200 t. Non-refrigerated spherical pressure vessels with capacities up to 1500 t ammonia have been used [833].

- In semi-refrigerated tanks where the ammonia is refrigerated–usually by a refrigeration unit with a single stage ammonia compressor–to a temperature often around 0 °C. The storage pressure corresponds to ammonia vapour pressure at the refrigeration temperature and may be as low as 2.5 kg/cm^2g. Semi-refrigerated storage vessels, normally spheres or in some cases special constructions referred to as "Horton Spheres", may have capacities up to about 3000 t.
- In fully refrigerated storage tanks (cylindrical, insulated tanks with flat bottom and domed roof) at essentially atmospheric pressure and − 33 °C. The refrigeration is provided by a refrigeration unit, most often with a two stage ammonia compressor. The refrigeration may also be integrated with the ammonia synthesis loop refrigeration system. The tanks, most often referred to as "Atmospheric Ammonia Storage Tanks"–may have capacities up to above 40000 t. Characteristic features of the three types of storage is given in Table 6.6 (based on [836, 837]).

The choice between the three types evidently depends on the capacity required and is made on the basis of economics in such a way that fully refrigerated tanks are used for the largest capacities and non-refrigerated vessels for the smallest capacities. Semi-refrigerated vessels are used in an intermediate capacity range, but according to [832, 833] they are becoming less popular.

In addition to the storage capacity, the rate of required supply of ammonia at different conditions (gaseous, warm liquid, cold liquid) must also be taken into account. This is discussed with reference to a specific situation in [844]. In many cases ammonia terminals–both for export from producing plants and for import from the distribution system–involve a combination of pressurized spheres and refrigerated storage tanks. As an example, Fig. 6.33 (from [831)] shows a system capable of delivering product ammonia from a manufacturing plant to a variety of distribution systems. The figure assumes that the refrigeration system for the refrigerated storage is integrated with the refrigeration for the ammonia synthesis loop, and that ammonia must be supplied at − 33 °C to a ship or barge and at ambient temperature to pipeline or to rail or road transport cars.

Table 6.6. Characteristic features of ammonia storage tanks

Type	Typical Pressure kg/cm^2g	Design Temperature °C	t Ammonia per t Steel	Capacity Range t ammonia	Refrigeration Compressor
Non-refrigerated	18	ambient	2.75	< 270	None
Semi-refrigerated	3–5	≈ 0	10	450–2700	Single-stage
Fully refrigerated	1.1–1.2	− 33	41–45	4500–45000	Two-stage

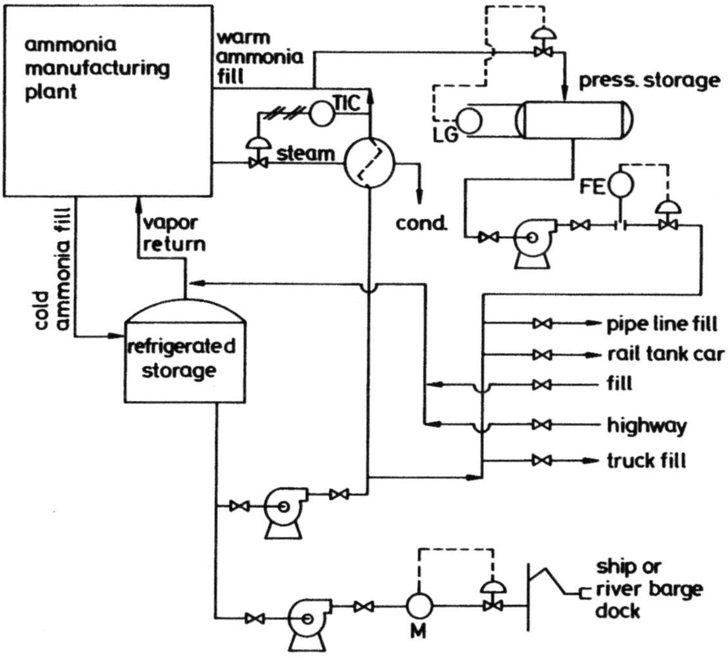

Fig. 6.33. Generalized scheme for supply of ammonia from manufacturing plant to various transport facilities (from [831])

6.6.1 Pressurized Storage (Non-Refrigerated and Semi-Refrigerated)

This type of storage, as mentioned above, is most often used for storage of relatively small quantities of ammonia as intermediate storage between the ammonia plant and ammonia consumers or transport systems accepting warm ammonia. Pressurized tanks are also used extensively in the distribution system for storage of small quantities of ammonia, for example in field tanks in connection with direct application of liquid ammonia as a fertilizer. Non-refrigerated pressurized storage tanks may be protected against overheating in hot weather by insulation, by water spray, or by circulating ammonia through a water cooler. The design pressure is typically around 18 bar.

The design of pressurized storage tanks and related safety aspects are reviewed in [845]. Stress corrosion cracking is a much debated risk in pressurized storage tanks. The mechanism of stress corrosion cracking (including the effect of oxygen and water) is discussed in [846–850, 937] and, in an authoritative review, in [938]. It has been generally accepted that the addition of water to ammonia inhibits stress corrosion cracking [832, 846], and in certain situations (transport of ammonia in containers made from certain specified types of steel) a minimum water content of 0.2 wt% is a requirement [851]. Further protection may be obtained by cathodic polarization, for example by aluminum or zinc metal spray [937, 938]. The most recent research [849, 938], however, has shown that water does not give complete protection. It has also been more or less accepted that stress corrosion cracking does not constitute a risk in atmospheric storage tanks. Recently, incidents have been reported, however, [852, 940, 941] in which stress corrosion cracking had occurred in large atmospheric storage tank. General surveys of cracking incidents are given in [835, 854]. Specific cases including examples of inspection and repair procedures are given in [855–858, 939, 942, 943].

6.6.2 Atmospheric Storage (Fully Refrigerated)

Atmospheric storage at − 33 °C is, as mentioned, used for large-capacity storage; single tanks with a capacity above 40000 t of ammonia are known. The storage tanks are cylindrical with a flat bottom and domed roof. The refrigeration is provided by recompression of boil-off, normally in a two-stage compressor. Incoming warm ammonia is flashed to − 33 °C before entering the tank, and the vapors from the flash vessel are also handled by the refrigeration compressor. A sketch of a typical storage unit is shown in Fig. 6.34 (from [832]).

Two types of tanks are used, single and double wall tanks. The single wall tank consists of a single shell designed to withstand the operating pressure. The shell is insulated; the insulation must be completely vapour-tight and requires a very high standard of construction of maintenance to avoid hazards.

Double wall tanks consist of an inner tank designed for the storage temperature and pressure surrounded by a second tank, with a gap between the two shells of normally 18 inches or more. The gap is filled with insulation material, and the main purpose of the outer tank is to contain the insulation. It is, however, becoming more and more normal to design the outer shell to the same standards as the inner shell. These so-called "double-integrity" tanks offer the added safety that the outer tank will hold the ammonia in case of failure of the inner tank. As a further safety measure, ammonia storage tanks are normally surrounded by dikes or placed in concrete basins with the capacity to contain the liquid ammonia in case of complete failure of the tank. Discussions of such secondary retainment are given in general and with reference to specific cases, in [859–863]. Owing to the refrigeration, there is a risk of ice formation under the tank. This must be avoided since it could lead to movements that could damage the bottom of the tank, and the foundation of the tank is therefore normally provided with heating, or the tank is constructed on piles with ventilation under the bottom. Problems relating to tank foundations are discussed with reference to specific cases in [864–866, 944]. Details about the design of storage tanks are given in [832, 867, 868]. The construction of a new high-integrity tank is discussed in [869].

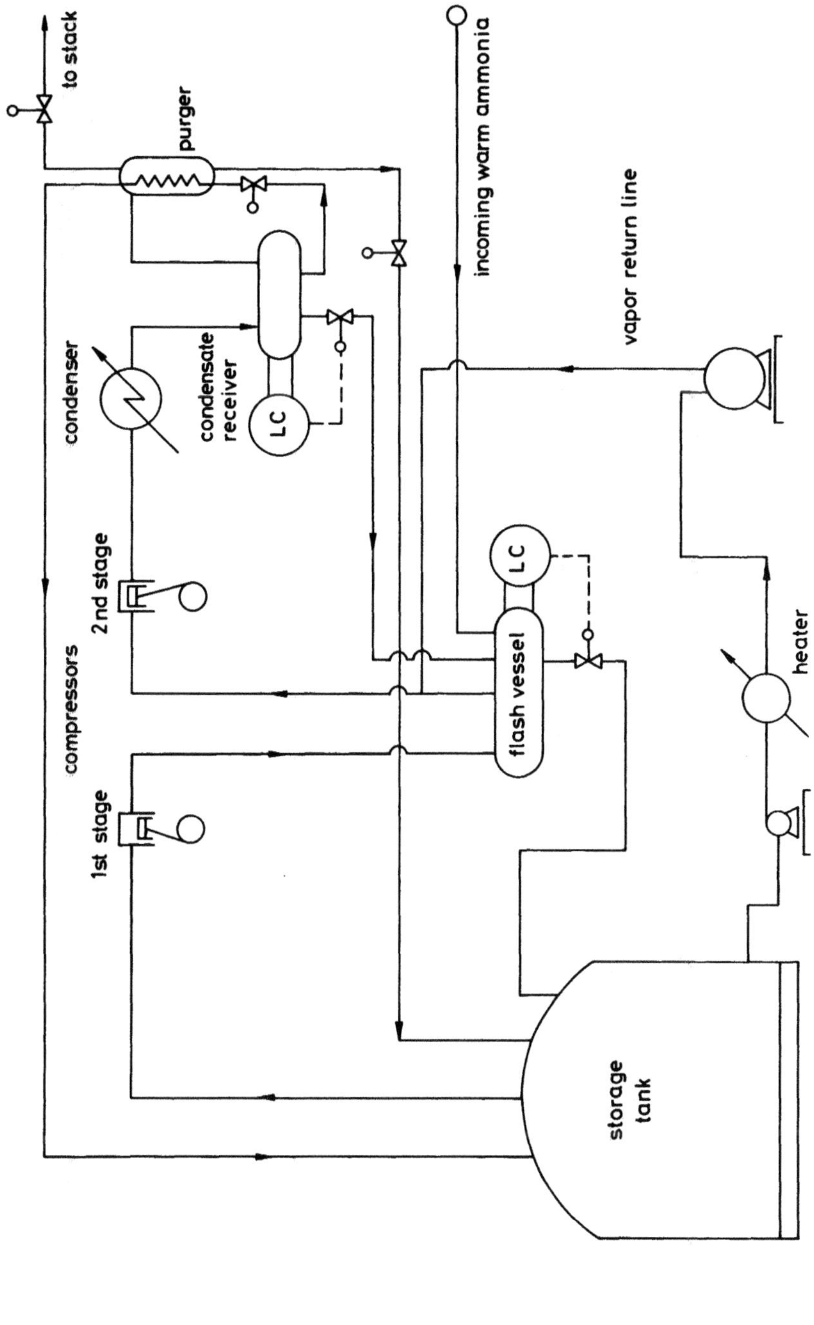

Fig. 6.34. Typical fully refrigerated ammonia storage unit. Simplified process flow diagram (from [832])

Models for tank behaviour and potential hazards during tank loading are described in [945].

Retrofit of existing tanks for added safety is discussed in [870]. Special design and operating considerations relating to the influence of climatic conditions are reported in [871]. Safety aspects including maintenance and inspection procedures are reviewed with reference to specific cases in [872–875, 946–948]. For risk of stress corrosion cracking in atmospheric ammonia storage tanks see Sect. 6.6.1.

6.7 References

1. Harding AJ (1959) Ammonia manufacture and uses, Oxford University Press, London
2. Vancini CA (1961) La Sintesi dell'Ammoniaca. Ulrico Hoepli, Milano
3. Sauchelli V, Stevenson FJ, Timm B, Danz W, Axelrod LC, O'Hare TE, Hansen HJ (1964) In: Sauchelli V (ed) Fertilizer nitrogen its chemistry and technology. Reinhold, New York
4. Vancini CA (1971) Synthesis of ammonia. Macmillan, London
5. LeBlanc JR Jr, Madhavan S, Porter RE (1978) In: Kirk–Othmer Encyclopedia of Chemical Technology, 3rd ed. Wiley, New York, vol 2, p 470
6. Slack AV, James GR (eds) Ammonia. Marcel Dekker, New York, vol. 1 (1973) vol 2 (1974) vol 3 (1977) vol 4 (1979)
7. Strelzoff S (1981) Technology and manufacture of ammonia. Wiley, New York
8. Bakemeier H, Huberich T, Krabetz R, Liebe W, Schunck M, Mayer D (1985) In: Ullman's Encyclopedia of Industrial Chemistry, 5th edn, vol A2, VCH, Weinheim, p 143
9. Noyes R (1964) Ammonia and synthesis gas 1964. Noyes Development Coporation, New York
10. Noyes R (1967) Ammonia and synthesis gas 1967. Noyes Development Coporation, New York
11. Brykowski FJ (ed) (1981) Ammonia and synthesis gas recent and energy-saving processes, Noyes Data Corporation, Park Ridge, p 1
12. Mittasch A (1951) Geschichte der Ammoniaksynthese. Verlag Chemie, Weinheim
13. Timm B (1960) Chem Ind (London) 12: 274
14. Timm B (1963) Chem Ing Tech 35: 817
15. Timm B, Danz W (1964) In: Sauchelli V (ed) Fertilizer nitrogen its chemistry and technology. Reinhold, New York, p 40
16. Appl M (1976) Nitrogen 100: 47
17. Slack AV (1973) In: Slack AV, James GR (ed) Ammonia, Marcel Dekker, New York 1: 5
18. Waeser B (1939) Chem Fabr 12(31, 32): 370
19. Brown CO (1954) Chem Eng Prog 50(11): 556
20. Nitrogen (1962) 16: 35
21. Strelzoff S, Vasan S, (1963) Chem Eng Prog 59(11): 60
22. Bresler SA, James GR (1965) Chem Eng (NY) 72(13): 109
23. Allen JB (1965) Chem Process Eng (London) 46(9): 473
24. Mayo HC, Finneran JA (1968) Erdöl Kohle, Erdgas, Petrochem 21(7): 404
25. Quartulli J, Wagener D (1973) Erdöl Kohle, Erdgas, Petrochem Brenns Chem 26(4): 192
26. Lyon SD (1975) Chem Ind (London) 17: 731
27. Quartulli OJ , Buividas LJ (1976) Nitrogen 100: 60
28. Pachaiyappan V (1979) Fert News (August) 41
29. CEER (1979) Chem Econ Eng Rev 11(5): 24
30. Zardi U, Antonini A (1979) Nitrogen 122: 33
31. Honti GD (1981) In: More AI (ed) Fert nitrogen: Proc Br Sulphus Corp Int Conf Fert Technol, 4th, part 1, 1
32. Saviano F, Laganà V, Bisi P (1981) Hydrocarbon Process 60(7): 99
33. Rai HS (1982) CEER Chem Eng World 17(1): 69
34. Brown FC (1983) Proc Fert Soc London 218: 1
35. Dybkjaer I, (1983) ECN Eur Chem News Fertilizers '83 Supplement (February 21): 15
36. Picciotti M, Pocini CA (1984) Chim Ind (Milan) 66(2): 97

37. Agarwal MR (1982) CEW Chem Eng World 17(1): 55
38. James GR, Stokes KJ (1984) Chem Eng Prog 80(6): 33
39. Blanken JM (1987) Proc Fert Soc London 259: 1
40. Blanken JM (1989) Ammonia Plant Saf 29: 273
41. Fertilizer Focus (1987) 4(10): 22
42. Fokdal A, Hansen HJ (1955) Ingeniøren 47: special issue 902
43. Catalyst handbook (1970) Wolfe Scientific Books, London
44. Twiggs MV (ed) (1989) Catalyst handbook, 2nd ed. Wolfe Publishing Ltc, London
45. Hansen HJ (1964) In: Sauchelli V (ed) Fertilizer nitrogen its chemistry and technology. Reinhold Publishing Corporation, New York, p 89
46. Rankin JD (1978) Proc. Fert Soc London 168: 1
47. Badische Anilin- & Soda-Fabrik, Ludwigshafen (1908) GER 235421, CA (1911) 5: No. 3137
48. Haber F (1909) GER 238450, CA (1912) 6: No. 1663
49. Slack AV (1973) In: Slack AV, James GR (eds) Ammonia, vol 1. Marcel Dekker, New York, p 20
50. Gmelins Handbuch der Anorganischen Chemie (1935) Stickstoff. Verlag Chemie, Berlin, p 295
51. Habermehl R, Long ND (1974) In: Slack AV, James GR (eds) Ammonia vol 2. Marcel Dekker, New York, p 333
52. Kjeldgaard K (1974) In: Slack AV, James GR (eds) Ammonia, vol 2. Marcel Dekker, New York, p 341
53. Phillipson JJ (1970) In: Catalyst handbook. Wolfe Scientific Books, London, p 46
54. Carnell JH (1989) In: Twiggs MV (ed) Catalyst handbook, 2nd ed. Wolfe Publishing Ltd, London, p 199
55. Nitrogen (1971) 72: 34
56. Livingston JY (1971) Hydrocarbon Process 50(1): 126
57. Lee MNY, Collins JJ (1969) Ammonia Plant Saf 11: 59
58. Burklow BW, Coleman RL (1977) Ammonia Plant Saf 19: 21
59. Jensen PE, Søndergaard K (1984) Ammonia Plant Saf 24: 47
60. Carnell PJH, Denny PJ (1985) Ammonia Plant Saf 25: 99
61. Carnell PJH, Denny PJ, Clark DN (1987) Ammonia Plant Saf 27: 99
62. Chatterjee A, Rajagopal KR (1982) CEW Chem Eng World 17(1): 63
63. Polasek J, Bullin J (1984) Energy Prog 4(3): 146
64. Rostrup-Nielsen JR (1975) Steam reforming catalysts. Teknisk Forlag A/S, Copenhagen
65. Rostrup-Nielsen JR (1984) In: Anderson JR, Boudart M (eds) Catalysis science end technology vol 5. Springer, Berlin
66. Vancini CA (1961) La Sintesi dell'Ammoniace. Ulrico Heopli, Milano, p 160
67. Noyes R (1967) Ammonia and synthesis gas 1967. Noyes Development Corporation, Park Ridge, p 29
68. Atwood K, Knight CB (1974) In: Slack AV, James GR (eds) Ammonia. Marcel Dekker, New York, vol 2 p 145
69. Strelzoff S (1981) Technology and manufacture of ammonia. Wiley, New York, p 101
70. Brykowski FJ (ed) (1981) Ammonia and synthesis gas recent and energy-saving processes. Noyes Data Corporation, Park Ridge, p 20
71. Gard NR (1966) Nitrogen 39: 25
72. Appl M, Gössling H (1972) Chem Ztg 96(3): 135
73. Nitrogen (1962) 18: 24
74. Voogd J, Tielrooy J (1963) Hydrocarbon Process Pet Refin 42(3): 144
75. Borgars DJ (1964) Chim Ind (Paris) 91(3): 276
76. Young RJ (1964) Chem Age India 15(6): 741
77. Bridger GW, Wyrwas W (1967) Chem Process Eng (London) 48(9): 101
78. Marschner F, Renner HJ (1982) Hydrocarbon Process 61(4): 176
79. Rostrup-Nielsen JR (1973) Ammonia Plant Saf 15: 82
80. Agarwal MR, Bajpai T (1977) Chem Age India 28(113): 995
81. Tottrup PB, Nielsen B (1982) Hydrocarbon Process 61(3): 89
82. Bridger GW (1966) Chem Process Eng (London) 47(12): 39
83. Bridger GW (1976) Oil Gas J 74(7): 73
84. Rounthwaite DP (1983) Plant/Oper Prog 2(2): 127
85. Nitrogen (1988) 174: 23
86. Congram GE (1976) Oil Gas J 74(7): 75
87. Rostrup-Nielsen JR (1982) In: Figueiredo JL (ed) Progress in catalyst deactivation. Martinus Nijhoff Publishers, The Hague, p 127

88. Rostrup-Nielsen JR (1982) In: Figueiredo JL (ed) Progress in catalyst deactivation. Martinus Nijhoff Publishers, The Hague, p 209
89. Davies J, Lihou DA (1971) Chem Process Eng (London) 52(4): 71
90. Marsch HD, Herbort HJ (1982) Hydrocarbon Process 61(6): 101
91. Nitrogen (1987) 166: 24
92. Nitrogen (1987) 167: 31
93. Leyel CP (1975) Ammonia Plant Saf 17: 41
94. Thuillier J, Pons F (1978) Ammonia Plant Saf 20: 89
95. Kawai T, Takemura K, Zaghloul MB (1984) Paper presented at International Plant Engr Conf, Bombay
96. Kawai T, Takemura K, Shibaksi T, Rump L, Danielsen B, Wrisberg J (1982) Paper presented at AIChE Ammonia Saf Symp. San Francisco
97. Muhlenforth CJ (1989) Finds 4(2): 14
98. Salot WJ (1972) Ammonia Plant Saf 14: 119
99. Salot WJ (1973) Ammonia Plant Saf 15: 1
100. Konoki K, Shinohara T, Shibata K, (1982) Plant/Oper Prog 1(2): 122
101. Imoto Y, Terada S, Maki K (1982) Plant/Oper Prog 1(2): 127
102. Kawai T, Takemura K, Shibasaki T, Mohri T (1982) Plant/Oper Prog 1(3): 181
103. Kawai T, Takemura K, Shibasaki T, Mohri T (1980) Ammonia Plant Saf 22: 119
104. Kawai T, Mohri T, Takemura K, Shibasaki T (1984) Ammonia Plant Saf 24: 131
105. Cockerham RG, Percival G (1957/58) Trans Inst Gas Eng 107: 390
106. Jockel H, Triebskorn BE (1973) Hydrocarbon Process 52(1): 93
107. Rall W, (1967) Erdöl Kohle, Erdgas, Petrochem 20(5): 351
108. Davies HS, Humphries KJ, Hebden D, Percy DA (1967) IGE J (October): 708
109. Jockel H (1969) GWF Gas Wasserfach 110(21): 561
110. Ishiguro T (1968) Hydrocarbon Process 47(2): 87
111. Rostrup-Nielsen JR, Tottrup PB (1979) In: Proc Symp Sci Catal its Appl Ind, Sindri, p 379
112. Clark DN, Henson WGS (1988) Ammonia Plant Saf 28: 99
113. Mosley T, Stephens RW, Stweart KD, Wood J (1972) J Catal 24: 18
114. Stahl H, Rostrup-Nielsen J, Udengaard NR (1985) In: Fuel cell seminar 1985. Tuscon, Arizona, p 83
115. Udengaard NR, Christiansen LJ, Summers WA (1988) Endurance testing of a high-efficiency steam reformer for fuel cell power plants. EPRI AP-6071, Project 2192–1. Electric Power Research Institute, California.
116. Sederquist RA (1978) US 4071330; (1978) CA 89: No. 8711
117. Smith ASP, Doy RJ, Limbach APJ (1986) EP 194067, CA (1986) 105: No. 229260.
118. Ruziska PA (1984) EP 113198; CA 101: No. 113278
119. Pinto A (1988) US 4750986; see EP 124226 CA (1985) 102: No. 27497
120. Crawford DB, Becker CL, LeBlanc JR (1978) US 4079017; CA (1979) 91: No. 213715
121. Crawford DB, Becker CL, LeBlanc JR (1979) US 416229. see BE 856919 CA (1978) 89: No. 179566
122. Miyasugi T, Kosaka S, Kawai T, Suzuki A (1984) Ammoonia Plant Saf 24: 64
123. Pagani G, Brusasco G, Gramatica G (1981) In: More AI (ed) Fert nitrogen: Proc Br Sulphur Corp Int Conf Fert Technol, 4th, part 1, p 195
124. Nitrogen (1989) 179: 16
125. Bogart MJP (1973) US 3743488; CA (1973) 79: No. 106547
126. Bogart MJP (1974) US 3795485; CA (1974) 81: No. 15210
127. Nobles EJ (1973) In: Slack AV, James GR (eds) Ammonia, vol 1. Marcel Dekker, New York, p 275
128. Fuderer A (1987) US 4650651; CA (1987) 106: No. 179566
129. Herbort H-J, Marsch H-D (1987) (GB 2181740; see DE 3532413 CA (1987) 106: No. 158813
130. Marsch H-D, Thiagarajan N (1989) Ammonia Plant Saf 29: 195
131. Noyes R (1967) Ammonia and synthesis gas 1967. Noyes Development Corporation, Park Ridge, p 78
132. Nitrogen (1962) 17: 35
133. Chem Eng Int Ed (1962) 69(14): 89
134. Chem Eng Int Ed (1966) 73(1): 24
135. Hydrocarbon Process (1984) 63(4): 103
136. Flytzani-Stephanopoulos M, Voecks GE (1981) Energy Prog 1(1–4): 52
137. Reidel JC (1954) Oil Gas J 52(40): 60

138. Reidel JC (1954) Oil Gas J 53(5): 86
139. Foster AL (1946) Oil Gas J (June 22): 98
140. Tuttle HA (1952) Chem Eng Prog 48(6): 272
141. Mayland BJ, Comley EA, Reynolds JC (1954) Chem Eng Prog 50(4): 177
142. Blanken JM (1988) Procestechnologie 2: 37
143. Eickmeyer AG, Marshall WH, Jr (1955) Chem Eng Prog 51(9): 418
144. Chopey NP (1961) Chem Eng (NY) 68(8): 158
145. Weyermuller G, Ogden JW (1952) Chem Process (London) (February): 36
146. Fricke H (1972) Chem Ztg 96(3): 123
147. Noyes R (1967) Ammonia and synthesis gas 1967. Noyes Development Corporation, Park Ridge, p 69
148. Lee KT, (1973) In: Slack AV, James GR (eds) Ammonia vol 1. Marcel Dekker, New York, p 293
149. Strelzoff S (1981) Technology and manufacture of ammonia. Wiley, New York, p 165
150. Strelzoff S (1974) Hydrocarbon Process 53(12): 79
151. Nitrogen (1973) 83: 40
152. Shimizu Y (1978) CEER Chem Econ Eng Rev 10(7): 9
153. Jungfer H (1985) Linde Rep Sci Technol 40: 14
154. van den Berg GJ, Rijnaard P, Byrne DJ (1966) Hydrocarbon Process 45(5): 193
155. van den Berg GJ, Reinmuth E, Kapp E (1971) Chem Process Eng (London) 52(10): 49
156. ter Haar LW (1973) CEER Chem Econ Eng Rev 5(12): 22
157. ter Haar LW (1973) Het Ingenieursblad 21: 3
158. Reed CL, Kuhre CJ (1979) Hydrocarbon Process 58(9): 191
159. Banquy DL (1970) Fert News 15(11): 22
160. Banquy DL (1970) Proc Fert Soc London 117: 82
161. Becker PD, Hiller H, Hochgesand G, Sinclair AM (1971) Chem Process Eng (London) 52(11): 59
162. Haese E (1974) Chem Ind (Düsseldorf) 26(11): 732
163. Kammholz G, Müller G-A (1973) Erdöl Kohle, Erdgas, Petrochem Brenns Chem 26(12): 695
164. Butzert HE (1976) Chem Eng Prog 72(1): 56
165. Supp E, (1977) Nitrogen 109: 36
166. Morrison J (1969) Oil Gas J 67(8): 76
167. Morrison J (1969) Oil Gas Int 9(3): 78
168. Auer W (1971) Erdöl Kohle, Erdgas Petrochem Brenns Chem 24(3): 145
169. Balz DF, Gettert HF, Gruendler KH (1983) Plant/Oper Prog 2(1): 47
170. Happe U (1973) In: Slack AV, James GR (eds) Ammonia, vol 1. Marcel Dekker, New York, p 325
171. Forney AJ, Field JH (1973) In: Slack AV, James GR (eds) Ammonia, vol 1. Marcel Dekker, New York, p 347
172. Noyes R (1967) Ammonia and synthesis gas 1967. Noyes Development Corporation, Park Ridge p 94
173. Jüntgen H, Heek van KH (1979) GWF-Gas/Erdgas 120 (12): 559 and (1980) 121(1): 6
174. Vorres KS (1980) Energy Res 4: 109
175. Lacey JA, Roberts GFI (1984) In: Proc World Pet Congr 11th 1983, London, p 1
176. Von Gratkowski H-W (1975) Erdöl Kohle, Erdgas, Petrochem Brenns Chem 28(2): 81
177. Wurzbacher G (1977) Chem Tech 6(8): 317
178. Nitrogen (1980) 126: 32
179. Ammonia Coal Symp (1979) Tennessee Valley Authority, Muscle Shoals, AL USA
180. Teggers H, Jüntgen H (1984) Erdöl Kohle, Erdgas, Petrochem Brenns Chem 37(4): 163
181. Brown F (1977) Hydrocarbon Process 56(11): 361
182. Brown FC, Hargreaves HG (1979) Proc Fert Soc London 184: 1
183. Nitrogen (1986) 161: 23
184. Staege H (1979) Chem Eng (NY) 86(19): 106
185. Michaels HJ, Leonard HF (1978) Chem Eng Prog 74(8): 85
186. Staege H (1982) Hydrocarbon Process 61(3): 92
187. Sharpe RA (1976) Hydrocarbon Process 55(11): 171
188. Beck B, (1979) In: Ammonia Coal Symp. Tennessee Valley Authority, Muscle Shoals, AL USA, p 72
189. Engelbrecht AD, Partridge LJ (1979) In: Ammonia Coal Symp. Tennessee Valley Authority, Muscle Shoals, AL USA p 176

190. Patridge LJ (1976) Chem Eng Prog 72(8): 57
191. Engelbrecht AD, Partridge LJ (1981) Oil Gas J 79(6): 113
192. Chaurey KH (1981) In: FAI seminar 1980. The Fertiliser Association of India, New Delhi, TECH 1-2/1
193. Chaurey KH, Sharma KC, (1979) In: Ammonia Coal Symp. Tennessee Valley Authority, Muscle Shoals, AL USA, p 166
194. Völkel HK (1978) Energie 30(6): 196
195. Vogt EV, Burgt van der MJ (1980) Chem Eng Prog 76(3): 65
196. Burgt van der MJ (1979) Hydrocarbon Process 58(1): 161
197. Hydrocarbon Process (1984) 63(4): 96
198. IPG International Power Generation (1990) 13(4): 29
199. Hydrocarbon Process (1984) 63(4): 95
200. Schmidt-Traub H, Pohl HC (1983) Chem Ing Tech 55(11): 850
201. Schellberg W, Braun H (1988) Paper presented at symp on trends and developments in the fertilizer industry. Bursa, Turkey, p 1
202. Bailey E, Subramaniam TK, Anwer J, Bogner F (1979) In: Ammonia Coal Symp. Tennessee Valley Authority, Muscle Shoals, AL USA, p 51
203. Theis KA, Nitschke E (1982) Hydrocarbon Process 61(9): 233
204. Franke FH, Pattas E, Nitschke E, Keller J (1979) In: Ammonia Coal Symp. Tennessee Valley Authority, Muscle Shoals, AL USA, p 86
205. Hiller H (1975) Erdöl Kohle, Erdgas, Petrochem Brenns Chem 28(2): 74
206. Becker PD (1979) In: Ammonia Coal Symp. Tennessee Valley Authority, Muscle Shoals, Al USA, p 44
207. Serrurier R (1976) Hydrocarbon Process 72(9): 253
208. King WEH (1981) Fuel 60: 803
209. Child ET (1979) In: Ammonia Coal Symp. Tennessee Valley Authority, Muscle Shoals, AL USA, p 79
210. Schlinger WG (1980) Energy Res 4: 127
211. Cornils B, Hibbel J, Ruprecht P, Dürrfeld R, Langhoff J (1981) Hydrocarbon Process 60(1): 149
212. Konkol W, Ruprecht P, Cornils B, Dürrfeld R, Langhoff J (1982) Hydrocarbon Process 61(3): 97
213. Sueyama T, Tsujino T (1985) In: More AI (ed) Proc Int Conf Fertilizer 85 1 Br. Sulphur Corp., London, p 83
214. Jpn Chem Week (1984) (Nov. 15): 4.
215. Burnett JA, Jr (1979) In: Ammonia Coal Symp. Tennessee Valley Authority, Muscle Shoals, AL USA, p 27
216. Chem Eng News (1979) 57(23): 27
217. Alves GW, Waitzmann DA (1982) Erdöl Kohle, Erdgas, Petrochem Brenns Chem 35(2): 70
218. Watson JR, McClanahan TS, Weatherington RW (1985) Altern Energy Sources 3: 407
219. Lee RG (1981) In: FAI Seminar 1980, The Fertiliser Association of India, New Delhi, part II, TECH 1-3/1
220. Alpert SB, Gluckman MJ (1986) Ann Rev Energy 11: 315
221. Bohlbro H (1969) An investigation on the kinetics of the conversion of carbon monoxide with water vapour over iron oxide based catalysts, 2nd ed. Gjellerup, Copenhagen
222. Moe JM (1962) Chem Eng Prog 58(3): 33
223. Ting AP, Wan S-W (1969) Chem Eng (NY) 76(11): 185
224. Chem Eng News (1963) 41(7): 46
225. Habermehl R (1965) Chem Eng Prog 61(1): 57
226. Habermehl R, Atwood K (1965) Paper presented at United Nations Interregional Seminar in the Production of Fertilizers, Kiev, USSR
227. Nitrogen (1966) 40: 28
228. Allen D (1974) In: Slack AV, James GR (eds) Ammonia, vol 2. Marcel Dekker, New York, p 3
229. Mertzsch N (1979) Chem Tech Umsch 11(2): 5
230. Newsome DS (1980) Catal Rev Sci Eng 21(2): 275
231. Hawker PN (1982) Hydrocarbon Process 61(4): 183
232. Völter J, Berndt H, Lietz G (1976) Chem Tech 28(19): 606
233. Lloyd L, Twigg MV (1979) Nitrogen 118: 30
234. Nitrogen (1987) 168: 29
235. Lombard JF (1969) Hydrocarbon Process 48(8): 111

236. Campbell JS (1970) Ind Eng Chem Process Des Dev 9(4): 588
237. Young PW, Clark CB (1973) Chem Eng Prog 69(5): 69 and Ammonia Plant Saf 15: 18
238. Rostrup-Nielsen JR, Højlund Nielsen PE (1985) In: Düdar J, Wise H (eds.) Deactivation and poisoning of catalysts. Marcel Dekker, New York, p 259
239. Kitchen D (1988) In: Nitrogen 88 Br Sulphur's 12th Int Conf, Geneva, p 127
240. Pedersen PS (1988) In: Nitrogen 88 Br Sulphur's 12th Int Conf Geneva, p 111
241. Lundberg WC (1979) Chem Eng Prog 75(6): 81 and Ammonia Plant Saf 21: 105
242. Lorenz E (1967) Ind Chim Belge 32: (Spec No.) (Pt. 2) 377
243. Auer W, Lorenz E, Gründler KH (1971) Paper presented at The 68th National Meeting of the American Institute of Chemical Engineers, Houston, USA
244. Dybkjaer I (1979) In: Ammonia Coal Symp. Tennessee Valley Authority, Muscle Shoals, AL USA, p 133
245. Dybkjaer I, Bohlbro H, Aldridge CL, Riley KL (1979) Ammonia Plant Saf 21: 145
246. Dybkjaer I (1981) In: More AI (ed) Fert nitrogen: Proc Br Sulphur Corp Int Conf Fert Technol, 4th, part II, p 503
247. Højlund-Nielsen PE, Bøgild-Hansen J (1982) J Mol Catal 17: 183
248. Højlund-Nielsen PE, Bøgild-Hansen J (1981) Paper presented at Communicacao ao Coloquio Nacional de Catalise Industrial, Lisboa
249. Hansen JB, Carstensen JH, Pedersen PS (1989) Ammonia Plant Saf 29: 204
250. Kitchen D, Pinto A, Praag H van (1989) Ammonia Plant Saf 29: 212
251. Carstensen JH, Bøgild-Hansen J, Pedersen PS (1990) Ammonia Plant Saf 30: 139
252. Buckthorp CM (1978) Nitrogen 113: 34
253. Bonacci JC, Otchy TG (1978) Ammonia Plant Saf 20: 165
254. Colby JH, White GA, Notwick PN Jr (1979) Ammonia Plant Saf 21: 138
255. Krishnaswami KR, Neelakantan PS (1989) Fert. News (December) 31
256. Cover AE, Hubbard DA, Jain SK, Shah KV, Koneru PB, Won EW (1985) Review of selected gas removal processes for SNG production. Gas Research Institute, Contract No. 5082-222-0754, p 293
257. Christensen KG, Stupin WJ (1978) FE-2240-49, Report to the United States Department of Energy and Gas Research Institute. Contract No. EX-76-C-01-2240
258. Kohl AL, Riesenfeld FC (1979) Gas Purification, 3rd ed. Gulf Publishing Company, Houston
259. Schmidt HW, Henrici H-J (1972) Chem Ztg 96(3): 154
260. Goar BG (1971) Oil Gas J 69(28): 75, 69(29): 84
261. Hochgesand G (1968) Chem Ing Tech 40(9/10): 432
262. Werner D (1981) Chem Ing Tech 53(2): 73
263. Christensen KG, Stupin WJ (1978) Hydrocarbon Process 57(2): 125
264. Tennyson RN, Schaaf RP (1977) Oil Gas J 75(2): 78
265. Schmidt HW (1968) Chem Ing Tech 40(9/10): 425
266. Thirkell H (1974) In: Slack AV, James GR (eds) Ammonia, vol 2. Marcel Dekker, New York, p 117
267. Strelzoff S (1981) Technology and manufacture of ammonia. Wiley, New York, p 193
268. Strelzoff S (1975) Chem Eng (NY) 82(19): 115
269. Stokes KJ (1981) In: More AI (ed) Fert nitrogen: Proc Br Sulphur Corp Int Conf Fert Technol 4th, part II, 525
270. Stokes KJ (1980) Ammonia Plant Saf 22: 178
271. Stokes KJ (1981) Nitrogen 131: 35
272. Brown FC, Leci CL (1982) Proc Fert Soc London 210: 1
273. Jackson JM (1974) In: Slack AV, James GR (eds) Ammonia, vol 2. Marcel Dekker, New York, p 183
274. Field JH (1974) In: Slack AV, James GR (eds) Ammonia, vol 2. Marcel Dekker, New York, p 153
275. Butwell KF, Kubek DJ (1977) Hydrocarbon Process 56(10): 173
276. Nitrogen (1975) 96: 33
277. Nitrogen (1976) 102: 40
278. Butwell KF, Kubek DJ, Sigmund PW (1979) Chem Eng Prog 75(2): 75; Ammonia Plant Saf 21: 156
279. Pierce JD, Chao EI (1985) Paper presented at Ammonia Saf Symp, Seattle paper 30c, p 1
280. Linsmayer S (1972) Chem Tech 24(2): 74
281. Elberling K, Gabriel W (1977) Chem Tech 29(1): 43

282. Maddox RN, Burns MD (1968) Oil Gas J 66(4): 91
283. Giammarco G (1974) In: Slack AV, James GR (eds) Ammonia, vol 2. Marcel Dekker, New York, p 171
284. Hydrocarbon Process (1982) 61(4): 95
285. Eickmeyer AG (1974) In: Slack AV, James GR (eds) Ammonia, vol 2. Marcel Dekker, New York, p 165
286. Benson HE, Parrish RW (1974) Hydrocarbon Process 53(4): 81
287. Bartoo RK, Ruzicka SJ (1983) In: Fertilizer 83, Br Sulphur Corp 7th Int Conf, London, p 129
288. Benson HE (1974) In: Slack AV, James GR (eds) Ammonia, vol 2. Marcel Dekker, New York, p 159
289. Grover BS, Holmes ES (1986) In: Proc Nitrogen 86 Amsterdam, Br Sulphur Corp, London p 101
290. van Hecke F (1977) US 4035166; CA (1977) 86: No. 173649
291. Crabs LE, Pouilliart R, van Hecke FC (1980) Ammonia Plant Saf 22: 185
292. Bögner F (1974) In: Slack AV, James GR (eds) Ammonia, vol 2. Marcel Dekker, New York, p 205
293. Zarker KE (1974) In: Slack AV, James GR (eds) Ammonia, vol 2. Marcel Dekker, New York, p 219
294. Sehrt B, Polster P (1980) Chem Tech 32(7): 345
295. Hochgesand G (1970) Ind Eng Chem 62(7): 37
296. Haase H (1970) Chem Anlagen Verfahren 10: 59
297. Linde W (1974) In: Slack AV, James GR (eds) Ammonia, vol 2. Marcel Dekker, New York, p 233
298. Chem Eng (NY) (1979) 86(26): 88
299. Lewis JL, Truby HA, Pascoo MB (1974) Oil Gas J 72(25): 120
300. Bucklin RW, Schendel RL (1984) Energy Prog 4(3): 137
301. Cook JP (1974) In: Slack AV, James GR (eds) Ammonia, vol 2. Marcel Dekker, New York, p 249
302. Dailey LW (1974) In: Slack AV, James GR (eds) Ammonia, vol 2. Marcel Dekker, New York, p 243
303. Kohrt H-U, Thormann K, Bratzler K (1963) Erdöl Kohle, Erdgas, Petrochem 16(2): 96
304. Judd DK (1978) Hydrocarbon Process 57(4): 122
305. Swanson CG Jr. (1979) Ammonia Plant Saf 21: 152
306. Swanson CG Jr, Burkhard FC (1984) Ammonia Plant Saf 24: 16
307. Shah VA (1988) Energy Prog 8(2): 67
308. Shah VA, McFarland J (1988) Hydrocarbon Process 67(3): 43
309. Hernandez RJ, Huurdeman TL (1988) In: Nitrogen 88 Br Sulhur's 12th Int Conf, Geneva p 147
310. Fertilizer Focus (1988) 5(11): 27
311. Volkamer K, Wagner E, Schubert F (1982) Plant/Oper Prog 1(2): 134
312. Strelzoff S (1981) Technology and manufacture of ammonia. J Wiley, New York, p 203
313. Strelzoff S (1974) In: Slack AV, James GR (eds) Ammonia, vol 2. Marcel Dekker, New York, p 201
314. Meissner RE, Wagner U (1983) Oil Gas J 81(5): 55
315. Volkamer K, Wagner U (1983) In: Fertilizer 83, Br Sulphyr Corp, 7th Int Conf, London, p 139
316. Meissner RE, Parsons RM (1984) Energy Prog 4(1): 17
317. Daviet GR, Sundermann R, Donnelly ST, Bullin JA (1984) Hydrocarbon Process 63(5): 79
318. Nobles JE, Shay LN, Stover JC (1988) In: Nitrogen 88 Br Sulphur's 12th Int Conf Geneva p 243
319. Gerhardt W, Hefner W (1989) Ammonia Plant Saf 29: 73
320. Meissner H, Hefner W (1990) Paper presented at European Conference on Energy Efficient Production of Fertilizers, Bristol UK
321. Allgood HY (1974) In: Slack AV, James GR (ed) Ammonia, vol 2. Marcel Dekker, New York, p 289
322. Yeandle WW, Klein GF (1952) Chem Eng Prog 48(7): 349
323. Egalon R, Vanhille R, Willemyns M (1955) Ind Eng Chem 47(5): 887
324. Grotz BJ Jr (1969) (US 3442613: CA (1967) 67: No. 66185
325. Nitrogen (1983) 144: 30
326. Grotz BJ (1967) Hydrocarbon Process 46(4): 197

423. Slack AV (1977) In: Slack AV, James GR (eds) Ammonia, vol 3. Marcel Dekker, New York, p 324
424. Vancini CA (1971) Synthesis of ammonia. The Macmillan Press Ltd, London and Basingstoke, p 241
425. Fertilizer Focus (1987) 4(10): 36
426. Topsøe HFA, Poulsen HF, Nielsen A (1967) Chem Eng Prog 63(10): 67
427. Nitrogen (1964) 31: 22
428. Vancini CA (1971) Synthesis of ammonia. The Macmillan Press Ltd, London and Basingstoke, p 245
429. Finneran JA, Mayo HC (1967) US 3350170; CA (1968) 68: No. 4558
430. Chem Eng (NY) (1967) 74(24): 112
431. Chakley DE, Turner W (1967) Oil Gas Int 7(10): 49
432. Axelrod LC, Quartulli OJ, Turner W (1966) Europe Oil 5(3): 58
433. Pet Process (1956) 10: 167
434. Slack AV (1977) In: Slack AV, James GR (eds) Ammonia, vol 3. Marcel Dekker, New York, p 303
435. Slack AV (1977) In: Slack AV, James GR (eds) Ammonia, vol 3. Marcel Dekker, New York, p 310
436. Slack AV (1977) In: Slack AV, James GR (eds) Ammonia, vol 3. Marcel Dekker, New York, p 319
437. Slack AV (1977) In: Slack AV, James GR (eds) Ammonia, vol 3. Marcel Dekker, New York, p 341
438. Slack AV (1977) In: Slack AV, James GR (eds) Ammonia, vol 3. Marcel Dekker, New York, p 350
439. Spielman M, Baumann GP, Hering B (1968) US 3388968; CA (1968) 69: No. 44970
440. Vancini CA (1971) Synthesis of ammonia. The Macmillan Press Ltd, London and Basingstoke, p 234
440. Vancini CA (1971) Synthesis of ammonia. The Macmillan Press Ltd, London and Basingstoke, p 237
442. Slack AV (1977) In: Slack AV, James GR (eds) Ammonia, vol 3. Marcel Dekker, New York, p 313
443. Mundo K-J, (1972) Chem Anlagen Verfahren 6: 49
444. Slack AV (1977) In: Slack AV, James GR (eds) Ammonia, vol 3. Marcel Dekker, New York, p 331
445. Vancini CA (1971) Synthesis of ammonia. The Macmillan Press Ltd, London and Basingstoke, p 248
446. Shearon WH Jr, Thompson HL (1952) Ind Eng Chem 44(2): 254
447. Thompson HL, Guillaumeron P, Updegraff NC (1952) Chem Eng Prog 48(9): 468
448. Nitrogen (1987) 169: 31
449. Neth N, Puhl H, Liebe W (1982) Chem Eng Prog 78(7): 69
450. Landrum LH (1984) Ammonia Plant Saf 24: 22
451. Vancini CA (1971) Synthesis of ammonia. The Macmillan Press Ltd, London and Basingstoke, p 122
452. Bendix H, Lenz L (1989) Ammonia Plant Saf 29: 221
453. Ripps DL (1977) In: Slack AV, James GR (eds) Ammonia, vol 3. Marcel Dekker, New York, p 235
454. Kjaer J (1966) Chem Tech (Berlin) 18(3): 138
455. Almasy GA, Hay JJ, Pallai IM (1966) Br Chem Eng 11(3): 188
456. Almasy G, Hay J, Jedlovszky P, Pallai, I (1966) Int Chem Eng 6(2): 233
457. Rase HF (1977) Chemical reactor design for process plants, vol 2. Wiley, New York, p 61
458. Kjaer J (1985) Comput Chem Eng 9(2): 153
459. Dybkjaer I (1986) In: De Lasa HI (ed) Chemical Reactor Design and Technology, Nasa ASI Series, Series E: Applied Sciences No. 110. Martinus Nijhoff Publishers, Dordrecht, p 795
460. Dybkjaer I, Gam EA (1985) Ammonia Plant Sat 25: 15
461. Dybkjaer I, Gam EA (1984) CEER Chem Econ Eng Rev 16(9): 29
462. Jarvan JE (1978) Oil Gas J 76(5): 178
463. Kjaer J (1979) Measurement and calculation of temperature and conversion in fixed-bed catalytic reactors, 2nd ed Haldor Topsøe A/S, Vedbaek, Denmark
464. Hlavacek V (1970) Ind Eng Chem 62(7): 8

377. Arnold JH, Dixon WT, (1955) Oil Gas J 54(30): 90
378. Czuppon TA, Lee JM (1989) In: Fertilizer Latin America Int Conf, vol 2. Br Sulphur Corp, Caracas, p 389
379. Fertilizer Focus (1989) 6(7): 46
380. Bögner F (1958) Chem Ing Tech 30(6): 382
381. Blaskowski HJ (1968) US 3404957; CA (1969) 70: No. 5585
382. Huebler J, Johnson JL, Schora C Jr, Tarman PB, Panos PS (1970) US 3539292; CA (1971) 74: No. 14671
383. Buskies U, Summers F (1978) Erdöl Kohle, Erdgas, Petrochem 31(10): 474
384. Sauchelli V (ed) (1964) Fertilizer nitrogen its chemistry and technology. Reinhold Publishing Corporation, New York, p 80
385. Hein LB (1952) Chem Eng Prog 48(8): 412
386. Vancini CA (1961) La Sintesi dell'Ammoniaca, Ulrico Hoepli, Milano, p 497
387. Kennedy JL (1967) Oil Gas J 65(46): 105, 65(48): 95, 65(1): 72, 65(51): 76, (1968) 66(2): 65, (1968) 66(4): 76
388. Fraschetti F, Filippini U, Ferrara PL (1967) Quad. Pignone 9: 5
389. Begg GAJ (1968) Chem Process Eng (London) 49(1): 58
390. Viti A (1971) Het Ingenieursblad 40(21): 619
391. Ferrara PL, Tesei A (1978) Quad Pignone 25: 131
392. Salviani J, Fraschetti F, Ferrara PL, Filippini U, Wickl R, Hile HB, Zech WA, Purcell VP (1977) In: Slack AV, James GR (ed) Ammonia, vol 2. Marcel Dekker, New York, p 1
393. Nitrogen (1968) 55: 37
394. Labrow S (1968) Chem Process Eng (London) 49(1): 55
395. Förster H (1971) Chem Tech 23(2): 93; (3): 157
396. Plötner W (1972) Chem Tech 24(6): 324
397. Nitrogen (1989) 181: 23
398. Nobles JE, Stover JC (1984) Chem Eng Prog 80(1): 81
399. Caplow SD, Bresler SA (1967) Chem Eng (NY) 74(7): 103
400. Stokes KJ (1979) Chem Eng Prog 75(7): 88
401. Twist DR, Standbridge DW (1968) GB 1134621; CA (1969) 70: No. 59416
402. Uji S, Ikeda M (1981) Hydrocarbon Process 60(7): 94
403. Bakemeier H, Huberich T, Krabetz R, Liebe W, Schunck M (1985) In: Ullman's Encyclopedia of industrial chemistry, vol 42, 5th ed VCH Verlagsgesellschaft mbH, Weinheim, p 187
404. Quartulli OJ, Fleming JB, Finneran JA (1968) Hydrocarbon Process. 47(11): 153
405. Quartulli OJ, Fleming JB, Finneran JA (1969) Nitrogen 58: 25
406. Vancini CA (1971) Synthesis of ammonia. The Macmillan Press Ltd, London and Basingstoke, p 231
407. Sauchelli V (ed) (1964) Fertilizer nitrogen its chemistry and technology. Reinhold Publishing Corporation, London, p 81.
408. Slack AV (1977) In: Slack AV, James GR (eds) Ammonia, vol 2. Marcel Dekker, New York, p 326
409. Wilson KC, Grotz BJ, Richez J (1986) In: Proc Nitrogen 86 Amsterdam, Br Sulphur Corp London, p 63
410. Nitrogen (1973) 82: 34
411. Glover W, Yoars JP (1973) Hydrocarbon Process 52(4): 165
412. Glover WA, Yoars JP (1973) Ammonia Plant Saf 15: 77
413. Process Eng (London) (1983) 64(3): 15
414. Chem Eng (London) (1983) 390: 30
415. Trotter SG, Pinto A, (1987) Fert News (India) (December) 29
416. Livingstone JG, Pinto A (1983) Chem Eng Prog 79(5): 62
417. Nitrogen (1986) 162: 27
418. Fertilizer Focus (1987) 4(10): 40
419. Fertilizer Focus (1987) 4(10): 24
420. Brown FC (1981) In: More AI (ed) Fert nitrogen: Proc Br Sulphur Corp Int Conf Fert Technol, 4th part I, p 39
421. LeBlanc JR Jr, Madhavan S, Porter RE (1978) In: Kirk–Othmer Encyclopedia of chemical technology, vol 2 3rd ed Wiley, New York, p 495
422. Finneran JA, Quartulli OJ (1977) In: Slack AV, James GR (ed) Ammonia, vol 3. Marcel Dekker, New York, p 159

423. Slack AV (1977) In: Slack AV, James GR (eds) Ammonia, vol 3. Marcel Dekker, New York, p 324
424. Vancini CA (1971) Synthesis of ammonia. The Macmillan Press Ltd, London and Basingstoke, p 241
425. Fertilizer Focus (1987) 4(10): 36
426. Topsøe HFA, Poulsen HF, Nielsen A (1967) Chem Eng Prog 63(10): 67
427. Nitrogen (1964) 31: 22
428. Vancini CA (1971) Synthesis of ammonia. The Macmillan Press Ltd, London and Basingstoke, p 245
429. Finneran JA, Mayo HC (1967) US 3350170; CA (1968) 68: No. 4558
430. Chem Eng (NY) (1967) 74(24): 112
431. Chakley DE, Turner W (1967) Oil Gas Int 7(10): 49
432. Axelrod LC, Quartulli OJ, Turner W (1966) Europe Oil 5(3): 58
433. Pet Process (1956) 10: 167
434. Slack AV (1977) In: Slack AV, James GR (eds) Ammonia, vol 3. Marcel Dekker, New York, p 303
435. Slack AV (1977) In: Slack AV, James GR (eds) Ammonia, vol 3. Marcel Dekker, New York, p 310
436. Slack AV (1977) In: Slack AV, James GR (eds) Ammonia, vol 3. Marcel Dekker, New York, p 319
437. Slack AV (1977) In: Slack AV, James GR (eds) Ammonia, vol 3. Marcel Dekker, New York, p 341
438. Slack AV (1977) In: Slack AV, James GR (eds) Ammonia, vol 3. Marcel Dekker, New York, p 350
439. Spielman M, Baumann GP, Hering B (1968) US 3388968; CA (1968) 69: No. 44970
440. Vancini CA (1971) Synthesis of ammonia. The Macmillan Press Ltd, London and Basingstoke, p 234
440. Vancini CA (1971) Synthesis of ammonia. The Macmillan Press Ltd, London and Basingstoke, p 237
442. Slack AV (1977) In: Slack AV, James GR (eds) Ammonia, vol 3. Marcel Dekker, New York, p 313
443. Mundo K-J, (1972) Chem Anlagen Verfahren 6: 49
444. Slack AV (1977) In: Slack AV, James GR (eds) Ammonia, vol 3. Marcel Dekker, New York, p 331
445. Vancini CA (1971) Synthesis of ammonia. The Macmillan Press Ltd, London and Basingstoke, p 248
446. Shearon WH Jr, Thompson HL (1952) Ind Eng Chem 44(2): 254
447. Thompson HL, Guillaumeron P, Updegraff NC (1952) Chem Eng Prog 48(9): 468
448. Nitrogen (1987) 169: 31
449. Neth N, Puhl H, Liebe W (1982) Chem Eng Prog 78(7): 69
450. Landrum LH (1984) Ammonia Plant Saf 24: 22
451. Vancini CA (1971) Synthesis of ammonia. The Macmillan Press Ltd, London and Basingstoke, p 122
452. Bendix H, Lenz L (1989) Ammonia Plant Saf 29: 221
453. Ripps DL (1977) In: Slack AV, James GR (eds) Ammonia, vol 3. Marcel Dekker, New York, p 235
454. Kjaer J (1966) Chem Tech (Berlin) 18(3): 138
455. Almasy GA, Hay JJ, Pallai IM (1966) Br Chem Eng 11(3): 188
456. Almasy G, Hay J, Jedlovszky P, Pallai, I (1966) Int Chem Eng 6(2): 233
457. Rase HF (1977) Chemical reactor design for process plants, vol 2. Wiley, New York, p 61
458. Kjaer J (1985) Comput Chem Eng 9(2): 153
459. Dybkjaer I (1986) In: De Lasa HI (ed) Chemical Reactor Design and Technology, Nasa ASI Series, Series E: Applied Sciences No. 110. Martinus Nijhoff Publishers, Dordrecht, p 795
460. Dybkjaer I, Gam EA (1985) Ammonia Plant Sat 25: 15
461. Dybkjaer I, Gam EA (1984) CEER Chem Econ Eng Rev 16(9): 29
462. Jarvan JE (1978) Oil Gas J 76(5): 178
463. Kjaer J (1979) Measurement and calculation of temperature and conversion in fixed-bed catalytic reactors, 2nd ed Haldor Topsøe A/S, Vedbaek, Denmark
464. Hlavacek V (1970) Ind Eng Chem 62(7): 8

465. Froment GF (1974) Chem Ing Tech. 46(9): 374
466. Jarvan JE (1970) Ber Bunsen Ges Phys Chem 74(2): 141
467. Kjaer J (1972) Computer methods in catalytic reactor calculations. Haldor Topsøe, Vedbaek Denmark
468. Jarvan JE (1978) Oil Gas J 76(3): 51
469. Hay JJ, Pallai IM (1963) Br Chem Eng 8(3): 171
470. Bakemeier H, Detzer H, Krabetz R (1965) Chem Ing Tech 37(4): 429
471. Fodor L (1971) Chim Ind Genie Chim 104(8): 1002
472. Singh CPP, Saraf DN (1979) Ind Eng Chem Process Des Dev 18(3): 364
473. Lucas K, Gelbin D (1962) Br Chem Eng 7(5): 336
474. Shah MJ (1967) Ind Eng Chem 59(1): 72
475. Gaines LD (1977) Ind Eng Chem Process Des Dev 16(3): 381
476. Gaines LD (1979) Chem Eng Sci 34: 37
477. Reddy KV, Husain A (1982) Ind Eng Chem Process Des Dev 21(3): 359
478. Shipman LM, Hickman JB (1968) Chem Eng Prog 64(5): 59
479. Siminiceanu I, Petrila C, Pop A (1983) Chem Tech 35(12): 628
480. Baddour RF, Brian PLT, Logeais BA, Eymery JP (1965) Chem Eng Sci 20: 281
481. Brian PLT, Baddour RF, Eymery JP (1965) Chem Eng Sci 20: 297
482. Almasy GA, Jedlovszky P, Pallai IM (1967) Br Chem Eng 12(8): 1219
483. Porubszky I, Simonyi E, Ladanyi G (1969) Br Chem Eng 14(4): 495
484. Murase A, Roberts HL, Converse AO (1970) Ind Eng Chem Process Des Dev 9(4): 503
485. Balzer D, Iijin B, Reinig G, Scholz V (1971) Chem Tech 23(9): 513
486. Lewin DR, Lavie R (1984) Ind Chem Eng Symp Ser 87: 393
487. Nielsen A (1953) In: Advances in catalysis, vol V. Academic Press Inc, New York, p 1
488. Nielsen A (1968) An investigation on promoted iron catalysts for the synthesis of ammonia, 3rd ed. Jul. Gjellerups Forlag, Copenhagen
489. Nielsen A (1981) Catal Rev Sci Eng 23(1,2): 17
490. Zardi U (1982) Hydrocarbon Process 61(8): 129
491. Nitrogen (1972) 75: 33
492. Nitrogen (1982) 140: 30
493. Strelzoff S (1981) Technology and manufacture of ammonia. Wiley, New York, p 23
494. Bakemeier H, Huberich T, Krabetz R, Liebe W, Schunck M (1985) In: Ullman's Encyclopedia of industrial chemistry, vol A2. 5th ed. VCH Verlagsgesellschaft mbH, Weinheim, p 190
495. Worn GA (1962) US 3041150; CA (1963) 58(8): No. 10997
496. Slack AV, Allgood HY, Maune HE (1953) Chem Eng Prog 49(8): 393
497. Vancini CA (1961) La Sintesi dell'Ammoniaca. Ulrico Hoepli, Milano, p 951
498. Vancini CA (1971) Synthesis of ammonia. Macmillan London p 239
499. Slack AV (1977) In: Slack AV, James GR (eds) Ammonia, vol 3. Marcel Dekker, New York, p 353
500. Slack AV (1977) In: Slack AV, James GR (eds) Ammonia, vol 3. Marcel Dekker, New York, p 333
501. de Rycker HGEL, Mathieu MFVG (1959) US 2887365; CA (1959) 53: No. 12766
502. Vek V (1988) Paper presented at 7th Int Symp Large Chemical Plants 1988. Bruges, Belgium, p 77
503. Gmelins Handbuch der Anorganischen Chemie (1935) Stickstoff, nitrogen system 4, vol 2. Verlag Chemie, Berlin, p 405
504. Vancini CA (1971) Synthesis of ammonia. Macmillan London p 235
505. Vancini CA (1971) Synthesis of ammonia. Macmillan London p 249
506. Vancini CA (1971) Synthesis of ammonia. Macmillan London p 232
507. Gmelins Handbuch der Anorganischen Chemie (1935) Stickstoff, nitrogen system 4, vol 2. Verlag Chemie, Berlin, p 405
508. Pinto A, Ward SA (1980) US 4215099; see GB 1565824 CA (1980) 93: No. 134577
509. Schober R (1962) US 3031274; CA (1961) 55: No. 6962
510. Dohnalek R, Hribal J (1965) NE 6400361; CA (1966) 64: No. 10809
511. Breus YV (1965) SU 170917; CA (1965) 63: No. 9516
512. Christensen A, Rayfiel RD (1958) US 2953371; CA (1959) 53: No. 6707
513. Christensen A (1962) US 3041151; CA (1962) 56: No. 12706

514. Christensen A (1962) US 3050377; CA (1962) 57: No. 9623
515. Worn GA (1958) US 2861873; CA (1959) 53: No. 3797
516. Vilceanu M, Bors C (1976) US 3932139; see FR 2141488 CA (1973) 79: No. 94152
517. Hennel W (1964) Br Chem Eng 9(5): 290
518. Slack AV (1977) In: Slack AV, James GR (eds) Ammonia, vol 3. Marcel Dekker, New York, p 358
519. Takagi T (1968) Jpn Chem Q 4(4): 47
520. Ohsaki K, Zanma J, Watanabe H (1984) DE 3334775; CA (1984) 101: No. 40286
521. Papillon A, Lesur P, Faury C, Badoual C, Lafleur G (1989) EP 314550 CA (1989) 111: No 80790
522. Lee JA (1945) Chem Metall Eng (December) 94
523. Vancini CA (1971) Synthesis of ammonia. Macmillan London p 240
524. Vancini CA (1971) Synthesis of ammonia. Macmillan London p 244
525. Slack AV (1977) In: Slack AV, James GR (eds) Ammonia, vol 3. Marcel Dekker, New York, p 312
526. Vancini CA (1971) Synthesis of ammonia. Macmillan London p 236
527. Eschenbrenner GP, Honigsberg CA (1969) US 3475136; see FR 1528951 CA (1969) 71: 14502
528. Strelzoff S (1981) Technology and manufacture of ammonia. Wiley, New York, p 34
529. Lesur P (1977) Nitrogen 108: 29
530. Kubec J, Saroch V (1974) US 3784362; see FR 2046522 CA (1971) 75: No. 131039
531. Kubec J, Saroch V (1972) US 3653846; see DE 1927850 CA (1971) 72: No. 68564
532. Eschenbrenner GP, Wagner III GA (1972) Chem. Eng. Prog. 68(1): 62; Ammonia Plant Saf (1971) 14: 51
533. Axelrod LC, Eagle RS, Fleming JB, Kahn GM, Quartulli OJ (1971) US 3567404; CA (1971) 74: No. 143654
534. Wett T (1971) Oil Gas J 69(41): 70
535. Chem Eng (NY) (1971) 78(24): 90
536. Quartulli OJ, Wagner II GA, (1978) Hydrocarbon Process 57(12): 115
537. Fawcett R, Smith AW, Westwood D (1972) US 3694169; see DE 2025486 CA (1971) 74: No. 113691
538. Ridler DE (1972) Ammonia Plant Saf 14: 57
539. Mehta DD, Miller EJ (1972) US 3663179; see DE 2119127 CA (1972) 76: No. 26712
540. Nielsen A (1969) Chim Ind (Milan) 51(10): 1052
541. Vek V (1973) Chem Ing Tech 45(9, 10): 608
542. Vek V (1977) Ind Eng Chem Process Des Dev 16(3): 412
543. Pet Int (1974) 14(11): 72
544. Hansen HJ (1968) US 3372988; CA (1965) 63: No. 14703
545. Chem Week (1966) 98(9): 97
546. Nielsen A (1972) Ammonia Plant Saf 14: 46
547. Slack AV (1977) In: Slack AV, James GR (eds) Ammonia, vol 3. Marcel Dekker, New York, p 345
548. Slack AV (1977) In: Slack AV, James GR (eds) Ammonia, vol 3. Marcel Dekker, New York, p 354
549. Hinrichs H, Lehner H, Wagner A, Pesl J, Niedetzky J, Faschinger F (1973) US 3754078, see FR 2089755 CA (1972) 77: No. 93392
550. Kohn HB, Friedman G (1975) US 3918918; CA (1976) 84: No.166677
551. Zardi U (1983) US 4372920; see DE 3026199 CA (1981) 94: No. 123623
552. Zardi U, Pagani G (1988) EP 287765; CA (1989) 110: No. 40965
553. Zardi U, Commandini E, Gallazzi C (1981) In: More AI (ed) Fert nitrogen: Proc Br Sulphur Corp Int Fert Technol, 4th part I, p 173
554. Fauser G (1959) US 2898183; CA (1959) 53: No. 19489
555. Vancini CA (1971) Synthesis of ammonia. Macmillan London p 233
556. Strelzoff S (1981) Technology and manufacture of ammonia. Wiley, New York, p 27
557. Hinrichs H, Niedtzky J (1962) Chem Ing Tech 34: 88
558. Graeve HW (1981) Chem Eng Prog 77(10): 54
559. Graeve HW (1981) Ammonia Plant Saf 23: 78
560. Ind Eng Chem (1962) 54(5): 11

561. Nitrogen (1962) 17: 23
562. Nitrogen (1966) 42: 40
563. Nitrogen (1973) 81: 29
564. Hinrichs H, Lehner H (1972) Chem Anlagen Verfahren 6: 55
565. Oesterreichische Stickstoffwerke A-G (1966) US 3440021; CA (1977) 66: No. 12354
566. Förster F (1980) Chem Eng 87(9): 62
567. Strelzoff S (1981) Technology and manufacture of ammonia. Wiley, New York, p 26
568. Slack AV (1977) In: Slack AV, James GR (eds) Ammonia, vol 3. Marcel Dekker, New York, p 315
569. Hennel W, Sobolewski C, Bartoski Z (1975) US 3892535; see DE 2323678 CA (1974) 80: No. 147234
570. Handman SE, Leblanc JR Jr (1983) Chem Eng Prog 79(5): 56
571. Peterson R, Finello R, Denavit G (1985) EP 134260; CA (1985) 102: No. 187456
572. Wright LE, Pickfor AE (1973) US 3721532; CA (1973) 79: No. 33234
573. Wright LE, Pickford AE (1974) US 3851046: CA (1975) 82: No. 142248
574. Grotz BJ (1988) US 4744966; CA (1988) 109: No. 57548
575. Grotz BJ (1988) EP 268469; CA (1988) 109: No. 57554
576. Wilson KC, Grotz BJ (1990) J Tech Dev 14: 54
577. Ward SA (1979) US 4180543; see DE 2815856 CA (1979) 90: No. 123588
578. LeBlanc JR, Peterson RB (1981) US 4298589; CA (1982) 96: No. 37843
579. Nitrogen (1976) 101: 42
580. Topsoe HFA, Gam EA (1980) US 4181701; see DE 2710247 CA (1978) 88: No. 52614
581. Haldor Topsoe A/S (1981) CA 1112847; see DE 2710247 CA (1978) 88: No. 52614
582. Zardi U, Commandini E (1983) US 4405562; see DE 3146778 CA (1983) 98: No. 56154
583. Zardi U (1987) EP 222069; CA (1987) 107: No. 98608
584. Weicken R (1986) Chem Ind (Berlin) 8: 674
585. Uhde Engineering News (1987) 1–87: 1
586. Hakmann R, Nübel W, Weicken R, Lehner H (1986) Ammonia Plant Saf 26: 44
587. Eagle RA, Patel VA (1980) US 4230669; see EP 7743 CA (1980) 93: No. 28739
588. Becker CL, LeBlanc JR (1980) JS 4230680; see EP 7743 CA (1980) 93: No. 28739
589. Pagani G (1978) US 4101281; see DE 2631898 CA (1977) 86: No. 123879
590. Nitrogen (1987) 165: 27
591. Shannahan N (1989) Hydrocarbon Process 68(1): 60
592. Peterson RB (1984) US 4482523; CA (1985) 102: 26997
593. Nitrogen (1986) 162: 35
594. Howerton RG, Noe SA (1989) Ammonia Plant Saf 29: 157
595. Dybkjaer I (1989) In: Fertilizer Latin America International Conference, vol 1. Br Sulphur Corp, Caracas, p 77
596. Novacek D, Newland RL, Horsey JF (1990) Ammonia Plant Saf 30: 161
597. Fert News (1987) (December): 19
598. Brown FC, Zardi U, Pangani G (1988) In: Nitrogen 88 Br Sulphur's 12th Int Conf Geneva, p 159
599. Nitrogen (1987) 169: 33
600. Comandini E, Zardi U (1989) In: Fertilizer Latin America Int Conf vol 1. Br Sulphur Corp, Caracas, p 113
601. Sutherland L, Wallace B (1989) Ammonia Plant Saf 29: 167
602. Pagani G, Zardi U (1989) In: FAI Seminar 1988. The Fertiliser Society of India, New Delhi, SII/1–1/9
603. Newland RL, Pierce JD, Borzik DM (1988) Ammonia Plant Saf 28: 53
604. Axelrod LC, O'Hare TE (1964) In: Sauchelli V (ed) Fertilizer nitrogen its chemistry and technology. Reinhold Publishing Corporation, New York, p 78
605. Hinrichs H (1962) Ind Chem (January): 7
606. Hinrichs H (1963) Ind Chem (October): 519
607. Bögner F, Karsch D (1965) Nitrogen 37: 21
608. Lachmann H, Fromm D (1988) Ammonia Plant Saf 28: 91
609. Prescott GR, Podhorsky M, Blommaert P (1988) Ammonia Plant Saf 28: 135
610. Podhorsky M, Prescott GR, Blommaert P, Grisola L (1989) Ammonia Plant Saf 29: 253
611. Stahl H (1982) Proc Fert Soc London 206: 61
612. Weber H, Graeve HW, Herbort H-J, Marsch H-D (1984) Chem Ing Tech 56(5): 356
613. Johnson DL, O'Connor TM (1987) Ammonia Plant Saf 27: 27
614. Cocchi V, Mezzedimi V (1983) Quad Pignone 36: 17
615. Nitrogen (1988) 176: 25

616. ECN Eur Chem News (1966) (August 19): 34
617. Silberring L (1985) Nitrogen 157: 33
618. Vancini CA (1961) La Sintesi dell'Ammoniaca. Ulrico Hoepli, Milano, p 769
619. Nitrogen (1962) 16: 48
620. Steels for hydrogen service at elevated temperatures and Pressures in petroleum refineries and petrochemical plants. API publication 941, (1990) 4th ed, American Petroleum Institute
621. Bonner WA (1977) Hydrocarbon Process 56(5): 165
622. Kobrin G, Kopecki ES (1978) Chem Eng (NY) 85(28): 115
623. Prescott GR, Badger FW (1980) Ammonia Plant Saf 22: 146
624. Prescott GR (1982) Plant/Oper Prog 1(2): 94
625. Prescott GR, Blommaert P, Grisola L (1986) Ammonia Plant Saf 26: 228
626. Marsch HD (1982) Plant/Oper Prog 1(3): 152
627. Sathe SY, O'Connor T (1988) Ammonia Plant Saf 28: 115
628. van Grieken CA (1989) Ammonia Plant Saf 29: 145
629. Zeis LA (1975) Ammonia Plant Saf 17: 56
630. Ueda S, Onishi H, Okubo M, Takegawa T (1978) Ammonia Plant Saf 20: 98
631. van Grieken CA (1979) Ammonia Plant Saf 21: 18
632. Phillips BR (1979) Ammonia Plant Saf 21: 21
633. Orbons HG, Huurdemann TL (1985) Ammonia Plant Saf. 25: 89; Plant/Oper Prog (1985) 4(1): 49
634. Cermak J, Burex Z (1967) Br Chem Eng 12(9): 1372
635. Bresler SA (1977) US 4056603; CA (1978) 88: No. 25032
636. Lipman M (1979) Hydrocarbon Process 58(1): 153
637. Bogart MJP (1981) In: More AI (ed) Fert nitrogen: Proc Br Sulphur Corp Int Fert Technol, 4th part I, p 141
638. Bogart MJP (1982) Plant/Oper Prog 1(3): 147
639. Pagani G, Zardi U (1972) Hydrocarbon Process 51(7): 106
640. Lavie R (1985) Chem Eng Sci 40(11): 2019
641. Lavie R (1987) Plant/Oper Prog 6(2): 118
642. Lavie R (1985) US 4537760; see FR 2511666 CA (1983) 98: No. 218143
643. Brubaker DW, Kammermeyer K (1954) Ind Eng Chem 46(4): 733
644. Laciak DV, Pez GP (1988) US 4758250; CA (1988) 109: No. 152435
645. Pez GP, Laciak DV (1988) US 4762535; CA (1988) 109: No. 131707
646. Otto W (1972) Chem Tech 24(8): 489
647. Parrish RW (1985) Ammonia Plant Saf 25: 162
648. Wang SI, Nicholas DM, Dimartino SP (1984) Oil Gas J 82(6): 111
649. Merlo M, Saviano F (1985) Fert Ind 63
650. Isalski WH (1984) Nitrogen 152: 100
651. Oil Gas J (1979) 77(10): 182
652. Banks R (1977) Chem Eng (NY) (October 10): 90
653. Isalski WH (1979) Chem Age (London) (November 30): 19
654. Haslam A, Brook P, Isalski H (1976) Hydrocarbon Process 55(1): 103
655. Nitrogen (1976) 102: 35
656. Haslam A, Isalski WH (1975) Ammonia Plant Saf 17: 80
657. Harmon R (1981) In: More AI (ed) Fert Nitrogen: Proc Br Sulphur Corp Int Fert Technol, 4th part I, p 113
658. Finn A (1988) Nitrogen 175: 25
659. Fabian R, Tillmann D (1986) Linde Ber Tech Wiss 59: 6
660. Harmon R, Isalski WH (1981) Ammonia Plant Saf 23: 39
661. Banks R (1978) Ammonia Plant Saf. 20: 79
662. Combs CA (1981) Ammonia Plant Saf 23: 32
663. Berry RI (1979) Chem Eng (NY) 86(15): 62
664. Overmann L, Staude E (1987) Erdöl Kohle, Erdgas, Petrochem Brenns Chem 40(10): 427
665. Sculz G, Michele H, Werner U (1982) Chem Ing Tech 54(4): 351
666. Fritzsche AK, Narayan RA (1987) CEER Chem Econ Eng Rev 19(1,2,3): 19
667. Knieriem H Jr (1980) Hydrocarbon Process 59: 65
668. Rosenzweig MD (1981) Chem Eng (NY) 88(24): 62
669. Nitrogen (1982) 136: 29
670. Chae YC, Legendre GS, van JM Gelder (1981) In: More AI (ed) Fert Nitrogen: Proc Br Sulphur Corp Int Fert Technol, 4th part I, p 457

671. MacLean DI, Prince CE, Chae YC (1980) Ammonia Plant Saf 22: 1
672. Schendel RL, Mariz CL, Mak JY (1983) Hydrocarbon Process 62(8): 58
673. Chem Eng (London) (1979) 345: 395
674. Sheridan III JJ, Eisenberg FG, Greskovich EJ, Sandrock GD, Huston EL (1983) J Less Common Met 89: 447
675. Santangelo JG, Chen GY (1983) CHEMTECH 13(10): 621
676. Duff BS (1955) Chem Eng Prog 51(1): 12-J
677. Mundo K-J (1973) Chem Eng Tech 45(10a): 632
678. Strelzoff S (1974) Hydrocarbon Processes 53(10): 133
679. Chem Eng (NY) (1974) 81(23): 52
680. Buividas LJ, Finneran JA, Quartulli OJ (1974) Chem Eng Prog 70(10): 21
681. Rothman SN, Frank ME (1975) Ammonia Plant Saf 17: 19
682. Hess M (1976) Hydrocarbon Process 55(11): 97
683. Netzer D, Moe J (1977) Chem Eng (NY) 84(23): 129
684. Waitzman DA (1978) Chem Eng (NY) 85(3): 69
685. Czuppon TA, Buividas LJ (1979) Hydrocarbon Process 58(9): 197
686. Nichols D, Blouin GM (1979) CHEMTECH 9(9): 512
687. Buividas LJ (1981) Chem Eng Prog 77(5): 44; Ammonia Plant Saf (1981) 23: 67
688. Axelrod L (1981) Catal Rev Sci Eng 23(1,2): 53
689. Omori T (1982) CEER Chem Econ Eng Rev 14(9): 7
690. Staege H (1982) Tech Mitt Krupp Werksber 40(1): 1
691. Stratton A, Teper M (1984) The economics of producing ammonia and hydrogen. IEA Coal Research, London, p 1
692. Holroyd R (1967) Chem Ind (August 5): 1310
693. Bögner H (1965) Nitrogen 34: 32
694. Axelrod L, Daze RE, Wickham HP (1968) Chem Eng Prog 64(7): 17
695. Quartulli OJ, Turner W (1972) Nitrogen 80: 28; (1973) 81: 32
696. Nitrogen (1976) 100: 77
697. Ennis R, Lesur PF (1977) Hydrocarbon Process 56(12): 121
698. Arkley KI, Pinto A (1987) In: FAI Seminar 1986. The Fertiliser Society of India, New Delhi, p SI/2
699. Pinto A, Trotter SG (1987) Proc Fert Soc London 260
700. Arkley KI, Halstead JM, Pinto A (1988) Ammonia Plant Saf 28: 58
701. Dybkjaer I (1984) In: IFA Technical Conf, Paris, 13-1
702. Denbigh KG (1956) Chem Eng Sci 6(1): 1
703. Riekert L (1974) Chem Eng Sci 29: 1613
704. Gaggioli RA, Petit PJ (1977) CHEMTECH 7(7): 496
705. Gaggioli RA (ed) (1980) Thermodynamics second law analysis ACS Symp Ser 122
706. Sussman MV (1980) Chem Eng Prog 76(1): 37
707. Christiansen LJ (1987) In: CEF 87 Proc XVIII Cong. EFCE, Sicily, p 217
708. Vaclavek V, Duy Phi D, Kleinova N (1987) In: CEF 87 Proc XVIII Congr EFCE, Sicily, p 721
709. Silberring L (1971) Chem Ing Tech 43(12): 711
710. Pinto A, Rogerson PL (1977) Chem Eng Prog 73(7): 95
711. Cremer H (1980) In: Gaggioli RA (ed) Thermodynamics second law analysis ACS Symp Ser 122: 111
712. Nimmo NM (1967) Chem Eng (London) 210: CE 156
713. Topsøe H (1966) In: World Power Conf. Tokyo, paper 135 IIIA
714. Finneran JA, Mayo HC, Multharp RH, Smith RB (1969) US 3441393; CA (1969) 71: No. 23310
715. Silberring L (1969) Chem Process Eng Heat Transfer Surv (August) Suppl: 74
716. Slack JB (1972) Chem Eng (NY) 79(2): 107
717. Renker W (1972) Chem Tech 24(10): 596
718. Silberring L (1979) Nitrogen 120: 35
719. van Campagne WL (1981) Hydrocarbon Process 60(8): 117; (1982) 61(7): 145; (1983) 67(7): 57
720. Tasrif A, Madsen J (1989) In: Proc Fertilizer Asia Conf and Exhibition, Manila 1989, Br Sulphur Corp, p 155
721. Quartulli OJ (1975) Hydrocarbon Process 54(10): 94
722. Buividas LJ (1980) Proc Environ Symp, p 365
723. Bignon M (1970) In: Proc Fert Soc London 117: 98
724. Hinchley P (1979) Chem Eng (NY) 86(17): 120
725. Sault RA (1968) Chem Eng Prog 64(3): 57

726. Camps JA, Gentry JF (1987) Ammonia Plant Saf 27: 36
727. Badame PJ (1989) Ammonia Plant Saf 29: 179
728. Barnes KB (1949) Oil Gas J (August 4): 40
729. Chem Eng (NY) (1951) 58(8): 174
730. Reidel JC (1953) Oil Gas J 51(36): 82
731. Resen FL (1954) Oil Gas J 53(32): 120
732. Reidel JC (1954) Oil Gas J 52(44): 86
733. Pet Refin (1954) 33: (11) 168/170
734. Reidel JC (1956) Oil Gas J 54(47): 106
735. Reynolds W (1965) Proc Fert Soc London 89: 1
736. Guccione E (1965) Chem Eng (NY) 72(24): 124
737. Topsoe H (1965) Chim Ind 94(1): 1
738. Mayo HC, Finneran JA (1968) Oil Gas J 66(12): 78
739. Bakemeier H, Huberich T, Krabetz R, Liebe W, Schunk M (1985) In: Ullman's Encyclopedia of industrial Chemistry, vol A2, 5th ed. VCH Verlagsgesellschaft mbH, Weinheim, p 204
740. Brown FC (1976) Nitrogen 100: 65
741. Mundo KJ (1978) Erdöl Kohle, Erdgas, Petrochem Brenns Chem 31(2): 74
742. Nitrogen (1970) 65: 32
743. Fangel PB (1970) Nitrogen 67: 35
744. Daigre III LC, Nieman GR (1974) Ammonia Plant Saf 16: 45
745. Tijssen P (1977) Ammonia Plant Saf 19: 155
746. Weems S, Ball DH, Griffin DE (1979) Chem Eng Prog 75(5): 64
747. Yost CC, Curtis CR, Ryskamp CJ (1980) Ammonia Plant Saf 22: 200
748. Collier G, Voelkers JD, Griffin DE (1980) Ammonia Plant Saf 22: 206
749. Yazhari F (1982) Hydrocarbon Process 61(5): 187
750. Huurdeman TL (1989) Ammonia Plant Sat 29: 234
751. Madhavan S (1984) Plant/Oper Prog 3(1): 14
752. Solomon SM (1985) Ammonia Plant Saf 25: 129
753. Moore SC, Piper TM, Chen CC (1986) Ammonia Plant Saf 26: 56
754. Pathe DC (1986) Oil Gas J 84(14): 49
755. Nitrogen (1987) 169: 35
756. Jarvan JE (1987) Fert Ind Annu Rev India 10: 8
757. Fert Int (1988) 267: 25
758. Mani S, Shoor SK, Pedersen HS (1989) Ammonia Plant Saf 29: 244
759. Christensen KG, Gosnell JH, Grotz BJ (1989) In: Fertilizer Latin American Int Conf, Br Sulphur Corp, Caracas p 99; Nitrogen (1989) 181: 31
760. Setoyama I, Wada E, Funakoshi Y (1974) Ammonia Plant Saf 21: 82
761. Grotz BJ, Gosnell JH, Grisola L (1985) Ammonia Plant Saf 25: 38
762. Grisolia L, Johnstone, WD, Grotz BJ, Crowley MR (1988) In: Nitrogen 88 Br Sulphur's 12th Int Conf Geneva, p 11
763. Bizzotto V (1986) In: FAI Seminar 1986, The Fertiliser Society of India, New Delhi p S I/1-1
764. Pinto A (1981) US 4298588; see JP 54060298 CA (1979) 91: No. 142758
765. Pinto A, Johnson BH (1987) US 4695442; see EP 157480 CA (1985) 103: No 217689
766. LeBlanc JR (1985) Ammonia Plant Saf 25: 29
767. LeBlanc JR (1984) Hydrocarbon Process 63(7): 69
768. LeBlanc JR (1985) Energy Prog 5(1): 4
769. Becker CL (1979) US 4153673; CA (1979) 91: No. 23422
770. Rudbeck P (1982) J Tech Dev 7(4): 24
771. Bajpai T (1988) In: Nitrogen 88 Br Sulphur's 12th Int Conf Geneva, p 39
772. Hariharan TS (1987) J Tech Dev 12(2): 42
773. Sahore SR, Krishnan TS (1989) Fert News (December): 15
774. Dybkjaer I (1990) Fertilizer Industry Annu Rev XIII: 42A
775. Kalthoff P (1989) Fert News (December): 21
776. Ruziska PA, Song CC, Wilkinson RA, Unrch W (1985) Ammonia Plant Saf 25: 22; Plant/Oper Prog (1985) 2: 79
777. Ruziska PA, Dranse P, Song CC (1986) In: Nitrogen 86 Amsterdam, Br Sulphur Corp, London, p 45
778. Bogart MJP (1978) Hydrocarbon Process 57(4): 145
779. Banquy DL (1981) US 4296085; see EP 32096 CA (1981) 95: No. 189496
780. Brook PH (1983) In: Fertilizer 83, Br Sulphur Corp 7th Int Conf London

781. Banquy D (1984) Ammonia Plant Saf 24: 8
782. Brown FC (1978) Paper presented at Process Dev Tech Symp 1978, The Institution of Chemical Engineers, p 1
783. Winter CL (1984) GB 2126208; CA (1984) 101: No. 9578
784. Brown FC (1982) ECN Eur Chem News Process Rev (November 15): 8
785. Nitrogen (1983) 141: 36
786. Brown FC, Topham C (1990) In: Proceedings European Conf on Energy Efficient Production of Fertilizers, Bristol UK, p 1
787. Orphanides P, Polychronides C (1984) In: IFA Technical Conf, Paris, p 15-1
788. Taffe P (1978) Chem Age (London) (December 15): 8
789. Ghosal SR, Karkun K (1982) Fert News (December): 47
790. Karkun K, Ghosal SR (1985) IN 156125; CA (1986) 104: No. 170928
791. Chatterjee B (1980) Fert News (December): 19
792. Thompson DC (1981) In: More AI (ed) Fert Nitrogen: Proc Br Sulphur Corp Int Conf Fert Technol, 4th part II, p 545
793. Madhavan S, Landry B (1988) Ammonia Plant Saf 28: 1
794. Becker CL (1981) In: More AI (ed) Fert Nitrogen: Proc Br Sulphur Corp Int Conf Fert Technol 4th, part II, p 537
795. Low G (1984) Nitrogen 147 Suppl: 1
796. Nielsen A, Hansen JB, Houken J, Gam EA (1982) Plant/Oper Prog 1(3): 186
797. Graeve HW (1988) In: Nitrogen 88 Br Sulphur's 12th Int Conf Geneva, p 77
798. Wang SI, Patel NM (1984) Plant/Oper Prog 3(2): 101; Ammonia Plant Saf (1984) 24: 1
799. Dark AM, Stallworthy EA (1985) Nitrogen 153: 25
800. Singh D (1986) Process Plant Eng (Jan/March): 73
801. LeBlanc JR, Moore DO, Schneider III RV (1982) Oil Gas J 80(38): 115
802. LeBlanc JR (1986) Hydrocarbon Process 65(8): 39
803. Darajat R, LeBlanc JR (1988) In: Nitrogen 88 Br Sulphur's 12th Int Conf, Geneva, p 29
804. MacLean DL, Prince CE, Chae YC (1980) Ammonia Plant Saf 22: 1
805. Howerton RG (1980) Ammonia Plant Saf 22: 9
806. Patel NM, Wang SI, Kittelstad KJ (1989) Ammonia Plant Saf 29: 33
807. Reidel JC (1954) Oil Gas J 53(13): 78
808. Skinner GF, Kowal WM (1983) US 4409196; CA (1984) 100: No. 9464
809. Skinner GF (1981) In: More AI Ied) Fert nitrogen: Proc Br Sulphur Corp Int Conf Fert Technol, 4th, part II, p 491
810. Gent CW, Ward SA (1980) Chem Eng (London) 353: 85
811. Partridge LJ (1980) Chem Eng (London) 353: 88
812. Tippmer K (1986) Erdöl Kohle, Erdgas, Petrochem Brenns Chem 39(8): 361
813. Brown F (1977) Hydrocarbon Process 56(11): 361
814. Nurse TW (1982) In: Proc 5th Int Symp Large Chem Plants, Antwerpen, p 131
815. Nitrogen (1986) 160: 39
816. Hargreaves HG, Kirk RJ (1980) Inst Chem Eng Symp Ser 62: s1
817. Stewart RM, Marqueen TJ (1978) Coal Process Technol 4: 201
818. CEER Chem Econ Eng Rev (1980) 12(6-7): 33
819. Dyekjaer Hansen E (1980) Chem Age India 31(12) : DEV 3/1
820. Keens D (1982) Inst Chem Eng Symp Ser 74: 55
821. Nitrogen (1966) 43: 26
822. Oil Gas J (1966) 64(33): 48
823. Nitrogen (1970) 64: 17
824. Hsu H, Jaeschke WC (1967) US 3349126; CA (1968) 68: No. 12509
825. Quartulli OJ, Axelrod LC, Randall R (1971) Het Ingenieursblad 40(21): 642
826. Lagana V, Zardi U (1977) Proc Fert Soc London 167: 1
827. Lagana V (1978) Chem Eng (NY) 85(1): 37
828. Farinola G, Lagana V (1979) Hydrocarbon Process 58(9): 202
829. Van Dijk CP (1989) US 4869887; CA (1990) 112: No. 39261
830. Hale CC, Lichtenberg WH (1988) Ammonia Plant saf 28: 76
831. Hale CC (1988) In: Nitrogen 88 Br Sulphur's 12th Int Conf, Geneva, p 161
832. Reed JD (1979) In: Slack AV, James GR (eds) Ammonia, vol 4. Marcel Dekker, New York, p 1
833. Bakemeier H, Huberich T, Krabetz R, Liebe W, Schunck M (1985) In: Ullman's Encyclopedia of industrial chemistry, vol A2, 5th ed. VCH Verlagsgesellschaft mbH, Weinheim, p 221
834. Hale CC (1974) Ammonia Plant Saf 16: 23

835. Hale CC (1979) Ammonia Plant Saf 21: 61
836. Hale CC (1979) Nitrogen 119: 30
837. Hale CC (1984) Plant/Oper Prog 3(3): 147
838. Hale CC (1984) Ammonia Plant Saf 24: 181
839. Hale CC (1984) Nitrogen 150: 27
840. Swaminathan BV, Jain BK (1986) Fert News (May): 35
841. Safety recommendations for the large scale storage of anhydrous ammonia (1985) IFA, International Fertilizer Industry Association Ltd, Paris
842. Hale CC, Lichtenberg WH (1980) Ammonia Plant Saf 22: 35
843. Parmar AS (1986) Fert News (May): 65
844. Huberich T (1982) Plant/Oper Prog 1(2): 117
845. Suri IK, Bhola RK (1986) Fert News (May): 52
846. Phelps EH (1972) Ammonia Plant Saf 14: 109
847. Arup H (1977) Ammonia Plant Saf 19: 73
848. Lunde L (1984) Ammonia Plant Saf 24: 154
849. Lunde L, Nyborg R (1987) Ammonia Plant Saf 27: 49; Plant/Oper Prog (1987) 6(1): 11
850. Stephens JD, Vidalin F (1988) Ammonia Plant Saf 28: 9
851. Olsen EA (1969) Ammonia Plant Saf 11: 46
852. Byrne JR, Moir FE, Williams RD (1989) Ammonia Plant Saf 29: 122
853. Nielsen A (1971) Ammonia Plant Saf 13: 103
854. Blanken JM (1984) Ammonia Plant Saf 24: 140
855. KA van Grieken (1976) Ammonia Plant Saf 18: 15
856. Cracknell A (1980) Ammonia Plant Saf 22: 63
857. Brown RS (1982) Plant/Oper Prog 1(2): 97
858. Guth DC, Clark DA (1985) Ammonia Plant Saf 25: 60
859. Guth DC, Clark DA (1985) Plant/Oper Prog 4(1): 16
860. Comeau ET (1972) Ammonia Plant Saf 14: 69
861. Hendriks NA (1979) Ammonia Plant Saf 21: 69
862. Aarts JJ, Morrison DM (1981) Ammonia Plant Saf 23: 124
863. Al-Abdulally F, Al-Shuwaib S, Gupta BL (1987) Ammonia Plant Saf 27: 153
864. Comeau ET, Weber ML (1977) Ammonia Plant Saf 19: 63
865. Vick KA, Withaus JB, Mayo HC (1980) Ammonia Plant Saf 22: 54
866. Martinez OA, Madhavan S, Kellett DJ (1987) Ammonia Plant Saf 27: 162
867. Code of Practice for Large Scale Storage of Fully Refrigerated Anhydrous Ammonia in the UK (1975). Chemical Industries Association, London
868. Shah JM (1982) Plant/Oper Prog 1(2): 90
869. Thompson JR, Carnegie RN (1989) Ammonia Plant Saf 29: 116; Plant/Oper Prog (1989) 8(2): 92
870. Hale CC, Josefson AJ, Mattick DE (1985) Ammonia Plant Saf 25: 172
871. Hale CC (1982) Plant/Oper Prog 1(2): 107
872. Arunachalam PN (1986) Fert News (May): 57
873. Dayasagan I, (1986) Fert News (May): 60
874. Briggs PP, Richards III JM, Fiesinger EG (1986) Ammonia Plant Saf 26: 89
875. Prasek F (1988) Ammonia Plant Saf 28: 66
876. Development and Transfer of Technology Series, No. 13 (1980). Fertilizer Manual. United Nations, New York, p 49
877. Dybkjær I (1991) In: Nitrogen 91 Preprints, Br Sulphur Corp, London, p 59; Indian Fertilizer Scene Annual p 1
878. Wilhelm SM (1991) Ammonia Plant Saf 31: 20
879. Orieux A, Barthel Y (1992) Hydrocarbon Techn Int 3: 113
880. Nitrogen (1992) 195: 22
881. Storgaard L (1991) In: Nitrogen 91 Preprints, Br Sulphur Corp, London, p 97
882. Cromarty BJ (1991) In: Nitrogen 91 Preprints, Br Sulphur Corp, London, p 115
883. Hansen J-H, Bak Storgaard L, Pedersen PS (1992) Ammonia Plant Saf 32: 52
884. Wanger ES, Froment GF (1992) Hydrocarbon Process (July): 69
885. Salot WJ (1991) Ammonia Plant Saf 31: 121
886. Cromarty BJ (1992) Ammonia Plant Saf 32: 197
887. Cromarty BJ, Crewdson BJ (1991) Nitrogen 191: 30
888. Vannby R, Madsen SEL Winter (1992) Ammonia Plant Saf 32: 122
889. Elkins KJ, Jeffery IC, Kitchen D, Pinto A (1991) In: Nitrogen 91 Preprints, Br Sulphur Corp, London, p 83

890. Schneider RV III: (1991) FAI Seminar 1990, The Fertiliser Association of India, New Delhi, p SII-1
891. Shiris PJ, Cassata JR, Mandelik BG, van Dijk CP (1984) US 4479925
892. Degand P, Schurmans JP, Julemont V (1992) Ammonia Plant Saf 32: 129
893. Quintana ME, Skinner GF (1991) FAI Seminar 1990. The Fertiliser Association of India, New Delhi, p SII-2
894. Kumar R, Srinivasan V (1991) Indian Fertilizer Annual Scene p 17
895. Hauser N, Bayens CA (1992) Ammonia Plant Saf 32: 268
896. Nitrogen (1990) 186: 21
897. Roos H, Wanjek H, Sprague M (1990) Ammonia Plant Saf 30: 187
898. Kitchen D, Henson WGS, Madsen JK (1990) Ammonia Plant Saf 30: 105
899. Carstensen JH, Bøgild-Hansen J, Pedersen PS (1991) Ammonia Plant Saf 31: 113
900. Bartoo RK, Gemborys TM, Wolf CW (1991) In: Nitrogen 91 Preprints. Br Sulphur Corp London, p 127
901. Nitrogen (1992) 198: 26
902. Song CC, Wohlgeschaffen KR, Verduijn WD, Qadir A (1992) Ammonia Plant Saf 32: 187
903. Shiyong Z (1991) In: Nitrogen 91 Preprints, Br Sulphur Corp London, p 141
904. Shah VA, Huurdeman TL (1990) Ammmonia Plant Saf 30: 216
905. Hefner W, Herion C, Meissner H, Appl M (1991) FAI Seminar, 1990. The Fertiliser Association of India, New Delhi, p SII-3
906. Meissner H, Wammes W, Hefner W (1991) In: Nitrogen 91 Preprints. Br Sulphur Corp London, p 151
907. Nitrogen (1992) 197: 18
908. Søgaard-Andersen P, Hansen O (1992) Ammonia Plant Saf 32: 177
909. Rall W (1991) Ammonia Plant Saf 31: 247
910. Rasmussen CV (1991) Nitrogen 194: 30
911. Peterson RB, Finello R, Denavit GA (1984) US 4452760; CA (1985) 102: No. 187456t
912. Grotz BJ (1986) WO Patent Application 86/06058; CA (1987) 106: No. 69646q
913. Clayton KA, Shannahan N, Wallace B (1991) Ammonia Plant Saf 31: 212
914. Campbell JD, Rawlinson RL, WIlson KC (1992) Ammonia Plant Saf 32: 1
915. Huurdeman TL, Schrijen HC (1992) Ammonia Plant Saf 32: 10
916. Prescott GR (1992) Ammonia Plant Saf 32: 217
917. Heuser A (1992) Ammonia Plant Saf 32: 243
918. Wagner GH, Heuser A, Heinke G (1992) Ammonia Plant Saf 32: 252
919. Timbres DH, van Moorsel WH, Johnson DL (1990) Ammonia Plant Saf 30: 200
920. Madsen J (1991) Ammonia Plant Saf 31: 227
921. Patel NM, Kittelstad KJ (1990) Ammonia Plant Saf 30: 208
922. Allen RL Jr, Moser GA (1992) Ammonia Plant Saf 32: 170
923. Grossmann G, Dejaeger J (1992) Ammonia Plant Saf 32: 164
924. Kitchen D, Pinto A (1991) FAI Seminar 1990. The Fertiliser Association of India, New Delhi, p SIII-1
925. Hicks TC, Pinto A, Moss MJS (1990) Ammonia Plant Saf 31: 219
926. Kitchen D, Pinto A (1991) Ammonia Plant Saf 31: 152
927. Armitage PM, Elkins KJ, Kitchen D, Pinto A (1992) Ammonia Plant Saf 32: 111
928. LeBlanc JR, Shires PJ (1992) Hydrocarbon Technology Int 3: 141
929. Dybkjær I (1991) FAI Seminar 1990. The fertiliser Assoication of India, New Delhi, p S III-2
930. Hakmann R (1991) FAI Seminar 1990. The Fertiliser Assoication of India, New Delhi, p SIII 3
931. Jungr M (1991) Nitrogen 191: 42
932. Sharma SP, Bhaduri AK (1992) FAI Seminar 1991. The Fertiliser Assoication of India, New Delhi, p SII/1-1
933. Tsujimoto M, Bendix H, Lenz L, Johannes D (1990) Ammonia Plant Saf 30: 167
934. Tusjimoto M, Johannes D (1991) In: Nitrogen 91 Preprints. Br Sulphur Corp London, p 69
935. Brown F (1992) Finds, First Quarter, p 32
936. Hale CC, Lichtenberg WH (1990) Ammonia Plant Saf 30: 225
937. Lunde L, Nyborg R (1990) Ammonia Plant Saf 30: 60
938. Lunde L, Nyborg R (1991) Proc No. 307. The Fertiliser Society, London, p 1
939. Hewerdine S (1991) Proc No. 308. The Fertiliser Society, London, p 1
940. Appl M, Fässler K, Fromm D, Gebhard H, Porte H (1990) Ammonia Plant Saf. 30: 22
941. Selva RA, Heuser AH, (1990) Ammonia Plant Saf 30: 39
942. Burke BG, Moore DE (1990) Ammonia Plant Saf 30: 91

943. Conley MJ, Angelsen S, WIlliams D (1991) Ammonia Plant Saf. 31: 159
944. Thompson JH (1990) Ammonia Plant Saf 30: 241
945. Tilton JN, Squire RH, Saffle CS, Atkins CR (1992) Ammonia Plant Saf 32: 63
946. Wiltzen RCA (1991) Ammonia Plant Saf 31: 28
947. Squire RH (1991) Ammonia Plant Saf 31: 131
948. Ali SB, Smallwood RE (1991) Ammonia Plant Saf 31: 142
949. Appl M (1992) Nitrogen 199: 46; (1992) 200: 27; (1992) 202: 44
950. Tomasi L (1992) Nitrogen 199: 35
951. Elkins KJ, Gow AJ, Kitchen D, Pinto A (1992) Proc No. 319. The Fertiliser Society, London, p 1
952. Grotz BJ, Grisolia L (1992) Nitrogen 199: 39
953. Nitrogen (1989) 180: 20
954. Nitrogen (189) 182: 25
955. Karthikeyan S, Neelakantan PS (1991) IFA-FADINAP Regional Fertilizer Conf for Asia and the Pacific. New Delhi, December 1991, p 3
956. Simanjuntak R, Eko NB (1992) Asiafab 1: 18
957. Asiafab (1992) 1: 15
958. Grotz BJ, Stupin WJ (1990) Fertilizer Latin American International Conf (British Sulphur). Caracas, Venezuela (October 1990) p 1
959. Dybkjær I. (1990) IFA Technical Conference, October 1–5, 1990. Venice, Italy, p 1
960. LeBlanc JR (1986) CEER Chem Econ Eng Rev 18(5): 22
961. Rostrup-Nielsen JR, Nielsen PEH, Sørensen NK, Carstensen JH (1993) Ammonia Plant Saf. 33: 184
962. Mohri T, Takemura K, Shibasaki T (1993) Ammonia Plant Saf 33: 86
963. Verduijn WD (1993) Ammonia Plant Saf 33: 165
964. March HD, Thiagarajan N (1993) Ammonia Plant Saf 33: 108
965. Deshmukh D, Raagaard S, Chawes L, Olesen L (1993) Ammonia Plant Saf 33: 278
966. Dybkjær I (1992) IFA Technical Conference, October 5–8, 1992; The Hague, The Netherlands, p 1
967. Dybkjær I (1992) IFA–FADINAP Regional Conference for Asia and the Pacific, Bali, Indonesia, November 30–December 2, 1992, p 1
968. Lee JM, Cialkowski EJ (1993) Ammonia Plant Saf 33: 53

Chapter 7

Ammonia Storage and Transportation-Safety

Anders Nielsen
Haldor Topsøe A/S Copenhagen, Denmark

Contents

Nielsen (Ed.), Ammonia

7.1 General Aspects

This section deals with certain physical and chemical data, which have particular reference to the safety in storage, transportation and use of ammonia.

At atmospheric temperature and pressure, ammonia is a colorless gas with a sharp and pungent odor. Ammonia also exists as a colorless liquid. Its vapor pressure at −33.6 °C is 1 atm, at +4.7 °C it is 5 atm, at 20 °C it is 8 atm, and at 25.7 °C it is 10 atm. At 50.1 °C, the vapor pressure is 20 atm. Ammonia vapor at the boiling point of −33 °C will have a vapor density of approximately 70% of the density of ambient air. However, ammonia and air can, under certain conditions, form mixtures that are denser than ambient air [1], as a result of the mixture being at a lower temperature caused by the evaporation of cold ammonia. An ammonia cloud formed by an accidental release of ammonia may contain a mist of liquid ammonia and its density may be greater than that of air. For the density of an ammonia cloud containing liquid ammonia as it is diluted with dry air, see [2].

Several literature sources list different values for the flammability limits of ammonia mixtures with air, but 16–27% ammonia by volume appears typical. The flammability limits of ammonia in oxygen are 15 to 79% [3]. Limits of flammability for various oxygen/nitrogen ammonia mixtures of various temperatures as well as the flammability characteristics of ammonia/water vapor/air mixtures are given in [4]. Flammability limit measurements were carried out at three temperatures with a supporting atmosphere containing from 21% oxygen (air) down to 11% oxygen. All experiments were carried out at atmospheric pressure. The data show that the flammability limits widen as temperature increases. At 400 °C the flammability limits have widened to 11 to 37% in air [4].

The high solubility of NH_3 in H_2O and the high heat of mixing must be fully considered in the estimation of its behavior in accidents.

7.2 Toxicity and Contact Injuries

The threshold of perception of ammonia varies with the individual and may also depend on atmospheric conditions. A lower limit of perception in the range of 0.4 to 2 wt ppm is given in [5]. In this report it is also concluded that levels of ammonia at the perception threshold may cause changes in the biopotentials of the brain.

Plant surveys [6] report concentrations of 9 to 45 ppm for various areas. Such ammonia concentrations initially are irritating to the eyes or nose, but the exposed persons rather quickly became accustomed to the ammonia. Another report [7] lists recommendations as to the limits of ammonia concentrations in air to which the public may be exposed for short periods of time as follows:

10 min	100 ppm
30 min	75 ppm
60 min	50 ppm

The threshold limit value of the American Conference of Governmental Industrial Hygienists is 25 ppm (or 18 mg/m^3) [8].

In the USA, the above Time Weighted Average Threshold Value has been supplemented with a Threshold Limit Value-Short Term Exposure of 35 ppm in air (or 27 mg/m^3). The Short Term Exposure refers to an average of 15 minutes.

For symptomatology at various exposure limits, see [8, 9]. Medical management of anhydrous ammonia emergencies is thoroughly discussed in [9], distinguishing between the patient mildly affected, those moderately affected, and those with severe ocular lesions, skin lesions, and gastro-intestinal lesions. The prehospital phase, hospital care, long-term complications and accident follow-up are discussed. Based on [8–10], the effects of high concentrations of ammonia on health may roughly be indicated as follows. There are obviously individual differences:

400 ppm: Throat irritation.
700 ppm: Eye injury.
1700 ppm: Coughing appears and labored breathing may be present for hours. Some patients may suffer momentary inability to breath.
2500 to 4500 ppm: May be fatal after short exposure.
5000 ppm and higher: Death results from respiratory arrest.

According to [9], the most frequent sources of exposure are: (1) the spraying of liquid ammonia, (2) being caught in a cloud of ammonia vapor, and (3) by ingestion. Six less common means of exposure are also listed in [9].

Both ammonia spray and liquid are freezing cold and cause frostbite. Temperatures of liquid ammonia down to − 50 °C may be reached as a result of rapid evaporation of ammonia released at − 33 °C [11]. Lessenger [9] lists decontamination as the first treatment of victims, pointing out, however, that clothing may be frozen to the skin, in which case much water must be administered before clothing can be removed. In the prehospital phase, eye injuries are flushed with water for at least 15 minutes. Eyelids must be held open while a constant stream of water flushes out the ammonia [9].

The metabolism and pharmacokinetics of ammonia as well as the health effects of chronic, subchronic and acute exposure to ammonia is discussed in [81]. It is concluded that due to its chemical properties and interactions within the respiratory tract ammonia does not have the potential to cause chronic damage from long term, low level exposure.

7.3 Explosion and Fire Hazards

Accidents in which fire and/or explosion of ammonia-air mixtures have been reported to cause the death or injury of personnel are known [77, 78], but are not very frequent. Such accidents have occurred in cold storage warehouses in which ammonia was released in an enclosed space and ignition was caused accidentally. One reason that explosions and fire accidents have not been frequent is that upon release of ammonia from atmospheric or pressurized tanks, rapid evaporation of ammonia cools down the liquid and thereby reduces the vapor pressure of ammonia over the liquid. As noted above, even momentary fatal concentrations of ammonia are much lower than the lower explosion limit in air. Also the fairly high ignition temperature and the low normal flame velocity of an ammonia/air mixture (0.07 to 0.08 m/s) are important [12]. For protection against fire and explosion in ammonia storage or transportation vessels see [13].

Another explosion hazard associated with ammonia is related to the reaction of ammonia with mercury in the presence of trace amounts of water leading to the formation of an explosive compound [14]. This kind of explosion has been encountered when mercury manometers, after service for several years in an atmosphere of ammonia, explode at the moment of cleaning [15]. In this connection it should be mentioned that natural gas from a number of sources contain elemental mercury in significant quantity [82].

7.4 Safety in Plant Storage and User Storage

Storage facilities for ammonia are discussed in this volume in the chapter by Dybkjaer. Here we will discuss storage facilities with reference to associated safety aspects. Three principally different types of ammonia storage tanks are used in producing plants and distribution terminals. Pressurized (unrefrigerated) storage tanks operate at ambient temperature and are typically designed for a pressure of 18.25 bar. Storage spheres are refrigerated typically by a single-stage refrigeration compressor, operated at temperatures from − 1 to + 2 °C, and are designed for pressure of 3.8 to 5.15 bar. The fully-refrigerated storage tank is serviced by two-stage refrigeration, operated at a temperature of − 33 °C, and is typically designed for a pressure of 1.117 bar [16]. It is important to also design the tank for a certain underpressure. A capacity of pressurized storage tanks up to 270 t, refrigerated spheres from 450 to 2750 t, and fully-refrigerated storage tanks up to 45 000 t is reported in [16]. Agricultural-use tanks range from less than one ton to a few tons in size, the larger tanks being used as applicator tanks as well as nurse tanks for smaller tanks.

Liquid ammonia has a high thermal coefficient of expansion, and rules and regulations exist in various countries governing the maximum filling of stationary and transportation tanks. A guide for developing a training program for anhydrous ammonia workers has been issued by the U.S. National Institute for Occupational Safety and Health, Cincinnati [18].

A further reference to actual design, practice and experience of refrigerated ammonia storage in North America is reported in [16], while an ammonia storage terminal safety program discusses, among other subjects, safety training and accident prevention [13].

While the many thousands of small tanks used in distribution and local storage present their own safety problems, of which examples will be listed below, very large storage facilities represent a potential for very large ammonia spills. One effort to reduce the potential hazard is to use a double-walled tank or to protect the tank by dikes, retention pits or concrete walls. Dikes, and particularly concrete walls, will limit the surface of the pool of liquid ammonia which could be formed in the case of the failure of a large storage tank. In the case of a tank surrounded by a dike, an initial high rate of vaporization would be controlled primarily by heat transfer from the ground and therefore the initial flash depends upon the condition of the ground within the dike [19].

The initial vaporization can create toxic and fire hazards a considerable distance downwind from the tank for a period of time, which is determined largely by atmospheric conditions. Subsequent steady state vaporization presents a smaller fire hazard and a more moderate toxicity hazard, and is controlled by the surface area of the spill and atmospheric conditions.

Stress corrosion cracking has been a serious problem in Europe and also in the U.S. Early observations came from Denmark [20]. Quenched and tempered steels appear particularly vulnerable. Inhibition with a minimum of 0.2% water is used to reduce the risk of cracking. Oxygen (air) plays an important role in the corrosion. Since stress corrosion cracking is the result of chemical processes, it is much less likely and would occur much slower in atmospheric storage tanks containing liquid ammonia at $-33\,°C$ [20]. According to T. Hallan [94] crack initiation is very infrequent at $-33\,°C$ and crack growth rate is approximately three times slower at $-33\,°C$ than at $18\,°C$. Cf. also Nyborg, Lunde and Conley [95] concerning ammonia storage vessel life prediction and Nyborg and Lunde concerning measures for reducing stress, corrosion cracking [96].

For a long time it was believed in the industry that fully-refrigerated ammonia storage tanks would not suffer stress corrosion cracking. In 1987 [80] in the U.K., a fully-refrigerated ammonia storage tank was decommissioned for inspection after almost ten years of service. In the inspection, stress corrosion cracking was found. The water content of ammonia storaged in this tank had been typically 0.02 wt %, which had been considered sufficient to inhibit stress corrosion cracking. The oxygen content of the liquid or vapor had not been determined, and there was a possibility, particularly during the commissioning of the tank, that some oxygen could have been present and played a role in crack initiation. The structural integrity of the tank was analyzed and verified by fracture mechanics [83]. Certain precautions were taken during the recommissioning of the tank.

A new disclosure of stress corrosion cracking in a fully-refrigerated ammonia storage tank was reported a year later [89]. Two identical tanks were inspected of which one showed a significantly higher level of attack by SCC than the other which could not be correlated with differences in the oxygen or water content of the ammonia.

A detailed study of stress corrosion cracking has been carried out at "Institutt for Energiteknikk" Kjeller, Norway financed by ammonia producers and safety authorities. The experimental results and conclusions as to the influence of tank environment, electrochemical potential, temperature, steel properties and types of welding electrodes are reported and recommendations given [92]. Application of cathodic protection in a large atmospheric ammonia storage tank is described in [97].

A description of leakage in the floor of an ammonia storage tank (a particularly critical point of damage) and the repair operations are given in [21].

7.5 Ammonia Transportation and Related Accidents

Liquid ammonia is transported in tankers, barges, by railcars and by various means of road transportation. A newer way is the use of pipelines for ammonia transportation.

7.5.1 Pipelines

There are three major ammonia pipelines; two in the U.S. and one in the USSR. One of the American pipelines (MidAmerica Pipeline System) transports ammonia from the Texas Panhandle to points in Kansas, Nebraska and Iowa. Its total peak capacity delivered to a number of points is 8000 t/day [22]. The other U.S. pipeline, the Gulf Central Pipeline, transports anhydrous ammonia from major producers along the Texas and Louisiana Gulf Coast to points in Iowa, Illinois, Nebraska, Indiana and Missouri [23]. Further information on the maintenance and repair of the MidAmerica Pipeline System is given in [24]. The total weight of ammonia in the MidAmerica Pipeline System, when full, is approximately 20,000 tons. Lock valves are 10 miles apart so it would be possible that 400 tons could get out between lock valves [22]. It is known from [12] that in one accident 700 tons of liquid ammonia leaked from the pipeline in the USSR. In the USSR pipeline accident, the aerosol cloud of ammonia covered a forested territory about 40 km^2 in area; however, all residents in the region were evacuated in time. Pipeline operations in the U.S. are governed by the national Gas Pipeline Safety Act of 1968, as amended in 1979 [17].

Beside the main distribution lines, a number of smaller pipelines between ammonia plants and adjacent chemical plants and loading facilities exist. A rupture of such a small pipeline resulting in the release of 50 t of ammonia has been described [46].

7.5.2 Tankers and Barges

Large quantities of ammonia are transported by tankers and barges. In 1991 a total of about 10 million tons of anhydrous ammonia was transported by ocean going vessels (IFA Ammonia Statistics, yearly issue). According to a Battelle report [17] U.S. Coast Guard regulations limit the capacity of tankers to 25 000 tons and of barges to 2500 tons. The Safety Features and Loading Rates of Refrigerated Barges is described in [25]. There have been a number of accidents during loading and unloading of vessels. The ammonia loading hoses or lines appear a central part of the facilities. A tanker's loading line ruptured and 180 tons of ammonia leaked out and a cloud of ammonia gas moved over the vessel. After the ammonia cloud had dispersed, 2 men of the 7-man crew were found dead and the other 5 were rescued [26]. Another accident, occurred during loading of a large vessel [27]. No fatalities occurred in this accident. However, a dockman received first and second degree burns.

7.5.3 By Rail

Ammonia is extensively transported by rail car. The type of tank cars used in the U.S. for transportation of anhydrous ammonia as well as their safety aspects, inspections and safety record are discussed in [28]. Here also two accidents involving ammonia tank cars are dealt with. The detailed description of the accident which occurred in 1976 at Glen Ellyn, Illinois, is described in [29, 49]. This accident was due to a derailment as two trains passed each other. The tank head of an ammonia car was punctured by a coupler of an adjacent car. One of the conclusions of the analysis is that if the tank car had been provided with head shields, the tank head would not have been punctured during the derailment. Another accident in which a railroad tank car carrying anhydrous ammonia suffered a puncture is described in [30]. This tank car contained about 75 t of anhydrous ammonia at a pressure of 4 to 5 atm. The handling of the railroad derailments involving anhydrous ammonia and other hazardous materials is discussed in [31]. Computer-aided ammonia rail car loading is described in [32].

A first-hand account of the efforts of an emergency response team at the Pensacola ammonia accident, which was due to a train derailment, is given in [48]. A special feature in some ammonia tank car accidents is the appearance of delayed tank car failures. This is reviewed in [50]. In this article, delayed tank car ruptures which occurred in accidents at Cummings, Iowa and Crestview, Fla. are discussed. A detailed survey of damages from ammonia spills has been given by Markham [79].

For transportation accidents by railroad, see also [47]. This report gives a survey of damage during 177 anhydrous ammonia highway incidents and 570 rail incidents in the U.S. for the years 1971 to 1982. It lists the identification number for anhydrous ammonia as UN1005 and the required labeling as a poisonous gas.

7.5.4 By Road

Various road vehicles are used to transport ammonia. The U.S. Department of Transportation has limited the use of certain high stress steels only for ammonia which has a minimum content of water of 0.2 wt %, or a purity at least 99.995% [33]. A more recent reference [17] mentions that when ammonia is shipped in containers constructed of quenched and tempered steel it must contain a minimum of 0.2 wt % water. Various countries specify maximum percentage filling of tanks to provide a cushion. The European Council of Chemical Manufacturer's Federation (CEFIC) Zürich has issued a safety card for road transportation of anhydrous ammonia, containing certain information and recommendations which have been used in some European countries in connection with local recommendations and lists of laws and rules relevant to transportation and storage of anhydrous ammonia [34].

Two road vehicle accidents are particularly well known. In one, a semi-trailer truck ruptured suddenly in the yard of a factory at Lievin (France). The accident and the analysis of the cause of the accident is reported in [35]. The almost instantaneous escape of 19 tons of ammonia caused severe injuries to 20 persons, 5 of whom died. Another serious accident occurred in Houston, Texas, when a tractor semi-trailer tank transporting approximately 27 m^3 of anhydrous ammonia struck and then penetrated a bridge rail on a ramp connecting two highways. The tractor and trailer left the ramp and fell on to a highway. The anhydrous ammonia was released and 78 persons were hospitalized. Six persons died as a result of the accident. Since the accident happened on a highway, most of the victims were motorists [36, 37].

7.6 Accidents with Fixed Tanks

The quantities of ammonia that may be released in the manufacturing facilities are typically smaller than quantities that may be released from storage or transportation tanks. However, under unfavorable conditions even a small quantity may present a danger to life.

One fatal accident is described in [17]. A cylinder used in servicing air conditioning equipment and containing only 5.7 kg of ammonia was being transported in the cargo space of a van-truck. The cylinder ruptured as the truck was moving at approximately 60 mph. The driver succeeded in stopping the truck and opening the door, after which he fell out and died, either at the scene or on the way to the hospital.

For pressurized storage tanks one concern is stress corrosion cracking. Stress corrosion cracking in pressurized storage tanks was first observed in Denmark in 1964 [38] when a new 230 m^3 cylindrical tank started to leak through a long crack. A detailed review of stress corrosion of steels in anhydrous ammonia service and laboratory investigations into its cause and prevention is presented in [39, 40, 92].

Other accidents with pressurized storage vessels have occurred as a result of brittle fractures. In one accident in South Africa, a 50 t pressurized storage tank failed, releasing 38 t of anhydrous ammonia and killing 18 people. Analysis of the tank after the accident showed that the failure was a brittle fracture and it was felt that reasidual stresses from an earlier weld repair of some minor cracks may have contributed to the cause of the failure. After the welding, the tank had not been subsequently stress-relieved. A hydraulic pressure testing of the vessel would have contributed to the strain-aging process of the metal [41].

The large atmospheric storage tanks represent a significant potential hazard due to the large inventories. Stress corrosion cracking occurs much slower in these tanks because of the low temperature, − 33 °C, and pressure, and because they are newer and more information on the selection of steels for tanks and other related factors were available when they were built. The kind of incidents that have been reported are associated with settling of the ground below the tanks, loss of heating below, and improper welding techniques in the floor plate or the cylindrical part of the tank [42, 43, 44]. Cf also earlier discussion of stress corrosion found during tank inspections.

A serious accident occurred in a Lithuanian storage facility in 1989. The tank in question was a single-wall tank protected by a concrete wall. The cause of the failure, which was very dramatic with the tank opening up on one side at the bottom seam and part of the tank being dislodged from its foundation, smashing through the concrete wall and being displaced 40 m, is not entirely clear [86]. There is a possibility that the cause of the accident may have been the so-called "roll-over". This may be caused by a quantity of warm ammonia being fed into the bottom of a storage tank, partly filled with cold ammonia. The phenomena associated with the feeding of warm ammonia into a fully-refrigerated ammonia storage tank are discussed in [87].

An instance of an ammonia spill has been reported when a refrigerated ammonia storage tank was filled from a barge and the high level alarm and shutdown system failed to operate [17]. Also, instances of creating a vacuum in the tank, due to instrument failure as a result of freezing up, have been reported [45].

A general survey of refrigerated ammonia storage in North America, treating design practices and experience, tank locations relative to neighborhoods, and various technical questions, is given in [16].

7.7 Behavior of Ammonia Spills

Upon release into the atmosphere of an ammonia spill, a portion of it will flash off as a vapor. This portion is the initial flash followed by a period of evaporation until all of the ammonia has evaporated, been dissolved in water or otherwise disposed of. Some of the fundamentals of the behavior of ammonia released are discussed in [1, 51].

There is a significant difference in the release of ammonia from a refrigerated tank operating slightly above atmospheric pressure, where the amount of the initial flash is only a few tenths of a percent of the total, and the release from a pressurized storage system under a pressure of about 9 atm and 24 °C, in which case approximately 20% of the spilled ammonia would be flashed. Figures are given showing the amount of flash depending upon the initial pressure and temperature. Figures are also shown about the behavior of small and large pools regarding temperature development and evaporation rates at certain wind and temperature conditions. The importance of meteorological conditions is pointed out. There is a difference between the situation at night with an overcast condition, and the situation on a hot sunny day, where as much as 80% of the radiation may be absorbed and re-radiation is negligible due to the low temperature of the ammonia pool. In addition, the question of vapor dispersion in dealt with [51].

In [1], the various ways of releasing ammonia are discussed, and spills on land and on water are discussed in a qualitative manner. The question of liquid ammonia jets, their direction, the fraction of ammonia remaining airborne and falling to the ground, the formation of dense air ammonia mixtures, and the formation of fogs are discussed in [1].

A simple model has been established [52] for the release of anhydrous ammonia from pressurized containers. The flash off and the entrainment of air in ammonia vapor is discussed. It is pointed out that the density of the mixture depends on the airborne fraction of the liquid, the ambient temperature, the relative humidity and the mass of air entrained in a given sized ammonia release. Based on the model, the calculation of hazard ranges is discussed and two instances of releases from pressurized containers, that are fairly well described, are used for comparison. It is concluded that the mixture of ammonia and air resulting from a certain release from a pressurized tank is likely to be denser than air. A detailed account of the Houston, Texas anhydrous ammonia release following the crash of a tank semi-trailer in a highway accident is given in [37]. The report shows a map of the area and also photographs of the accident site and the cloud at 1, 2, 3 and 4 minutes after the crash.

A review of the production of dense gas mixtures from ammonia releases is given in [2]. The paper concentrates on issues in the modeling of accidental releases. It lists a number of major ammonia releases and discusses some differences between spillage of refrigerated liquid and release through smaller or larger holes from pressurized containers. It points out that in the spillage of refrigerated liquid, the boiling is likely to be gentle and without a large amount of liquid ammonia ejected into the air, except possibly for the very early stages of such spill.

The importance of containing a spill of refrigerated ammonia within a space with a fairly small liquid surface area so as to restrict heat transfer and boiling has been pointed out earlier. In the spillage of refrigerated ammonia, the initial pool will not remain at − 33 °C. Buoyancy of ammonia vapor will cause strong convection currents leading to evaporation and cooling down of the pool to − 50 °C or possibly lower. Atmospheric conditions are very important in relation to spills of refrigerated as well as hot ammonia.

In the case of one rail car failure, it was pointed out that a ground fog was reported, the wind was calm, and a temperature inversion had occurred [53]. In the Houston semi-trailer accident in which about 16 t of ammonia was released, witnesses reported that the white ammonia vapor cloud initially reached a height of about 30 m before being carried by the wind for a distance of approximately 800 m. Within 3 minutes of release the maximum width of the vapor cloud over the ground was about 300 m and after 5 minutes, when most of the liquid ammonia had boiled off, the cloud had dispersed [36]. The importance of wind direction and strength cannot be overemphasized. It is evident in the accounts from many accidents.

A specific question relates to the use of water on ammonia releases. If a stream of liquid water is directed into a liquid pool of anhydrous ammonia, there is an exothermic reaction and a large release of ammonia vapor results. On the other hand, water is a main tool to use to absorb and control ammonia vapors. Fog nozzles can be used to contain a release of ammonia and prevent the spreading of a cloud downwind. Rapid and wide rotation of fog nozzles increases the efficiency of ammonia absorption. When using the right procedures, as demonstrated in many ammonia vapor control exercises, a "capture" zone can be created downwind to prevent the cloud from spreading [54].

In the Pensacola accident, 50% of the contents of one ammonia rail car were released within 10 minutes, while the contents of another ammonia rail car were slowly vaporized over 12 hours. A lethal cloud engulfed a populated area within minutes, and before the residents could be evacuated. They were later escorted on foot for distances from 60 to 300 m through the ammonia cloud, until out of the affected area. One of the lessons learned in this accident and also in the Crete, Nebraska accident, in which 3 anhydrous ammonia tank cars were struck by a passing train and one tank ruptured, is that people who stay in their homes closing doors and windows survive, and that they can help themselves by stuffing towels in openings to make the house airtight [28, 55]. (The chance of survival in a house will depend on the time the house is engulfed in an ammonia cloud, the conditions of the cloud and the tightness of the house). In the Pensacola accident the cloud travelled almost 15 miles in a period of an hour before dissipating [55]. Fire fighters directed a hose into the liquid ammonia stream and backed this up with a water fog to knock down the vapors which at the time had reached about 40 m upwind from the spill [48]. A two-phase cloud that formed during release from a pressure tank is discussed in [56] as well as the atmospheric dispersion of ammonia.

It is reported that much of the information concerning ammonia aerosol clouds comes from investigators at the scene of the accidents rather than from controlled experiments. Risk models for the prediction of risk distances during release of toxic industrial chemicals, including ammonia, are discussed in [57]. Two sizes of releases are considered: one 15 to 25 t and one 100 to 150 t. The risk distances are given for two wind velocities.

7.8 Ammonia Spill Studies

A number of companies or groups of companies have carried out studies of ammonia spills. A number of French companies and organizations carried out a significant number of spill tests on land. A total of 14 spills were carried out involving quantities of ammonia between 80 kg and 1 t. Some liquid spills were vertically ejected into the air, others were parallel to the ground; also gaseous emissions were studied. Several spills of 100 to 300 kg of liquid ammonia, ejected vertically over 1 to 3 minutes from a 50 mm tubing bent upwards, were carried out. The atmospheric conditions are given. The wind velocity was from 0 to 30 m/s. It is found in all ejections that a cloud was formed right from the exit from the pipe. The liquid was ejected under a pressure of at least 6 atm and apparently a very large fraction appeared as microdroplets which formed an aerosol that was very stable due to the low wind velocity and the low temperature of the cloud. Measuring devices were installed at three distances from the emission point and there are several photographs of the clouds. Ammonia concentrations in the 1 to 2% range were measured at 40 m from the ejection point. The cloud went up to approximately 20 m and then fell back towards the ground where it stretched out and could be followed for about 100 m until it disappeared in a forest area. In some tests the analyzer furthest away from the emission test was 175 m distant. In several of the tests maximum concentrations in excess of 1000 ppm were observed at that distance [58].

A report on ammonia spills at Frenchman's Flat has been given [59, 60]. Four tests were carried out with ammonia quantities of 25 to 60 m^3 (10–41 tons). Test number four was particularly interesting since it involved

41 tons of ammonia and was conducted under the most stable atmospheric conditions. An extensive instrumentation system was set up to measure width and height of the ammonia cloud as it moved in the wind direction, particularly at 100 m and 800 m from the spill point. Further measurements were carried out by portable ground level stations at 1.4, 2.8 and 5.5 km downwind. Preliminary results of maximum ammonia concentrations measured in test four were as follows: 100 m 6.5%, 800 m 2.1% and 2800 m 0.5%. It is noteworthy that a fatal concentration of 0.5% ammonia was registered at a distance of 2800 m. Reference [93] includes a simulation of Test No. 1.

A study on the predicted hazards from ammonia spills on and under water has been reported [61–63]. An important aspect is the partition between ammonia dissolving in water and that being released to the atmosphere. This partition function is higher, close to 0.9, when liquid ammonia is released underwater, and lower, depending on release conditions and atmospheric conditions, when a large quantity is released instantaneously on the water surface where liquid ammonia will boil to produce saturated ammonia vapor, and the reaction between ammonia and water also contributes to the vapor formation. Another phenomenon is the aerosol formation as fine drops of ammonia are thrown into the air. Partition function ratios of approximately 0.56 have been found for such releases. It must be pointed out that this is an extrapolation from smaller spills of approximately 10 and 200 kg of ammonia, the first in a swimming pool and the second on the surface of a lake. Models have been established. The situation for release on a river such as from a barge carrying liquid ammonia are studied in detail. Two cases treated are a 200 t surface release and a 3000 t surface spill. It is pointed out that a liquid ammonia barge that sinks in water may eventually release liquid or gaseous ammonia. The partition between ammonia picked up by water and ammonia released to the atmosphere depends on many factors; an important one for underwater releases is the ratio of the water depth to the pipe diameter of release. The model predicts that if the depth is on the order of 20 outlet pipe diameters, the partition ratio may be about 0.9, and it may approach 1.0 for greater depths. One of the conclusions of the report is that vapor clouds formed from spills on water, although containing a fraction of aerosols, will have buoyant behavior depending on wind velocity.

The model has been used to describe the simulation of a number of situations of transport of hazardous cargo by ship [64]. One of the situations covered is a fictitious anhydrous ammonia casualty at Louisville, Kentucky. The greatest potential hazards in ammonia spills, on or below the surface of water, are obviously those associated with barge transport on rivers and in narrow coastal waters, and those associated with tanker transport in narrow streams and harbors.

7.9 Ammonia Storage Facilities

The largest ammonia storage facilities are located as part of ammonia producing plants, or at large distribution centers or terminals. A large number of smaller storage tanks are typically operated by ammonia distribution companies and thousands of small tanks are used in distribution and local storage. In Denmark where the use of liquid ammonia as a direct application fertilizer is wide spread, there are on the order of 10 000 small tanks [20].

A large amount of ammonia is stored in the transportation system in vessels, barges, tank cars and in ammonia pipelines. According to [65] the Gulf Central Pipeline holds about 70 000 tons of ammonia.

For transfer of − 33 °C ammonia, e.g., from an ocean going tanker, into a pressure or semi-refrigerated storage installation, an ammonia heating installation is required. The actual temperature and pressure in refrigerated storage tanks depends upon the location of the storage, whether at sea level or at a higher elevation.

The pressurized storage tanks are typically made for capacities of up to 270 tons, but are found in all sizes down to a few cubic meters in size and even smaller pressurized containers are widely used in industry down to sizes of one to a few kg capacities.

For maintaining safety around ammonia installations rules, and regulations exist in the various countries and even the smallest containers are subject to rigid and detailed specifications [17]. A detailed survey of refrigerated ammonia storage tanks in the U.S. and Canada, with information on location and design features of such tanks, is given by Hale [16]. A HAZOP study of a fully-refrigerated storage installation is described in [90]. Cf. also [91] on topics of risk analysis and emergency management.

Two underground ammonia facilities are described. One storage cavern of a capacity of 20 000 tons of ammonia has been operated by Dupont, and Norsk Hydro, is operating underground storage in Norway of approximately 50 000 tons in size [66].

Modern ammonia plants typically in the size of 1000–2000 t/d capacity, in most cases include one or more atmospheric storage tanks as part of the facility. Details of the design and construction of a high-integrity ammonia storage tank is given in [84]. Comeau [67] discussed precautions for ammonia storage tanks and particularly the design of surrounding protection dikes. The function of such dikes is to keep evaporation to a minimum in the event of a leak and to protect the tanks against external physical damage. Many consider the best protection of an atmospheric storage tank to be a high wall of prestressed concrete around the tank which would further limit evaporation if the tank fails.

One of the technical and safety aspects of storing anhydrous ammonia in atmospheric tanks is to protect the soil below the tank from freezing. This problem and experience with tank foundation heaters are discussed by Comeau and Weber [43]. Special problems arise in the shutdown and startup of refrigerated atmospheric ammonia storage tanks [68, 69]. One has to be particularly careful in certain situations not to create a vacuum inside the tank. An inspection outline for retrofitting refrigerated ammonia storage for improved safety and economy is given in [70].

The high integrity ammonia storage tank described in [84] is characterized by double-integrity steel walls and a double-integrity bottom. The bottom is supported on an elevated foundation above grade, and the tank top is provided with a suspended deck. Such a construction is obviously safer than a single-wall atmospheric pressure storage tank, supported on the soil with only a layer of insulation, a layer of sand and gravel and with an electrical heating system below the insulation to protect against freezing of the soil. However, in a risk assessment, not only the tank itself but the complete installation should be considered [88].

As a further safety dimension an ammonia detection system in an area around the storage facility may be used. Such a system would give indications of any leaks from the ammonia storage and record even a few ppm of ammonia in the atmosphere [85]. The knowledge learned on stress corrosion cracking, construction, safe operation, and maintenance has contributed greatly to the safety of atmospheric ammonia storage systems.

7.10 Ammonia Transportation Systems

Much of the ammonia manufactured in the world is transported for various distances. Plants located in the areas with abundant natural gas and low population density may transport close to 100% of the ammonia production or 40–60% if combined with some capacity to make finished fertilizers. In the U.S. approximately 25% of the ammonia produced or imported is subjected to transportation.

Ocean transportation of ammonia is by ammonia tankers; river and coastal transportation by barges; and overland transportation by pipelines, rail or road tankers. Ammonia tankers may be designed as semi-refrigerated or fully refrigerated vessels [25]. Semi-refrigerated carriers typically contain up to 15 000 m^3 while fully refrigerated tankers have capacities of up to 46 500 tons [65]. Ammonia barges may have two cylindrical fully refrigerated tanks [25]. Transportation of ammonia by pipeline is discussed in [65], [72]. In the U.S. the MidAmerica Pipeline System and the Gulf Central Pipeline transport ammonia from high production areas in Texas and Oklahoma, and Texas and Louisiana respectively to heavy agricultural-use areas in the Midwest. A small ammonia pipeline connects an ammonia plant in Northeastern France with a fertilizer industry in Western Germany. The longest ammonia pipeline is found in the USSR. Its total length is 2424 km connecting Togliatti, Gordlovka and

Grigor'evskii. The USSR pipeline has a throughput of 2.5 million t/year and crosses several major rivers.

According to [65] ammonia transportation by pipelines requires the ammonia to be heated up, at least, to 2 °C, which in most cases means that it must be warmed at the supply terminal and cooled again to − 33 °C at the receiving terminal. For transportation of ammonia, particularly by pipeline, the PVT properties of ammonia with a small content of water are important. Such data are found in [73]. The Interstate Commerce Commission requires that all anhydrous ammonia transported interstate by pipeline must contain a minimum of 0.2 wt % of water [73]. The water is added as a corrosion inhibitor and will, according to experience gained by a survey of ammonia storage spheres, limit the risk of stress corrosion cracking in surfaces exposed to ammonia liquid [74, 75]. It is also stated in [75] that contamination with air is the primary cause of stress corrosion cracking in ammonia; and [75] also discusses the phenomena that the oxygen content is higher in the gas phase of the tank while water is contained predominantly in the liquid, and thus a fairly complex picture for stress corrosion cracking is developed which is important for transportation of ammonia in any tank. In [17] is given a survey of the various codes and regulations, covering storage and handling of ammonia in the USA as regulated through OSHA. It is emphasized that additional municipal, country and state regulations exist in many locations. It is pointed out that shipping containers constructed of quenched and tempered steel must contain, at least, 0.2 wt % water, which is also a requirement for ammonia transported by pipeline [65].

Rail cars can typically hold 26 tons of ammonia, while larger jumbo cars may hold up to 80 tons and road tankers are, in the U.S., limited to 25 tons of ammonia. One of the minimum requirements in the U.S. is that containers such as road and rail tankers used for ammonia shipments must not be filled to more than a certain percentage by volume to provide a cushion for thermal expansion should its temperature increase [17]. The tanks are typically designed for 15.5 bar (vapor pressure of ammonia at 43 °C) and provided with pressure relief devices [17]. Detailed description of a tanker loading station for liquid ammonia is given in [76]. The facility described has the capacity to handle the loading of 12 tankers per day and to unload 15 tankers per day. Details of a computer-aided ammonia rail car station are given in [32] including a description of the computer hardware and software used in the facility.

7.11 References

1. Blanken JM (1980) Ammonia Plant Saf 22: 25
2. Griffiths RF, Kaiser GD (1982) J Hazard Mater 6: 197
3. Charp C (1966) Agric Anydrous Ammonia Technol Use Proc Symp, St. Louis, MO, p.21
4. Harris GFP, MacDermott PE (1977) Inst Chem Eng Symp Ser No 49: 29
5. Nuttonson MY (1972) PB-209478, 1

6. National Institute for Occupational Safety and Health (1974) Rockville, MD, PB-246669, p.1
7. National Research Council (1972) Committee on toxicology, Washington, DC, PB-244336, p 1
8. Legters L (1980) AD-A094501 p 1
9. Lessenger JE (1985) Plant Oper Progr 4:20
10. Barber JC (1978) Ammonia Plant Saf 20:5
11. Morgan GO, Reed JD (1965) Ammonia Plant Saf 7:38
12. Zakaznov VF, Kursheva LA, Upadyshev KL (1980) Khim Prom (Moscow) 12:361; Soviet Chem Ind (1980) 12:728
13. Hale CC, Lichtenberg WH (1980) Ammonia Plant Saf 22:35
14. Michels AMJF, Dumoulin EM, Gerver JH, (1957) Rec Trav Chim 76:5
15. Leleu J (1976) Inst Natl Rech Secur Paris Cah Notes Doc No. 1024–8476, 427
16. Hale CC (1984) Ammonia Plant Saf 24:181
17. Brenchley DL, Athey GF, Bomelburg HJ (1981) PNL-4006:1
18. Karches GJ, Froehlich PA, Bicknell RJ (1978) PB 80-189475:1
19. Husa WH, Bulkley WL (1965) Ammonia Plant Saf 7:41
20. Arup H (1977) Ammonia Plant Saf 19:73
21. Lichtenberg WH (1972) Ammonia Plant Saf 14:24
22. Rohleder GV (1969) Ammonia Plant Saf 11:35
23. Inkofer WA (1969) Ammonia Plant Saf 11:40
24. Ludddeke DE (1975) Ammonia Plant Saf 17:99
25. Briley GG (1967) Ammonia Plant Saf 9:10
26. Håkansson R (1977) Ammonia Plant Saf 19:119
27. Caserta LV (1972) Ammonia Plant Saf 14:31
28. Heller FJ (1981) Ammonia Plant Saf 23:132
29. National Transportation Safety Board (1976) Washington, DC PB 267939, p 1
30. Cato GA, Dobbs WF (1971) Ammonia Plant Saf 13:1
31. O'Driscoll JJ (1974) Abtr Papers 78th Nat Meeting AIChE Salt Lake City Paper 35D, p 1
32. Arseneaux AA (1985) Ammonia Plant Saf 25:150
33. Olsen EA (1969) Ammonia Plant Saf 11:46
34. Conseil Europeen des Federations de l'Industrie Chimique (1979) Eur Council Chem Manuf Fed Safety Card-CEFIC TEC-T-1-Rev 3
35. Medard L (1970) Ammonia Plant Saf 12:17
36. National Transportation Safety Board (1977) Washington, DC PB-268251, p 1
37. National Transportation Safety Board (1979) Washington, DC PB-80-144942, p 1
38. Nielsen A (1971) Ammonia Plant Saf 13:103
39. Hutchings J, Sanderson G, Davies DGS, Davies MAP (1972) Ammonia Plant Saf 14: 102
40. Phelps EH (1972) Ammonia Plant Saf 14:109
41. Lonsdale H (1975) Ammonia Plant Saf 17:126
42. Lichtenberg WH (1977) Ammonia Plant Saf 19:59
43. Comeau ET, Weber ML (1977) Ammonia Plant Saf 19:63
44. Esrig MI, Ahmad S, Mayo HC (1975) Ammonia Plant Saf 17:93
45. Winegar BW (1980) Ammonia Plant Saf 22:226
46. Sterling MB (1977) Ammonia Plant Saf 19:77
47. Transportation Research Board (1983) Washington, DC PB-84-143635, p 1
48. Stueben WJ, Ball WL (1979) Ammonia Plant Saf 21:76
49. Day BF (1978) Ammonia Plant Saf 20:30
50. Eiber RJ (1981) Ammonia Plant Saf 23:146
51. Ball WL (1970) Ammonia Plant Saf 12:1
52. Kaiser CD, Walker BC (1978) Atoms Environ 12:2289
53. National Transportation Safety Board (1969) Washington, DC, PB-198790, p 1
54. Greiner ML (1984) Ammonia Plant Saf 24:109
55. National Transportation Safety Board (1978) Washington DC, PB-28325, p 1
56. Kansa EJ, Ermak DL, Chan ST, Rodean HC (1983) UCRL-88649-Rev 2, p 1
57. Andersson JO, Broxvall Å, Karlsson E, Karlsson N, Nyrén K, Rejnus L, Winter S (1983) FOA-C-40183-C2
58. Resplandy A (1969) Chim Ind Genie Chim 102:691
59. Goldwire HC (1986) Chem Eng Prog 82:35
60. Goldwire HC et al. (1985) UCID-20562
61. Raj PK, Hagopian JH, Kalelkar AS (1974) CG-D-74-74, AD-779400, p 1
62. Raj PK, Hagopian JH, Kalelkiar AS, Cece J (1975) Ammonia Plant Saf 17:102

63. Raj PK, Reid RC (1978) Environ Sci Technol 12:1422
64. National Research Council Panel on Response to Casualties Involving Ship-Borne Hazardous Cargoes (1979) AD-A075203, p 1
65. Hignett TP (1979) Transportation and Storage of Ammonia. Fertilizer Industry Round Table, Washington, DC
66. Sherman JL (1970) Ammonia Plant Saf 12:8
67. Comeau ET (1972) Ammonia Plant Saf 14:69
68. Bell HV (1982) Chem Eng Prog 78:74
69. Vick KA, Witthaus JB, Mayo WC (1980) Ammonia Plant Saf 22:54
70. Hale CC, Josephson AJ, Mattick DE (1985) Ammonia Plant Saf 25:172
71. Schrader RF (1979) Nitrogen No. 117, 26
72. Kharlamov VV, Tsybal YM (1983) Zh Vses Khim Ob-va 28:103
73. Inkofer W, Wilson GM, Adams JE (1971) Ammonia Plant Saf 13:67
74. Blanken JM (1984) Ammonia Plant Saf 24:140
75. Lunde L(1984) Ammonia Plant Saf 24:154
76. Schlichthärle G, Huberich T (1983) Plant Oper Progr 2:165
77. Klem TJ, Saunders JC (1986) Ammonia Plant Saf 26:145
78. McRae MH (1987) Plant/Oper Progr 6:17
79. Markham RS (1987) Ammonia Plant Saf 27:136
80. Byrne JR, Moir FE, Williams RD (1989) Ammonia Plant Saf 29:122
81. Ryer-Powder JE (1991) Ammonia Plant Saf 31:93
82. Mark Wilhelm S (1991) Ammonia Plant Saf 31:20
83. Selva RA, Heuser AH (1990) Ammonia Plant Saf 30:39
84. Thompson JR, Carnegie RN (1989) Ammonia Plant Saf 29:116
85. Thompson JR, Sekula LJ, Whitson PE (1992) Ammonia Plant Saf 32:77
86. Andersson BO (1991) Ammonia Plant Saf 31:5
87. Tilton JN, Squire RH, Saffle CS, Atkins CR (1992) Ammonia Plant Saf 32:63
88. Squire RH (1991) Ammonia Plant Saf 31:131
89. Appl M, Fässler K, Fromm D, Gebhard H, Portl H (1990) Ammonia Plant Saf 30:22
90. Lawley HG, Shepherd JS (1987) Fertilizer Focus May :57
91. Ham JM, Gansevoort J (1992) Ammonia Plant Saf 32:86
92. Lunde L, Nyborg R (1991) Proc No. 307, The Fertilizer Society, London, p 1
93. Statharas JC, Bartzis JG, Venetsanos A, Würtz J (1993) Process Safety Prog 12:118
94. Hallan T (1994) Ammonia Plant Saf 34 to appear
95. Nyborg R, Lunde LR, Conley MJ (1991) Mater Perform 30:61
96. Nyborg R, Lunde L (1995) Ammonia Plant Saf 35 to appear
97. Bickel W, Fässler K, Geis B, Guns L, Reininghaus J, Walter M (1995) Ammonia Plant Saf 35 to appear

Printing: COLOR-DRUCK DORFI GmbH, Berlin
Binding: Buchbinderei Lüderitz & Bauer, Berlin

Springer-Verlag and the Environment

We at Springer-Verlag firmly believe that an international science publisher has a special obligation to the environment, and our corporate policies consistently reflect this conviction.

We also expect our business partners – paper mills, printers, packaging manufacturers, etc. – to commit themselves to using environmentally friendly materials and production processes.

The paper in this book is made from low- or no-chlorine pulp and is acid free, in conformance with international standards for paper permanency.